Lecture Notes in Computer Science 13961

The series Lecture Notes in Computer Science (LNCS), including its subseries Lecture Notes in Artificial Intelligence (LNAI) and Lecture Notes in Bioinformatics (LNBI), has established itself as a medium for the publication of new developments in computer science and information technology research, teaching, and education.

LNCS enjoys close cooperation with the computer science R & D community, the series counts many renowned academics among its volume editors and paper authors, and collaborates with prestigious societies. Its mission is to serve this international community by providing an invaluable service, mainly focused on the publication of conference and workshop proceedings and postproceedings. LNCS commenced publication in 1973.

Maribel Fernández · Christopher M. Poskitt
Editors

Graph Transformation

16th International Conference, ICGT 2023
Held as Part of STAF 2023
Leicester, UK, July 19–20, 2023
Proceedings

Editors
Maribel Fernández 🆔
King's College London
London, UK

Christopher M. Poskitt 🆔
Singapore Management University
Singapore, Singapore

ISSN 0302-9743   ISSN 1611-3349 (electronic)
Lecture Notes in Computer Science
ISBN 978-3-031-36708-3   ISBN 978-3-031-36709-0 (eBook)
https://doi.org/10.1007/978-3-031-36709-0

This Springer imprint is published by the registered company Springer Nature Switzerland AG
The registered company address is: Gewerbestrasse 11, 6330 Cham, Switzerland

Preface

This volume contains the proceedings of ICGT 2023, the 16th International Conference on Graph Transformation, held during 19–20 July, 2023, in Leicester, UK. ICGT 2023 was affiliated with STAF (Software Technologies: Applications and Foundations), a federation of leading conferences on software technologies. ICGT 2023 took place under the auspices of the European Association for Theoretical Computer Science (EATCS), the European Association of Software Science and Technology (EASST), and the IFIP Working Group 1.3, Foundations of Systems Specification.

The ICGT series aims at fostering exchange and collaboration of researchers from different backgrounds working with graphs and graph transformation, either by contributing to their theoretical foundations or by applying established formalisms to classic or novel areas. The series not only serves as a well-established scientific publication outlet but also as a platform to boost inter- and intra-disciplinary research and to stimulate new ideas. The use of graphs and graph-like structures as a formalism for specification and modelling is widespread in all areas of computer science as well as in many fields of computational research and engineering. Relevant examples include software architectures, pointer structures, state-space and control/data flow graphs, UML and other domain-specific models, network layouts, topologies of cyber-physical environments, quantum computing, and molecular structures. Often, these graphs undergo dynamic change, ranging from reconfiguration and evolution to various kinds of behaviour, all of which may be captured by rule-based graph manipulation. Thus, graphs and graph transformation form a fundamental universal modelling paradigm that serves as a means for formal reasoning and analysis, ranging from the verification of certain properties of interest to the discovery of fundamentally new insights.

ICGT 2023 continued the series of conferences previously held in Barcelona (Spain) in 2002, Rome (Italy) in 2004, Natal (Brazil) in 2006, Leicester (UK) in 2008, Enschede (The Netherlands) in 2010, Bremen (Germany) in 2012, York (UK) in 2014, L'Aquila (Italy) in 2015, Vienna (Austria) in 2016, Marburg (Germany) in 2017, Toulouse (France) in 2018, Eindhoven (The Netherlands) in 2019, online in 2020 and 2021, and in Nantes (France) in 2022, following a series of six International Workshops on Graph Grammars and Their Application to Computer Science from 1978 to 1998 in Europe andthe USA.

This year, the conference solicited research papers across three categories: research papers, describing new and unpublished contributions to the theory and applications of graph transformation; tool presentation papers, demonstrating the main features and functionalities of graph-based tools; and blue skies papers, reporting on new research directions or ideas that are at an early or emerging stage. From an initial 32 abstract announcements, 29 submissions were received, and were each reviewed by three Programme Committee members and/or additional reviewers. Following an extensive discussion phase, the Programme Committee selected 13 research papers, one tool presentation paper, and two blue skies papers for publication in these proceedings. The topics of the accepted papers cover a wide spectrum, including theoretical approaches to graph transformation, logic and verification for graph transformation, and model transformation, as well as the application of graph transformation in new areas such as bond graphs and graph neural networks.

In addition, we solicited proposals for 'journal-first' talks, allowing authors to present relevant (previously published) work to the community. We accepted three talks under this category. Finally, we were delighted to host invited talks by Dan Ghica (Huawei Research and University of Birmingham) and Mohammad Abdulaziz (King's College London).

We would like to thank everyone who contributed to the success of ICGT 2023, including the members of our Programme Committee, our additional reviewers, and our invited speakers. We are grateful to Reiko Heckel, the Chair of the ICGT Steering Committee, for his valuable suggestions; the organising committee of STAF 2023, for hosting and supporting ICGT 2023; conf.researchr.org, for hosting our website; and EasyChair, for supporting the review process.

May 2023   Maribel Fernández
                                                            Chris Poskitt

Organization

Program Committee Chairs

Maribel Fernández   King's College London, UK
Chris Poskitt   Singapore Management University, Singapore

Program Committee

Nicolas Behr	Université Paris Cité, CNRS, IRIF, France
Paolo Bottoni	Sapienza University of Rome, Italy
Andrea Corradini	Università di Pisa, Italy
Juan de Lara	Universidad Autónoma de Madrid, Spain
Rachid Echahed	CNRS and University of Grenoble, France
Maribel Fernández	King's College London, UK
Holger Giese	Hasso Plattner Institute at the University of Potsdam, Germany
Russ Harmer	CNRS, France
Reiko Heckel	University of Leicester, UK
Wolfram Kahl	McMaster University, Canada
Barbara König	University of Duisburg-Essen, Germany
Leen Lambers	BTU Cottbus - Senftenberg, Germany
Yngve Lamo	Western Norway University of Applied Sciences, Norway
Fernando Orejas	Universitat Politècnica de Catalunya, Spain
Detlef Plump	University of York, UK
Chris Poskitt	Singapore Management University, Singapore
Arend Rensink	University of Twente, The Netherlands
Leila Ribeiro	Universidade Federal do Rio Grande do Sul, Brazil
Andy Schürr	TU Darmstadt, Germany
Pawel Sobocinski	Tallinn University of Technology, Estonia
Gabriele Taentzer	Philipps-Universität Marburg, Germany
Kazunori Ueda	Waseda University, Japan
Jens Weber	University of Victoria, Canada

Additional Reviewers

Barkowsky, Matthias
Barrett, Chris
Courtehoute, Brian
Ehmes, Sebastian
Fritsche, Lars
Ghani, Mustafa
Hadzihasanovic, Amar
Kratz, Maximilian

Maximova, Maria
Sakizloglou, Lucas
Schneider, Sven
Stoltenow, Lars
Strecker, Martin
Söldner, Robert
Vastarini, Federico

Contents

Tool Presentation

Blue Skies

Theoretical Advances

A Monoidal View on Fixpoint Checks[*]

Paolo Baldan[1], Richard Eggert[2(✉)], Barbara König[2], Timo Matt[3],
and Tommaso Padoan[4]

[1] Università di Padova, Padova, Italy
[2] Universität Duisburg-Essen, Duisburg, Germany
richard.eggert@uni-due.de
[3] Universität Duisburg-Essen, Essen, Germany
[4] Università di Trieste, Trieste, Italy

Abstract. Fixpoints are ubiquitous in computer science as they play a
central role in providing a meaning to recursive and cyclic definitions.
Bisimilarity, behavioural metrics, termination probabilities for Markov
chains and stochastic games are defined in terms of least or greatest fixpoints. Here we show that our recent work which proposes a technique
for checking whether the fixpoint of a function is the least (or the largest)
admits a natural categorical interpretation in terms of gs-monoidal categories. The technique is based on a construction that maps a function
to a suitable approximation and the compositionality properties of this
mapping are naturally interpreted as a gs-monoidal functor. This guides
the realisation of a tool, called UDEfix that allows to build functions
(and their approximations) like a circuit out of basic building blocks and
subsequently perform the fixpoints checks. We also show that a slight
generalisation of the theory allows one to treat a new relevant case study:
coalgebraic behavioural metrics based on Wasserstein liftings.

1 Introduction

For the compositional modelling of graphs and graph-like structures it has proven
useful to use the notion of monoidal categories [17], i.e., categories equipped
with a tensor product. There are several extensions of such categories, such as
gs-monoidal categories that have been shown to be suitable for specifying term
rewriting (see e.g. [15,16]). In essence gs-monoidal categories describe graph-
like structures with dedicated input and output interfaces, operators for disjoint
union (tensor), duplication and termination of wires, quotiented by the axioms
satisfied by these operators. Particularly useful are gs-monoidal functors that
preserve such operators and hence naturally describe compositional operations.

We show that gs-monoidal categories and the composition concepts that come
with them can be fruitfully used in a scenario that – at first sight – might seem

Partially supported by the DFG project SpeQt (project number 434050016) and by
the Ministero dell'Università e della Ricerca Scientifica of Italy, under Grant No.
201784YSZ5, PRIN2017 - ASPRA, and by project iNEST, No. J43C22000320006.
[*]

M. Fernández and C. M. Poskitt (Eds.): ICGT 2023, LNCS 13961, pp. 3–21, 2023.
https://doi.org/10.1007/978-3-031-36709-0_1

quite unrelated: methods for fixpoints checks. In particular, we build upon [8] where a theory is proposed for checking whether a fixpoint of a given function is the least (greatest) fixpoint. The theory applies to a variety of fairly diverse application scenarios, such as bisimilarity [21], behavioural metrics [4,10,13,24], termination probabilities for Markov chains [3] and simple stochastic games [11].

More precisely, the theory above deals with non-expansive functions $f : \mathbb{M}^Y \to \mathbb{M}^Y$, where $\mathbb{M}$ is a set of values (more precisely, an MV-chain) and Y is a finite set. The rough idea consists in mapping such functions to corresponding approximations, whose fixpoints can be computed effectively and give insights on the fixpoints of the original function.

We show that the approximation framework and its compositionality properties can be naturally interpreted in categorical terms. This is done by introducing two gs-monoidal categories in which the concrete functions respectively their approximations live as arrows, together with a gs-monoidal functor, called #, mapping one to the other. Besides shedding further light on the theoretical approximation framework of [8], this view guided the realisation of a tool, called UDEfix that allows to build functions (and their approximations) like a circuit out of basic building blocks and subsequently perform the fixpoints checks.

We also show that the functor # can be extended to deal with functions $f : \mathbb{M}^Y \to \mathbb{M}^Y$ where Y is not necessarily finite, becoming a lax functor. We prove some properties of this functor that enable us to give a recipe for finding approximations for a special type of functions: predicate liftings that have been introduced for coalgebraic modal logic [19,22]. This recipe allows us to include a new case study for the machinery for fixpoint checking: coalgebraic behavioural metrics, based on Wasserstein liftings.

The paper is organized as follows: In Sect. 2 we give some high-level motivation, while in Sect. 3 we review the theory from [8]. Subsequently in Sect. 4 we introduce two (gs-monoidal) categories $\mathbb{C}$, $\mathbb{A}$ (of concrete and abstract functions), show that the approximation # is a (lax) functor between these categories and prove some of its properties, which are used to handle predicate liftings (Sect. 5) and behavioural metrics (Sect. 6). Next, we show that the categories $\mathbb{C}$, $\mathbb{A}$ and the functor # are indeed gs-monoidal (Sect. 7) and lastly discuss the tool UDEfix in Sect. 8. We end by giving a conclusion (Sect. 9). Proofs and further material can be found in the full version of the paper [5].

2   Motivation

We start with some motivations for our theory and the tool UDEfix, which is based on it, via a case study on behavioural metrics. We consider probabilistic transition systems (Markov chains) with labelled states, given by a finite set of states X, a function $\delta : X \to \mathcal{D}(X)$ mapping each state $x \in X$ to a probability distribution on X and a labelling function $\ell : X \to \Lambda$, where Λ is a fixed set of labels (for examples see Fig. 1). Our aim is to determine the behavioural distance of two states, whose definition is based on the so-called Kantorovich or Wasserstein lifting [25] that measures the distance of two probability distributions on

X, based on a distance $d: X \times X \to [0, 1]$. In more detail: given d, we define
$d^{\mathcal{D}}: \mathcal{D}(X) \times \mathcal{D}(X) \to [0, 1]$ as

$$d^{\mathcal{D}}(p_1, p_2) = \inf\{ \sum_{x_1, x_2 \in X} d(x_1, x_2) \cdot t(x_1, x_2) \mid t \in \Gamma(p_1, p_2)\}$$

where $\Gamma(p_1, p_2)$ is the set of couplings of p_1, p_2 (i.e., distributions $t: X \times X \to [0, 1]$ such that $\sum_{x_2 \in X} t(x_1, x_2) = p_1(x_1)$, $\sum_{x_1 \in X} t(x_1, x_2) = p_2(x_2)$). The Wasserstein lifting gives in fact the solution of a transport problem, where we interpret p_1, p_2 as the supply respectively demand at each point $x \in X$. Transporting a unit from x_1 to x_2 costs $d(x_1, x_2)$ and t is a transport plan (= coupling) whose marginals are p_1, p_2. In other words: $d^{\mathcal{D}}(p_1, p_2)$ is the cost of the optimal transport plan, moving the supply p_1 to the demand p_2.

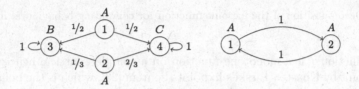

Fig. 1. Two probabilistic transition systems.

The behavioural metric is then defined as the least fixpoint of the function $f: [0, 1]^{X \times X} \to [0, 1]^{X \times X}$ where $f(d)(x_1, x_2) = 1$ if $\ell(x_1) \neq \ell(x_2)$ and $f(d)(x_1, x_2) = d^{\mathcal{D}}(\delta(x_1), \delta(x_2))$ otherwise. For instance, the best transport plan for the system on the left-hand side of Fig. 1 and the distributions $\delta(1), \delta(2)$ is t with $t(3, 3) = 1/3$, $t(3, 4) = 1/6$, $t(4, 4) = 1/2$ and 0 otherwise.

One can observe that the function f can be decomposed as

$$f = \max_\rho \circ (c_k + (\delta \times \delta)^* \circ \min_u \circ \tilde{\mathcal{D}}),$$

where $+$ stands for disjoint union and we use the functions given in Table 1.[1] More concretely, the types of the components and the parameters k, u, ρ are given as follows, where $Y = X \times X$:

- $c_k: [0, 1]^{\emptyset} \to [0, 1]^Y$ where $k(x, x') = 1$ if $\ell(x) \neq \ell(x')$ and 0 otherwise.
- $\tilde{\mathcal{D}}: [0, 1]^Y \to [0, 1]^{\mathcal{D}(Y)}$.
- $\min_u: [0, 1]^{\mathcal{D}(Y)} \to [0, 1]^{\mathcal{D}(X) \times \mathcal{D}(X)}$ where $u: \mathcal{D}(Y) \to \mathcal{D}(X) \times \mathcal{D}(X)$, $u(t) = (p, q)$ with $p(x) = \sum_{x' \in X} t(x, x')$, $q(x) = \sum_{x' \in X} t(x', x)$.
- $(\delta \times \delta)^*: [0, 1]^{\mathcal{D}(X) \times \mathcal{D}(X)} \to [0, 1]^Y$.
- $\max_\rho: [0, 1]^{Y+Y} \to [0, 1]^Y$ where $\rho: Y + Y \to Y$ is the obvious map from the coproduct (disjoint union) $Y + Y$ to Y.

In fact this decomposition can be depicted diagrammatically, as in Fig. 2.

---

[1] If the underlying sets are infinite, min, max can be replaced by inf, sup.

Table 1. Basic functions of type $M^Y \to M^Z$, $a\colon Y \to M$.

Function	c_k	g^*	$\min_u$	$\max_u$	$\mathrm{av}_D = \tilde{D}$
	$k\colon Z \to M$	$g\colon Z \to Y$	$u\colon Y \to Z$	$u\colon Y \to Z$	$M = [0,1]$, $Z = \mathcal{D}(Y)$
Name	constant	reindexing	minimum	maximum	expectation
$a \mapsto \dots$	k	$a \circ g$	$\lambda z.\min\limits_{u(y)=z} a(y)$	$\lambda z.\max\limits_{u(y)=z} a(y)$	$\lambda z.\lambda y. \sum\limits_{y \in Y} z(y) \cdot a(y)$

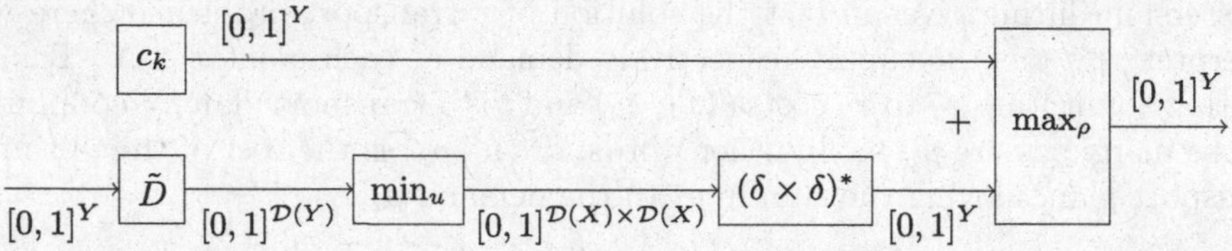

Fig. 2. Decomposition of the fixpoint function for computing behavioural metrics.

The function f is a monotone function on a complete lattice, hence it has a least fixpoint by Knaster-Tarski's fixpoint theorem [23], which is the behavioural metric. By giving a transport plan as above, it is possible to provide an upper bound for the Wasserstein lifting and hence there are strategy iteration algorithms that can approach a fixpoint from above. The problem with these algorithms is that they might get stuck at a fixpoint that is not the least. Hence, it is essential to be able to determine whether a given fixpoint is indeed the smallest one (cf. [2]).

Consider for instance the transition system in Fig. 1 on the right. It contains two states $1,2$ on a cycle. In fact these two states should be indistinguishable and hence $d(1,2) = d(2,1) = 0$ if $d = \mu f$ is the least fixpoint of f. However, the metric a with $a(1,2) = a(2,1) = 1$ (0 otherwise) is also a fixpoint and the question is how to determine that it is not the least.

For this, we use the techniques developed in [8] that require in particular that f is non-expansive (i.e., given two metrics d_1, d_2, the sup-distance of $f(d_1), f(d_2)$ is smaller or equal than the sup-distance of d_1, d_2). In this case we can associate f with an approximation $f_\#^a$ on subsets of $X \times X$ such that, given $Y' \subseteq X \times X$, $f_\#^a(Y')$ intuitively contains all pairs (x_1, x_2) such that, decreasing function a by some value δ over Y', resulting in a function b (defined as $b(x_1, x_2) = a(x_1, x_2) - \delta$ if $(x_1, x_2) \in Y'$ and $b(x_1, x_2) = a(x_1, x_2)$ otherwise) and applying f, we obtain a function $f(b)$, where the same decrease took place at (x_1, x_2) (i.e., $f(b)(x_1, x_2) = f(a)(x_1, x_2) - \delta$). More concretely, here $f_\#^a(\{(1,2)\}) = \{(2,1)\}$, since a decrease at $(1,2)$ will cause a decrease at $(2,1)$ in the next iteration. In fact the greatest fixpoint of $f_\#^a$ (here: $\{(1,2),(2,1)\}$) gives us those elements that have a potential for decrease (intuitively there is "slack" or "wiggle room") and form a vicious cycle as above. It holds that a is the least fixpoint of f iff the the greatest fixpoint of $f_\#^a$ is the empty set, a non-trivial result [6,8].

The importance of the decomposition stems from the fact that the approximation is in fact compositional, that is $f_{\#}^a$ can be built out of the approximations of $\max_\rho$, c_k, $(\delta \times \delta)^*$, $\min_u$, $\tilde{D} = \mathrm{av}_D$, which can be easily determined (see [8]). For general functors, beyond the distribution functor, the characterization is however still missing and will be provided in this paper.

We anticipate that in our tool UDEfix we can draw a diagram as in Fig. 2, from which the approximation and its greatest fixpoint is automatically computed in a compositional way, allowing us to perform such fixpoint checks.

3 Preliminaries

This section reviews some background used throughout the paper. This includes the basics of lattices and MV-algebras, where the functions of interest take values. We also recap some results from [8] useful for detecting if a fixpoint of a given function is the least (or greatest).

We will also need some standard notions from *category theory*, in particular categories, functors and natural transformations. The definition of (strict) gs-monoidal categories is spelled out in detail in Definition 7.1.

For sets X, Y, we denote by $\mathcal{P}(X)$ the powerset of X and $\mathcal{P}_f(X)$ the set of finite subsets of X. The set of functions from X to Y is denoted by Y^X.

A *partially ordered set* $(P, \sqsubseteq)$ is often denoted simply as P, omitting the order relation. The *join* and the *meet* of a subset $X \subseteq P$ (if they exist) are denoted $\bigsqcup X$ and $\bigsqcap X$. We write $x \sqsubset y$ when $x \sqsubseteq y$ and $x \neq y$.

A *complete lattice* is a partially ordered set $(\mathbb{L}, \sqsubseteq)$ such that each subset $X \subseteq \mathbb{L}$ admits a join $\bigsqcup X$ and a meet $\bigsqcap X$. A complete lattice $(\mathbb{L}, \sqsubseteq)$ always has a least element $\bot = \bigsqcap \mathbb{L}$ and a greatest element $\top = \bigsqcup \mathbb{L}$.

A function $f : \mathbb{L} \to \mathbb{L}$ is *monotone* if for all $l, l' \in \mathbb{L}$, if $l \sqsubseteq l'$ then $f(l) \sqsubseteq f(l')$. By Knaster-Tarski's theorem [23, Theorem 1], any monotone function on a complete lattice has a least fixpoint μf and a greatest fixpoint νf.

For a set Y and a complete lattice $\mathbb{L}$, the set of functions $\mathbb{L}^Y$, with pointwise order (for $a, b \in \mathbb{L}^Y$, $a \sqsubseteq b$ if $a(y) \sqsubseteq b(y)$ for all $y \in Y$), is a complete lattice.

We are also interested in the set of finitely supported probability distributions $\mathcal{D}(Y) \subseteq [0,1]^Y$, i.e., functions $\beta : Y \to [0,1]$ with finite support such that $\sum_{y \in Y} \beta(y) = 1$.

An *MV-algebra* [18] is a tuple $\mathbb{M} = (M, \oplus, 0, \overline{(\cdot)})$ where $(M, \oplus, 0)$ is a commutative monoid and $\overline{(\cdot)} : M \to M$ maps each element to its *complement*, such that for all $x, y \in M$ (i) $\overline{\overline{x}} = x$; (ii) $x \oplus \overline{0} = \overline{0}$; (iii) $\overline{(\overline{x} \oplus y)} \oplus y = \overline{(\overline{y} \oplus x)} \oplus x$.

We define $1 = \overline{0}$ and subtraction $x \ominus y = \overline{\overline{x} \oplus y}$.

MV-algebras are endowed with a partial order, the so-called *natural order*, defined for $x, y \in M$, by $x \sqsubseteq y$ if $x \oplus z = y$ for some $z \in M$. When $\sqsubseteq$ is total, $\mathbb{M}$ is called an *MV-chain*. We will write $\mathbb{M}$ instead of M.

The natural order gives an MV-algebra a lattice structure where $\bot = 0$, $\top = 1$, $x \sqcup y = (x \ominus y) \oplus y$ and $x \sqcap y = \overline{\overline{x} \sqcup \overline{y}} = x \ominus (x \ominus y)$. We call the MV-algebra *complete* if it is a complete lattice, which is not true in general, e.g., $([0,1] \cap \mathbb{Q}, \leq)$.

Example 3.1. *A prototypical MV-algebra is* $([0,1], \oplus, 0, \overline{(\cdot)})$ *where* $x \oplus y = \min\{x+y, 1\}$, $\overline{x} = 1 - x$ *and* $x \ominus y = \max\{0, x-y\}$ *for* $x, y \in [0,1]$. *The natural order is* $\leq$ *(less or equal) on the reals. Another example is* $K = (\{0, \dots, k\}, \oplus, 0, \overline{(\cdot)})$ *where* $n \oplus m = \min\{n+m, k\}$, $\overline{n} = k - n$ *and* $n \ominus m = \max\{n-m, 0\}$ *for* $n, m \in \{0, \dots, k\}$. *Both MV-algebras are complete and MV-chains.*

We next briefly recap the theory from [8] which will be helpful in the paper for checking whether a fixpoint is the least or the greatest fixpoint of some underlying endo-function. For the purposes of the present paper we actually need a generalisation of the theory which provides the approximation also for functions with an infinite domain (while the theory in [8] was restricted to finite sets). Hence in the following, sets Y and Z are possibly infinite.

Given $a \in \mathbb{M}^Y$ we define its *norm* as $\|a\| = \sup\{a(y) \mid y \in Y\}$. A function $f : \mathbb{M}^Y \to \mathbb{M}^Z$ is *non-expansive* if for all $a, b \in \mathbb{M}^Y$ it holds $\|f(b) \ominus f(a)\| \sqsubseteq \|b \ominus a\|$. It can be seen that non-expansive functions are monotone. A number of standard operators are non-expansive (e.g., constants, reindexing, max and min over a relation, average in Table 1), and non-expansiveness is preserved by composition and disjoint union (see [8]). Given $Y' \subseteq Y$ and $\delta \in \mathbb{M}$, we write $\delta_{Y'}$ for the function defined by $\delta_{Y'}(y) = \delta$ if $y \in Y'$ and $\delta_{Y'}(y) = 0$, otherwise.

Let $f : \mathbb{M}^Y \to \mathbb{M}^Y$, $a \in \mathbb{M}^Y$ and $0 \sqsubset \delta \in \mathbb{M}$. Define $[Y]^a = \{y \in Y \mid a(y) \neq 0\}$ and consider the functions $\alpha^{a,\delta} : \mathcal{P}([Y]^a) \to [a \ominus \delta, a]$ and $\gamma^{a,\delta} : [a \ominus \delta, a] \to \mathcal{P}([Y]^a)$, defined, for $Y' \in \mathcal{P}([Y]^a)$ and $b \in [a \ominus \delta, a]$, by

$$\alpha^{a,\delta}(Y') = a \ominus \delta_{Y'} \qquad \gamma^{a,\delta}(b) = \{y \in [Y]^a \mid a(y) \ominus b(y) \sqsupseteq \delta\}.$$

Here $[a, b] = \{c \in \mathbb{M}^Y \mid a \sqsubseteq c \sqsubseteq b\}$. In fact, for suitable values of δ, the functions $\alpha^{a,\delta}, \gamma^{a,\delta}$ form a Galois connection.

For a non-expansive function $f : \mathbb{M}^Y \to \mathbb{M}^Z$ and $\delta \in \mathbb{M}$, define $f_\#^{a,\delta} : \mathcal{P}([Y]^a) \to \mathcal{P}([Z]^{f(a)})$ as $f_\#^{a,\delta} = \gamma^{f(a),\delta} \circ f \circ \alpha^{a,\delta}$. The function $f_\#^{a,\delta}$ is antitone in the parameter δ and we define the *a-approximation* of f as

$$f_\#^a = \bigcup_{\delta \sqsupset 0} f_\#^{a,\delta}.$$

For finite sets Y and Z there exists a suitable value $\iota_f^a \sqsupset 0$, such that all functions $f_\#^{a,\delta}$ for $0 \sqsubset \delta \sqsubseteq \iota_f^a$ are equal. Here, the a-approximation is given by $f_\#^a = f_\#^{a,\delta}$ for $\delta = \iota_f^a$.

Intuitively, given some Y', the set $f_\#^a(Y')$ contains the points where a decrease of the values of a on the points in Y' "propagates" through the function f. The greatest fixpoint of $f_\#^a$ gives us the subset of Y where such a decrease is propagated in a cycle (so-called "vicious cycle"). Whenever $\nu f_\#^a$ is non-empty, one can argue that a cannot be the least fixpoint of f since we can decrease the value in all elements of $\nu f_\#^a$, obtaining a smaller prefixpoint. Interestingly, for non-expansive functions, it is shown in [8] that also the converse holds, i.e., emptiness of the greatest fixpoint of $f_\#^a$ implies that a is the least fixpoint.

Theorem 3.2 (soundness and completeness for fixpoints). *Let* $\mathbb{M}$ *be a complete MV-chain,* Y *a finite set and* $f : \mathbb{M}^Y \to \mathbb{M}^Y$ *be a non-expansive function. Let* $a \in \mathbb{M}^Y$ *be a fixpoint of* f. *Then* $\nu f^a_\# = \emptyset$ *if and only if* $a = \mu f$.

Using the above theorem we can check whether some fixpoint a of f is the least fixpoint. Whenever a is a fixpoint, but not yet the least fixpoint of f, it can be decreased by a fixed value in $\mathbb{M}$ (see [8] for the details) on the points in $\nu f^a_\#$ to obtain a smaller pre-fixpoint.

Lemma 3.3. *Let* $\mathbb{M}$ *be a complete MV-chain,* Y *a finite set and* $f : \mathbb{M}^Y \to \mathbb{M}^Y$ *a non-expansive function,* $a \in \mathbb{M}^Y$ *a fixpoint of* f, *and let* $f^a_\#$ *be the corresponding a-approximation. If* a *is not the least fixpoint and thus* $\nu f^a_\# \neq \emptyset$ *then there is* $0 \sqsubset \delta \in \mathbb{M}$ *such that* $a \ominus \delta_{\nu f^a_\#}$ *is a pre-fixpoint of* f.

The above theory can easily be dualised [8].

4   A Categorical View of the Approximation Framework

The framework from [8], summarized in the previous section, is not based on category theory, but – as we shall see – can be naturally reformulated in a categorical setting. In particular, casting the compositionality results into a monoidal structure (see Sect. 7) is a valuable basis for our tool. But first, we will show how the operation # of taking the a-approximation of a function can be seen as a (lax) functor between two categories: a concrete category $\mathbb{C}$ whose arrows are the non-expansive functions for which we seek the least (or greatest) fixpoint and an abstract category $\mathbb{A}$ whose arrows are the corresponding approximations.

More precisely, recall that given a non-expansive function $f : \mathbb{M}^Y \to \mathbb{M}^Z$, the approximation of f is relative to a fixed map $a \in \mathbb{M}^Y$. Hence objects in $\mathbb{C}$ are elements $a \in \mathbb{M}^Y$ and an arrow from $a \in \mathbb{M}^Y$ to $b \in \mathbb{M}^Z$ is a non-expansive function $f : \mathbb{M}^Y \to \mathbb{M}^Z$ required to map a into b. The approximations instead live in $\mathbb{A}$. Recall that the approximation is $f^a_\# : \mathcal{P}([Y]^a) \to \mathcal{P}([Z]^b)$. Since their domains and codomains are dependent again on a map a, we still employ elements of $\mathbb{M}^Y$ as objects, but functions between powersets as arrows.

Definition 4.1 (concrete and abstract categories). *The concrete category* $\mathbb{C}$ *has as objects maps* $a \in \mathbb{M}^Y$ *where* Y *is a (possibly infinite) set. Given* $a \in \mathbb{M}^Y$, $b \in \mathbb{M}^Z$ *an arrow* $f : a \dashrightarrow b$ *is a non-expansive function* $f : \mathbb{M}^Y \to \mathbb{M}^Z$, *such that* $f(a) = b$. *The abstract category* $\mathbb{A}$ *has again maps* $a \in \mathbb{M}^Y$ *as objects. Given* $a \in \mathbb{M}^Y$, $b \in \mathbb{M}^Z$ *an arrow* $f : a \dashrightarrow b$ *is a monotone (wrt. inclusion) function* $f : \mathcal{P}([Y]^a) \to \mathcal{P}([Z]^b)$. *Arrow composition and identities are the obvious ones.*

The lax functor $\# : \mathbb{C} \to \mathbb{A}$ *is defined as follows: for an object* $a \in \mathbb{M}^Y$, *we let* $\#(a) = a$ *and, given an arrow* $f : a \dashrightarrow b$, *we let* $\#(f) = f^a_\#$.

Note that abstract arrows are dashed ($\dashrightarrow$), while the underlying functions are represented by standard arrows ($\to$).

Lemma 4.2 (well-definedness). *The categories* $\mathbb{C}$ *and* $\mathbb{A}$ *are well-defined and* $\#$ *is a lax functor, i.e., identities are preserved and* $\#(f \circ g) \subseteq \#(f) \circ \#(g)$ *for composable arrows* f, g *in* $\mathbb{C}$.

It will be convenient to restrict to the subcategory of $\mathbb{C}$ where arrows are reindexings and to subcategories of $\mathbb{C}$, $\mathbb{A}$ with maps on finite sets.

Definition 4.3 (reindexing subcategory). *We denote by* $\mathbb{C}^*$ *the lluf[2] subcategory of* $\mathbb{C}$ *where arrows are reindexing, i.e., given objects* $a \in \mathbb{M}^Y$, $b \in \mathbb{M}^Z$ *we consider only arrows* $f : a \dashrightarrow b$ *such that* $f = g^*$ *for some* $g \colon Z \to Y$ *(hence, in particular,* $b = g^*(a) = a \circ g$*). We denote* $E \colon \mathbb{C}^* \hookrightarrow \mathbb{C}$ *the embedding functor.*

Definition 4.4 (finite subcategories). *We denote by* $\mathbb{C}_f$, $\mathbb{A}_f$ *the full subcategories of* $\mathbb{C}, \mathbb{A}$ *where objects are of the kind* $a \in \mathbb{M}^Y$ *for a finite set* Y.

Lemma 4.5. *The lax functor* $\# \colon \mathbb{C} \to \mathbb{A}$ *restricts to* $\# \colon \mathbb{C}_f \to \mathbb{A}_f$, *which is a (proper) functor.*

5   Predicate Liftings

In this section we discuss how predicate liftings [19,22] can be integrated into our theory. In this context the idea is to view a map in $\mathbb{M}^Y$ as a predicate over Y with values in $\mathbb{M}$ (e.g., if $\mathbb{M} = \{0, 1\}$ we obtain Boolean predicates). Then, given a functor F, a predicate lifting transforms a predicate over Y (a map in $\mathbb{M}^Y$), to a predicate over FY (a map in $\mathbb{M}^{FY}$). It must be remarked that every complete MV-algebra is a quantale[3] with respect to $\oplus$ and the inverse of the natural order (see [14]) and predicate liftings for arbitrary quantales have been studied, for instance, in [9].

First, we characterise which predicate liftings are non-expansive and second, derive their approximations. We will address both these issues in this section and then use predicate liftings to define behavioural metrics in Sect. 6.

The fact that there are some functors F, for which FY is infinite, even if Y is finite, is the reason why the categories $\mathbb{C}$ and $\mathbb{A}$ also include infinite sets. However note, that the resulting fixpoint function will be always defined for finite sets, although intermediate functions might not conform to this.

Definition 5.1 (predicate lifting). *Given a functor* $F \colon \mathbf{Set} \to \mathbf{Set}$, *a predicate lifting is a family of functions* $\tilde{F}_Y \colon \mathbb{M}^Y \to \mathbb{M}^{FY}$ *(where* Y *is a set), such that for* $g \colon Z \to Y$, $a \colon Y \to \mathbb{M}$ *it holds that* $(Fg)^*(\tilde{F}_Y(a)) = \tilde{F}_Z(g^*(a))$.

That is, predicate liftings must commute with reindexings. The index Y will be omitted if clear from the context. Such predicate liftings are in one-to-one

---

[2] A *lluf sub-category* is a sub-category that contains all objects.

[3] A quantale is a complete lattice with an associative operator that distributes over arbitrary joins.

correspondence to so called evaluation maps $ev\colon F\mathbb{M} \to \mathbb{M}$.[4] Given ev, we define the corresponding lifting to be $\tilde{F}(a) = ev \circ Fa\colon FY \to \mathbb{M}$, where $a\colon Y \to \mathbb{M}$.

In the sequel we will only consider well-behaved liftings [4,9], i.e., we require that (i) $\tilde{F}$ is monotone; (ii) $\tilde{F}(0_Y) = 0_{FY}$ where 0 is the constant 0-function; (iii) $\tilde{F}(a \oplus b) \sqsubseteq \tilde{F}(a) \oplus \tilde{F}(b)$ for $a, b\colon Y \to \mathbb{M}$; (iv) F preserves weak pullbacks.

We aim to have not only monotone, but non-expansive liftings.

Lemma 5.2. *Let $ev\colon F\mathbb{M} \to \mathbb{M}$ be an evaluation map and assume that its corresponding lifting $\tilde{F}\colon \mathbb{M}^Y \to \mathbb{M}^{FY}$ is well-behaved. Then $\tilde{F}$ is non-expansive iff for all $\delta \in \mathbb{M}$ it holds that $\tilde{F}\delta_Y \sqsubseteq \delta_{FY}$.*

Example 5.3. *We consider the (finitely supported) distribution functor $\mathcal{D}$ that maps a set X to all maps $p\colon X \to [0,1]$ that have finite support and satisfy $\sum_{x \in X} p(x) = 1$. (Here $\mathbb{M} = [0,1]$.) One evaluation map is $ev\colon \mathcal{D}[0,1] \to [0,1]$ with $ev(p) = \sum_{r \in [0,1]} r \cdot p(r)$, where p is a distribution on $[0,1]$ (expectation). It is easy to see that $\tilde{\mathcal{D}}$ is well-behaved and non-expansive. The latter follows from $\tilde{\mathcal{D}}(\delta_Y) = \delta_{\mathcal{D}Y}$.*

Example 5.4. *Another example is given by the finite powerset functor $\mathcal{P}_f$. We are given the evaluation map $ev\colon \mathcal{P}_f\mathbb{M} \to \mathbb{M}$, defined for finite $S \subseteq \mathbb{M}$ as $ev(S) = \max S$, where $\max \emptyset = 0$. The lifting $\tilde{\mathcal{P}}_f$ is well-behaved (see [4]) and non-expansive. To show the latter, observe that $\tilde{\mathcal{P}}_f(\delta_Y) = \delta_{\mathcal{P}_f(Y)\setminus\{\emptyset\}} \sqsubseteq \delta_{\mathcal{P}_f(Y)}$.*

Non-expansive predicate liftings can be seen as functors $\tilde{F}\colon \mathbb{C}^* \to \mathbb{C}^*$. To be more precise, $\tilde{F}$ maps an object $a \in \mathbb{M}^Y$ to $\tilde{F}(a) \in \mathbb{M}^{FY}$ and an arrow $g^*\colon a \dashrightarrow a \circ g$, , where $g\colon Z \to Y$, to $(Fg)^*\colon \tilde{F}a \dashrightarrow \tilde{F}(a \circ g)$.

Proposition 5.5. *Let $\tilde{F}$ be a (non-expansive) predicate lifting. There is a natural transformation $\beta\colon \#E \Rightarrow \#E\tilde{F}$ between (lax) functors $\#E, \#E\tilde{F}\colon \mathbb{C}^* \to \mathbb{A}$, whose components, for $a \in \mathbb{M}^Y$, are $\beta_a\colon a \dashrightarrow \tilde{F}(a)$ in $\mathbb{A}$, defined by $\beta_a(U) = \tilde{F}_\#^a(U)$ for $U \subseteq [Y]^a$.*

That is, the following diagrams commute for every $g\colon Z \to Y$ (on the left the diagram with formal arrows, omitting the embedding functor E, and on the right the functions with corresponding domains). Note that $\#(g) = g^{-1}$.

$$
\begin{array}{ccc}
\#(a) & \xdashrightarrow{\ \#(g^*)\ } & \#(a \circ g) \\
\beta_a \downarrow & & \downarrow \beta_{a \circ g} \\
\#(\tilde{F}a) & \xdashrightarrow{\ \#(\tilde{F}(g^*))\ } & \#(\tilde{F}(a \circ g))
\end{array}
\qquad
\begin{array}{ccc}
\mathcal{P}([Y]^a) & \xrightarrow{\ g^{-1}\ } & \mathcal{P}([Z]^{a \circ g}) \\
\tilde{F}_\#^a \downarrow & & \downarrow \tilde{F}_\#^{a \circ g} \\
\mathcal{P}([FY]^{\tilde{F}(a)}) & \xrightarrow{\ (Fg)^{-1}\ } & \mathcal{P}([FZ]^{\tilde{F}(a \circ g)})
\end{array}
$$

---

[4] This follows from the Yoneda lemma, see e.g. [17].

6   Wasserstein Lifting and Behavioural Metrics

In this section we show how the framework for fixpoint checking described before can be used to deal with coalgebraic behavioural metrics.

We build on [4], where an approach is proposed for canonically defining a behavioural pseudometric for coalgebras of a functor $F: \mathbf{Set} \to \mathbf{Set}$, that is, for functions of the form $\xi: X \to FX$ where X is a set. Intuitively ξ specifies a transition system whose branching type is given by F. Given such a coalgebra ξ, the idea is to endow X with a pseudo-metric $d_\xi: X \times X \to \mathbb{M}$ defined as the least fixpoint of the map $d \mapsto d^F \circ (\xi \times \xi)$ where $\_^F$ lifts a metric $d: X \times X \to \mathbb{M}$ to a metric $d^F: FX \times FX \to \mathbb{M}$. Here we focus on the so-called Wasserstein lifting and show how approximations of the functions involved in the definition of the pseudometric can be determined.

6.1   Wasserstein Lifting

Hereafter, F denotes a fixed endofunctor on $\mathbf{Set}$ and $\xi: X \to FX$ is a coalgebra over a finite set X. We also fix a well-behaved non-expansive predicate lifting $\tilde{F}$.

In order to define a Wasserstein lifting, a first ingredient is that of a coupling. Given $t_1, t_2 \in FX$ a *coupling* of t_1 and t_2 is an element $t \in F(X \times X)$, such that $F\pi_i(t) = t_i$ for $i = 1, 2$, where $\pi_i: X \times X \to X$ are the projections. We write $\Gamma(t_1, t_2)$ for the set of all such couplings.

Definition 6.1 (Wasserstein lifting). *The Wasserstein lifting* $\_^F: \mathbb{M}^{X \times X} \to \mathbb{M}^{FX \times FX}$ *is defined for* $d: X \times X \to \mathbb{M}$ *and* $t_1, t_2 \in FX$ *as*

$$d^F(t_1, t_2) = \inf_{t \in \Gamma(t_1, t_2)} \tilde{F}d(t)$$

For more intuition on the Wasserstein lifting see Sect. 2. Note that a coupling correspond to a transport plan. It can be shown that for well-behaved $\tilde{F}$, the lifting preserves pseudometrics (see [4,9]).

In order to make the theory for fixpoint checks effective we will need to restrict to a subclass of liftings.

Definition 6.2 (finitely coupled lifting). *We call a lifting* $\tilde{F}$ *finitely coupled if for all* X *and* $t_1, t_2 \in FX$ *there exists a finite* $\Gamma'(t_1, t_2) \subseteq \Gamma(t_1, t_2)$, *which can be computed given* t_1, t_2, *such that* $\inf_{t \in \Gamma(t_1,t_2)} \tilde{F}d(t) = \min_{t \in \Gamma'(t_1,t_2)} \tilde{F}d(t)$.

Observe that whenever the infimum above is a minimum, there is trivially a finite $\Gamma'(t_1, t_2)$. We however ask that there is an effective way to determine it.

The lifting in Example 5.4 (for the finite powerset functor) is obviously finitely coupled. For the lifting $\tilde{\mathcal{D}}$ from Example 5.3 we note that the set of couplings $t \in \Gamma(t_1, t_2)$ forms a polytope with a finite number of vertices, which can be effectively computed and $\Gamma'(t_1, t_2)$ consists of these vertices. The infimum (minimum) is obtained at one of these vertices [1, Remark 4.5].

6.2   A Compositional Representation

As mentioned above, for a coalgebra $\xi\colon X \to FX$ the behavioural pseudometric $d\colon X \times X \to \mathbb{M}$ arises as the least fixpoint of $\mathcal{W} = (\xi \times \xi)^* \circ (\_^F)$ where $(\_^F)$ is the Wasserstein lifting.

Example 6.3. *We can recover the motivating example from Sect. 2 by setting* $\mathbb{M} = [0,1]$ *and using the functor* $FX = \Lambda \times \mathcal{D}(X)$, *where* Λ *is a fixed set of labels. We observe that couplings of* $(a_1, p_1), (a_2, p_2) \in FX$ *only exist if* $a_1 = a_2$ *and – if they do not exist – the Wasserstein distance is the empty infimum, hence 1. If* $a_1 = a_2$, *couplings correspond to the usual Wasserstein couplings of* p_1, p_2 *and the least fixpoint of* $\mathcal{W}$ *equals the behavioural metrics, as explained in Sect. 2.*

Note that we do not use a discount factor to ensure contractivity and hence the fixpoint might not be unique. Thus, given some fixpoint d, the d-approximation $\mathcal{W}_\#^d$ can be used for checking whether $d = \mu \mathcal{W}$.

In the rest of the section we show how $\mathcal{W}$ can be decomposed into basic components and study the corresponding approximation.

The Wasserstein lifting can be decomposed as $\_^F = \min_u \circ \tilde{F}$ where $\tilde{F}\colon \mathbb{M}^{X \times X} \to \mathbb{M}^{F(X \times X)}$ is the predicate lifting – which we require to be non-expansive (cf. Lemma 5.2) – and $\min_u$ is the minimum over the coupling function $u\colon F(X \times X) \to FX \times FX$ defined as $u(t) = (F\pi_1(t), F\pi_2(t))$, which means that $\min_u\colon \mathbb{M}^{F(X \times X)} \to \mathbb{M}^{FX \times FX}$ (see Table 1).

We can now derive the corresponding d-approximation.

Proposition 6.4. *Assume that* $\tilde{F}$ *is finitely coupled. Let* $Y = X \times X$, *where* X *is finite. For* $d \in \mathbb{M}^Y$ *and* $Y' \subseteq [Y]^d$ *we have*

$$\mathcal{W}_\#^d(Y') = \{(x,y) \in [Y]^d \mid \exists t \in \tilde{F}_\#^d(Y'), u(t) = (\xi(x), \xi(y)),$$
$$\bar{F}d(t) = \min_{t' \in \Gamma(\xi(x), \xi(y))} \bar{F}d(t')\}.$$

Intuitively the statement of Proposition 6.4 means that the minimum must be reached in a coupling based on Y'.

For using the above result we next characterize $\tilde{F}_\#^d(Y')$. We rely on the fact that d can be decomposed into $d = \pi_1 \circ \bar{d}$, where the projection π_1 is independent of d and $\bar{d}$ is dependent on Y', and exploit the natural transformation in Proposition 5.5.

Proposition 6.5. *We fix* $Y' \subseteq Y$. *Let* $\pi_1\colon \mathbb{M} \times \{0,1\} \to \mathbb{M}$ *be the projection to the first component and* $\bar{d}\colon Y \to \mathbb{M} \times \{0,1\}$ *with* $\bar{d}(y) = (d(y), \chi_{Y'}(y))$ *where* $\chi_{Y'}\colon Y \to \{0,1\}$ *is the characteristic function of* Y'. *Then* $\tilde{F}_\#^d(Y') = (F\bar{d})^{-1}(\tilde{F}_\#^{\pi_1}((\mathbb{M}\backslash\{0\}) \times \{1\}))$.

Here $\tilde{F}_\#^{\pi_1}((\mathbb{M}\backslash\{0\}) \times \{1\}) \subseteq F(\mathbb{M} \times \{0,1\})$ is independent of d and has to be determined only once for every predicate lifting $\tilde{F}$. We will show how this set looks like for our example functors.

Lemma 6.6. *Consider the lifting of the distribution functor presented in Example 5.3 and let $Z = [0,1] \times \{0,1\}$. Then we have*

$$\tilde{D}_{\#}^{\pi_1}((0,1] \times \{1\}) = \{p \in \mathcal{D}Z \mid supp(p) \in (0,1] \times \{1\}\}.$$

This means intuitively that a decrease or "slack" can exactly be propagated for elements whose probabilities are strictly larger than 0.

Lemma 6.7. *Consider the lifting of the finite powerset functor from Example 5.4 and let $Z = \mathbb{M} \times \{0,1\}$. Then we have*

$$(\tilde{\mathcal{P}}_f)_{\#}^{\pi_1}((\mathbb{M}\backslash\{0\}) \times \{1\}) = \{S \in [\mathcal{P}_f Z]^{\tilde{\mathcal{P}}_f \pi_1} \mid \exists (s,1) \in S \; \forall (s',0) \in S : \; s \sqsupset s'\}.$$

The idea is that the maximum of a set S decreases if we decrease at least one its values and all values which are not decreased are strictly smaller.

Remark 8. Note that $\#$ is a functor on the subcategory $\mathbb{C}_f$, while some liftings (e.g., the one for the distribution functor) work with infinite sets. In this case, given a finite set Y, we actually focus on a finite $D \subseteq FY$. (This is possible since we consider coalgebras with finite state space and assume that all liftings are finitely coupled.) Then we consider $\tilde{F}_Y \colon \mathbb{M}^Y \to \mathbb{M}^{FY}$ and $e \colon D \hookrightarrow FY$ (the embedding of D into FY). We set $f = e^* \circ \tilde{F}_Y$. Given $a \colon Y \to \mathbb{M}$, we view f as an arrow $a \dashrightarrow \tilde{F}(a) \circ e$ in $\mathbb{C}$. The approximation in this subsection adapts to the "reduced" lifting, which can be seen as follows (cf. [5]: $\#$ preserves composition if one of the arrows is a reindexing):

$$f_{\#}^a = \#(f) = \#(e^* \circ \tilde{F}_Y) = \#(e^*) \circ \#(\tilde{F}_Y) = e^{-1} \circ \#(\tilde{F}_Y) = \#(\tilde{F}_Y) \cap D.$$

7   GS-Monoidality

We will now show that the categories $\mathbb{C}_f$ and $\mathbb{A}_f$ can be turned into gs-monoidal categories. This will give us a way to assemble functions and their approximations compositionally and this method will form the basis for the tool. We first define gs-monoidal categories in detail:

Definition 7.1. *A strict gs-monoidal category is a strict symmetric monoidal category, where $\otimes$ denotes the tensor and e its unit and symmetries are given by $\rho_{a,b} \colon a \otimes b \to b \otimes a$. For every object a there exist morphisms $\nabla_a \colon a \to a \times a$ (duplicator) and $!_a \colon a \to e$ (discharger) satisfying the axioms given below. (See also the visualizations as string diagrams in Fig. 3.)*

1. *functoriality of tensor:*
   - $(g \otimes g') \circ (f \otimes f') = (g \circ f) \otimes (g' \circ f')$
   - $id_{a \otimes b} = id_a \otimes id_b$
2. *monoidality:*
   - $(f \otimes g) \otimes h = f \otimes (g \otimes h)$
   - $f \otimes id_e = f = id_e \otimes f$

3. *naturality:*
   - $(f' \otimes f) \circ \rho_{a,a'} = \rho_{b,b'} \circ (f \otimes f')$
4. *symmetry:*
   - $\rho_{e,e} = id_e$
   - $\rho_{b,a} \circ \rho_{a,b} = id_{a \otimes b}$
   - $\rho_{a \otimes b, c} = (\rho_{a,c} \otimes id_b) \circ (id_a \otimes \rho_{b,c})$
5. *gs-monoidality:*
   - $!_e = \nabla_e = id_e$
   - *coherence axioms:*
     - $(id_a \otimes \nabla_a) \circ \nabla_a = (\nabla_a \otimes id_a) \circ \nabla_a$
     - $id_a = (id_a \otimes !_a) \circ \nabla_a$
     - $\rho_{a,a} \circ \nabla_a = \nabla_a$
   - *monoidality axioms:*
     - $!_{a \otimes b} = !_a \otimes !_b$
     - $(id_a \otimes \rho_{a,b} \otimes id_b) \circ (\nabla_a \otimes \nabla_b) = \nabla_{a \otimes b}$
       (or, equivalently, $\nabla_a \otimes \nabla_b = (id_a \otimes \rho_{b,a} \otimes id_b) \circ \nabla_{a \otimes b}$)

A functor $\#\colon \mathbb{C} \to \mathbb{D}$ is gs-monoidal if the following holds:

1. $\mathbb{C}$ *and $\mathbb{D}$ are gs-monoidal categories*
2. *monoidality:*
   - $\#(e) = e'$
   - $\#(a \otimes b) = \#(a) \otimes' \#(b)$
3. *symmetry:*
   - $\#(\rho_{a,b}) = \rho'_{\#(a),\#(b)}$
4. *gs-monoidality:*
   - $\#(!_a) = !'_{\#(a)}$
   - $\#(\nabla_a) = \nabla'_{\#(a)}$

where the primed operators are from the category $\mathbb{D}$, the others from $\mathbb{C}$.

In fact, in order to obtain strict gs-monoidal categories with disjoint union, we will work with the skeleton categories where every finite set Y is represented by an isomorphic copy $\{1, \ldots, |Y|\}$. This enables us to make disjoint union strict, i.e., associativity holds on the nose and not just up to isomorphism. In particular for finite sets Y, Z, we define disjoint union as $Y + Z = \{1, \ldots, |Y|, |Y| + 1, \ldots, |Y| + |Z|\}$.

Theorem 7.2. *The category $\mathbb{C}_f$ with the following operators is gs-monoidal:*

1. *The tensor $\otimes$ on objects $a \in \mathrm{M}^Y$ and $b \in \mathrm{M}^Z$ is defined as*

$$a \otimes b = a + b \in \mathrm{M}^{Y+Z}$$

where for $k \in Y + Z$ we have $(a + b)(k) = a(k)$ if $k \le |Y|$ and $(a + b)(k) = b(k - |Y|)$ if $|Y| < k \le |Y| + |Z|$.
On arrows $f\colon a \dashrightarrow b$ and $g\colon a' \dashrightarrow b'$ (with $a' \in \mathrm{M}^{Y'}$, $b' \in \mathrm{M}^{Z'}$) tensor is given by

$$f \otimes g\colon \mathrm{M}^{Y+Y'} \to \mathrm{M}^{Z+Z'}, \quad (f \otimes g)(u) = f(\bar{u}_Y) + g(\vec{u}_Y)$$

for $u \in \mathrm{M}^{Y+Y'}$ where $\bar{u}_Y \in \mathrm{M}^Y$ and $\vec{u}_Y \in \mathrm{M}^{Y'}$, defined as $\bar{u}_Y(k) = u(k)$ $(1 \le k \le |Y|)$ and $\vec{u}_Y(k) = u(|Y| + k)$ $(1 \le k \le |Y'|)$.

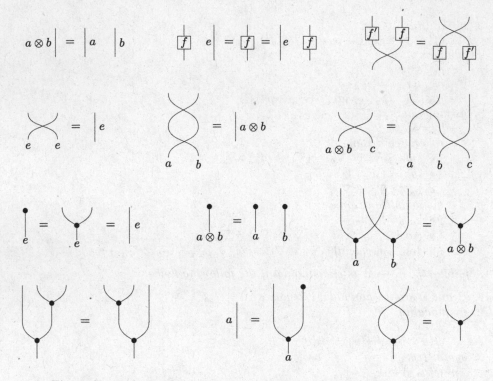

Fig. 3. String diagrams of the axioms satisfied by gs-monoidal categories.

2. *The symmetry $\rho_{a,b}: a \otimes b \dashrightarrow b \otimes a$ for $a \in \mathbb{M}^Y$, $b \in \mathbb{M}^Z$ is defined for $u \in \mathbb{M}^{Y+Z}$ as*

$$\rho_{a,b}(u) = \vec{u}_Y + \breve{u}_Y.$$

3. *The unit e is the unique mapping $e: \emptyset \to \mathbb{M}$.*
4. *The duplicator $\nabla_a: a \dashrightarrow a \otimes a$ for $a \in \mathbb{M}^Y$ is defined for $u \in \mathbb{M}^Y$ as*

$$\nabla_a(u) = u + u.$$

5. *The discharger $!_a: a \dashrightarrow e$ for $a \in \mathbb{M}^Y$ is defined for $u \in \mathbb{M}^Y$ as $!_a(u) = e$.*

We now turn to the abstract category $\mathbb{A}_f$. Note that here functions have as parameters sets of the form $U \subseteq [Y]^a \subseteq Y$. Hence, (the cardinality of) Y can not be determined directly from U and we need extra care with the tensor.

Theorem 7.3. *The category $\mathbb{A}_f$ with the following operators is gs-monoidal:*

1. *The tensor $\otimes$ on objects $a \in \mathbb{M}^Y$ and $b \in \mathbb{M}^Z$ is again defined as $a \otimes b = a + b$. On arrows $f: a \dashrightarrow b$ and $g: a' \dashrightarrow b'$ (where $a' \in \mathbb{M}^{Y'}$, $b' \in \mathbb{M}^{Z'}$ and $f: \mathcal{P}([Y]^a) \to \mathcal{P}([Z]^b)$, $g: \mathcal{P}([Y']^{a'}) \to \mathcal{P}([Z']^{b'})$ are the underlying functions), the tensor is given by*

$$f \otimes g: \mathcal{P}([Y + Y']^{a+a'}) \to \mathcal{P}([Z + Z']^{b+b'}), \quad (f \otimes g)(U) = f(\vec{U}_Y) \cup_Z g(\breve{U}_Y)$$

where $\vec{U}_Y = U \cap \{1, \dots, |Y|\}$ and $\vec{U}_Y = \{k \mid |Y| + k \in U\}$. Furthermore:

$$U \cup_Y V = U \cup \{|Y| + k \mid k \in V\} \quad \text{(where } U \subseteq Y)$$

2. *The symmetry $\rho_{a,b} \colon a \otimes b \dashrightarrow b \otimes a$ for $a \in \mathbb{M}^Y$, $b \in \mathbb{M}^Z$ is defined for $U \subseteq [Y + Z]^{a+b}$ as*

$$\rho_{a,b}(U) = \vec{U}_Y \cup_Z \vec{U}_Y \subseteq [Z + Y]^{b+a}$$

3. *The unit e is again the unique mapping $e \colon \emptyset \to \mathbb{M}$.*
4. *The duplicator $\nabla_a \colon a \dashrightarrow a \otimes a$ for $a \in \mathbb{M}^Y$ is defined for $U \subseteq [Y]^a$ as*

$$\nabla_a(U) = U \cup_Y U \subseteq [Y + Y]^{a+a}.$$

5. *The discharger $!_a \colon a \dashrightarrow e$ for $a \in \mathbb{M}^Y$ is defined for $U \subseteq [Y]^a$ as $!_a(U) = \emptyset$.*

Finally, the approximation $\#$ is indeed gs-monoidal, i.e., it preserves all the additional structure (tensor, symmetry, unit, duplicator and discharger).

Theorem 7.4. $\# \colon \mathbb{C}_f \to \mathbb{A}_f$ *is a gs-monoidal functor.*

8   UDEfix: A Tool for Fixpoints Checks

We exploit gs-monoidality as discussed before and present a tool, called UDEfix, where the user can compose his or her very own function $f \colon \mathbb{M}^Y \to \mathbb{M}^Y$ as a sort of circuit. Exploiting the fact that the functor $\#$ is gs-monoidal, this circuit is then transformed automatically and in a compositional way into the corresponding abstraction $f_{\#}^a$, for some given $a \in \mathbb{M}^Y$. By computing the greatest fixpoint of $f_{\#}^a$ and checking for emptiness, UDEfix can check whether $a = \mu f$.

In fact, UDEfix can handle all functions presented in Sect. 2, where for $\min_u$, $\max_u$ we also allow u to be a relation, instead of a function. Moreover, addition and subtraction by a fixed constant (both non-expansive functions) can be handled (see [7] for details). In addition to fixpoint checks, it is possible to perform (non-complete) checks whether a given post-fixpoint a is below the least fixpoint μf. The dual checks (for greatest fixpoint and pre-fixpoints) are implemented as well.

Building the desired function $f \colon \mathbb{M}^Y \to \mathbb{M}^Y$ requires three steps:

– Choosing the MV-algebra $\mathbb{M}$. Currently the MV-chains $[0, 1]$ and $\{0, \dots, k\}$ (for arbitrary k) are supported.
– Creating the required basic functions by specifying their parameters.
– Assembling f from these basic functions.

UDEfix is a Windows-Tool created in Python, which can be obtained from https://github.com/TimoMatt/UDEfix. The GUI of UDEfix is separated into three areas: Content area, Building area and Basic-Functions area. Under File the user can save/load contents and set the MV-algebra in Settings. Functions built in the Building area can be saved and loaded.

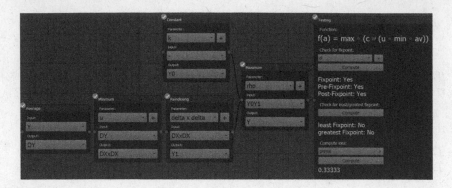

Fig. 4. Assembling the function f from Sect. 2.

Basic-Functions area: The Basic-Functions area contains the basic functions, encompassing those listed in Table 1 and additional ones. Via drag-and-drop (or right-click) these basic functions can be added to the Building area to create a Function box. Each such box requires three (in the case of $\hat{D}$ two) Contents: The Input set, the Output set and the required parameters. These Contents are to be created in the Content area. Additionally the Basic-Functions area contains the auxiliary function Testing which we will discuss in the next paragraph.

Building area: The user can connect the created Function boxes to obtain the function of interest. Composing functions is as simple as connecting two Function boxes in the correct order and disjoint union is achieved by connecting two boxes to the same box. We note that Input and Output sets of connected Function boxes need to match, otherwise the function is not built correctly. Revisiting the example in Fig. 1, we display in Fig. 4 how this function can be assembled.

The special box Testing is always required at the end. Here, the user can enter some mapping $a \colon Y \to \mathbb{M}$, test if a is a fixpoint/pre-fixpoint/post-fixpoint of the built function f and afterwards compute the greatest fixpoint of the approximation ($\nu f_{\#}^{a}$ if we want to check whether $\mu f = a$). If the result is not the empty set ($\nu f_{a}^{\#} \neq \emptyset$) one can compute a suitable value for decreasing a, needed for iterating to the least fixpoint from above (respectively increasing a for iterating to the greatest fixpoint from below). There is additional support for comparison with pre- and post-fixpoints.

In the left-hand system in Fig. 1, the function $d \colon Y \to [0, 1]$ with $d(3, 3) = 0$, $d(1, 1) = 1/2$, $d(1, 2) = d(2, 1) = d(2, 2) = 2/3$ and 1 for all other pairs is a fixpoint of f (d is not a pseudometric). By clicking Compute in the Testing-box, UDEfix displays that d is a fixpoint and tells us that d is in fact not the least and not the greatest fixpoint. It also computes the greatest fixpoints of the approximations step by step and displays the results to the user.

Content area: Here the user can create sets, mappings and relations which are used to specify the basic functions. Creating a set is done by entering a name for the new set and clicking on the plus ("+"). The user can create a variety

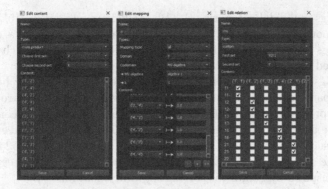

Fig. 5. Contents: Set Y, Mapping d, Relation ρ.

of different types of sets, for example the basic set $X = \{1, 2, 3, 4\}$ or the set $D = \{p_1, p_2, p_3, p_4\}$ which is a set of mappings resp. probability distributions.

Once, Input and Output sets are created we can define the required parameters (cf. Table 1). Here, the created sets can be chosen as domain and co-domain. Relations can be handled in a similar fashion: Given the two sets one wants to relate, creating a relation can be easily achieved by checking some boxes. Additionally the user has access to some useful in-built relations: "is-element-of"-relation and projections to the i-th component.

To ease the use, by clicking on the "+" in a Function box a new matching content with chosen Input and Output sets is created. The additional parameters (cf. Table 1) have domains and co-domains which need to be created or are the chosen MV-algebra. The Testing function d is a mapping as well.

See Fig. 5 for examples on how to create the contents Y (set), d (distance function) and ρ (relation).

Examples: There are pre-defined functions, implementing examples, that are shipped with the tool. These concern case studies on termination probability, bisimilarity, simple stochastic games, energy games, behavioural metrics and Rabin automata. See [7,8] for more details.

9 Conclusion, Related and Future Work

We have shown how our framework from [8] can be cast into a gs-monoidal setting, justifying the development of the tool UDEfix for a compositional view on fixpoint checks. In addition we studied properties of the gs-monoidal functor #, mapping from the concrete to the abstract universe and giving us a general procedure for approximating predicate liftings.

Related work: This paper is based on fixpoint theory, coalgebras, as well as on the theory of monoidal categories. Monoidal categories [17] are categories equipped with a tensor. It has long been realized that monoidal categories can have additional structure such as braiding or symmetries. Here we base our work on so called gs-monoidal categories [12,16], called s-monoidal in [15]. These are symmetric monoidal categories, equipped with a discharger and a duplicator. Note that "gs" originally stood for "graph substitution" and such categories were first used for modelling term graph rewriting.

We view gs-monoidal categories as a means to compositionally build monotone non-expansive functions on complete lattices, for which we are interested in the (least or greatest) fixpoint. Such fixpoints are ubiquitous in computer science, here we are in particular interested in applications in concurrency theory and games, such as bisimilarity [21], behavioural metrics [4,10,13,24] and simple stochastic games [11]. In recent work we have considered strategy iteration procedures inspired by games for solving fixpoint equations [7].

Fixpoint equations also arise in the context of coalgebra [20], a general framework for investigating behavioural equivalences for systems that are parameterized – via a functor – over their branching type (labelled, non-deterministic, probabilistic, etc.). Here in particular we are concerned with coalgebraic behavioural metrics [4], based on a generalization of the Wasserstein or Kantorovich lifting [25]. Such liftings require the notion of predicate liftings, well-known in coalgebraic modal logics [22], lifted to a quantitative setting [9].

Future Work: One important question is still open: we defined a lax functor #, relating the concrete category $\mathbb{C}$ of functions of type $\mathbb{M}^Y \to \mathbb{M}^Z$ – where Y, Z might be infinite – to their approximations, living in $\mathbb{A}$. It is unclear whether # is a proper functor, i.e., preserves composition. For finite sets functoriality derives from a non-trivial result in [8] and it is unclear whether it can be extended to the infinite case. If so, this would be a valuable step to extend the theory.

In this paper we illustrated the approximation for predicate liftings via the powerset and the distribution functor. It would be interesting to study more functors and hence broaden the applicability to other types of transition systems.

Concerning UDEfix, we plan to extend the tool to compute fixpoints, either via Kleene iteration or strategy iteration (strategy iteration from above and below), as detailed in [7]. Furthermore for convenience it would be useful to have support for generating fixpoint functions directly from a given coalgebra respectively transition system.

References

1. Bacci, G., Bacci, G., Larsen, K.G., Mardare, R.: On-the-fly exact computation of bisimilarity distances. Logic. Meth. Comput. Sci. **13**(2:13), 1–25 (2017)
2. Bacci, G., Bacci, G., Larsen, K.G., Mardare, R., Tang, Q., van Breugel, F.: Computing probabilistic bisimilarity distances for probabilistic automata. Logic. Meth. Comput. Sci. **17**(1) (2021)
3. Baier, C., Katoen, J.P.: Principles of Model Checking. MIT Press (2008)
4. Baldan, P., Bonchi, F., Kerstan, H., König, B.: Coalgebraic behavioral metrics. Logic. Meth. Comput. Sci. **14**(3) (2018). Selected Papers of the 6th Conference on Algebra and Coalgebra in Computer Science (CALCO 2015)
5. Baldan, P., Eggert, R., König, B., Matt, T., Padoan, T.: A monoidal view on fixpoint checks (2023). arXiv:2305.02957

6. Baldan, P., Eggert, R., König, B., Padoan, T.: Fixpoint theory - upside down (2023). arXiv:2101.08184
7. Baldan, P., Eggert, R., König, B., Padoan, T.: A lattice-theoretical view of strategy iteration. In: Proceedings of CSL 2023, vol. 252 of LIPIcs, pp. 7:1–7:19. Schloss Dagstuhl - Leibniz Center for Informatics (2023)
8. Baldan, P., Eggert, R., König, B., Padoan, T.: Fixpoint theory - upside down. In: Proceedings of FOSSACS 2021, pp. 62–81. Springer (2021). LNCS/ARCoSS 12650
9. Bonchi, F., König, B., Petrişan, D.: Up-to techniques for behavioural metrics via fibrations. In: Proceedings of CONCUR 2018, volume 118 of LIPIcs, pp. 17:1–17:17. Schloss Dagstuhl - Leibniz Center for Informatics (2018)
10. Chen, D., van Breugel, F., Worrell, J.: On the complexity of computing probabilistic bisimilarity. In: Birkedal, L. (ed.) FoSSaCS 2012. LNCS, vol. 7213, pp. 437–451. Springer, Heidelberg (2012). https://doi.org/10.1007/978-3-642-28729-9_29
11. Condon, A.: On algorithms for simple stochastic games. In: Advances in Computational Complexity Theory, volume 13 of DIMACS Series in Discrete Mathematics and Theoretical Computer Science, pp. 51–71 (1990)
12. Corradini, A., Gadducci, F.: An algebraic presentation of term graphs, via GS-monoidal categories. Appl. Categor. Struct. **7**, 299–331 (1999)
13. Desharnais, J., Gupta, V., Jagadeesan, R., Panangaden, P.: Metrics for labelled Markov processes. Theor. Comput. Sci. **318**, 323–354 (2004)
14. Di Nola, A., Gerla, B.: Algebras of Lukasiewicz's logic and their semiring reducts. In: Idempotent Mathematics and Mathematical Physics, vol. 377 of Proceedings of the AMS, pp. 131–144 (2005)
15. Gadducci, F.: On the Algebraic Approach To Concurrent Term Rewriting. PhD thesis, University of Pisa (1996)
16. Gadducci, F., Heckel, R.: An inductive view of graph transformation. In: Presicce, F.P. (ed.) WADT 1997. LNCS, vol. 1376, pp. 223–237. Springer, Heidelberg (1998). https://doi.org/10.1007/3-540-64299-4_36
17. Mac Lane, S.: Categories for the Working Mathematician. Springer-Verlag (1971)
18. Mundici, D.: MV-algebras. A short tutorial. http://www.matematica.uns.edu.ar/IXCongresoMonteiro/Comunicaciones/Mundici_tutorial.pdf
19. Pattinson, D.: Coalgebraic modal logic: soundness, completeness and decidability of local consequence. Theor. Comput. Sci. **309**(1), 177–193 (2003)
20. Rutten, J.J.M.M.: Universal coalgebra: a theory of systems. Theor. Comput. Sci. **249**, 3–80 (2000)
21. Sangiorgi, D.: Introduction to Bisimulation and Coinduction. Cambridge University Press (2011)
22. Schröder, L.: Expressivity of coalgebraic modal logic: the limits and beyond. Theor. Comput. Sci. **390**, 230–247 (2008)
23. Tarski, A.: A lattice-theoretical theorem and its applications. Pacific J. Math. **5**, 285–309 (1955)
24. van Breugel, F.: Probabilistic bisimilarity distances. ACM SIGLOG News **4**(4), 33–51 (2017)
25. Villani, C.: Optimal Transport - Old and New, volume 338 of A Series of Comprehensive Studies in Mathematics. Springer (2009)

Specification and Verification
of a Linear-Time Temporal Logic
for Graph Transformation

Fabio Gadducci[1], Andrea Laretto[2(⊠)], and Davide Trotta[1]

[1] Department of Computer Science, University of Pisa, Pisa, Italy
fabio.gadducci@unipi.it, trottadavide92@gmail.com
[2] Department of Software Science, Tallinn University of Technology, Tallinn, Estonia
andrea.laretto@taltech.ee

Abstract. We present a first-order linear-time temporal logic for reasoning about the evolution of directed graphs. Its semantics is based on the counterpart paradigm, thus allowing our logic to represent the creation, duplication, merging, and deletion of elements of a graph as well as how its topology changes over time. We then introduce a positive normal forms presentation, thus simplifying the actual process of verification. We provide the syntax and semantics of our logics with a computer-assisted formalisation using the proof assistant Agda, and we round up the paper by highlighting the crucial aspects of our formalisation and the practical use of quantified temporal logics in a constructive proof assistant.

Keywords: Counterpart semantics · Linear-time logics · Agda formalisation

1 Introduction

Among the many tools provided by formal methods, temporal logics have proven to be one of the most effective techniques for the verification of both large-scale and stand-alone programs. Along the years, the research on these logics focused on improving the algorithmic procedures for the verification process as well as on finding sufficiently expressive fragments of these logics for the specification of complex multi-component systems. Several models for temporal logics have been developed, with the leading example being transition systems, also known as Kripke structures. In a transition system, each state represents a configuration of the system and each transition identifies a possible state evolution. Often one is interested in enriching the states and transitions given by the model with more structure, for example by taking states as algebras and transitions as algebra homomorphisms. A prominent use case is that of graph logics [9,10,12], where states are specialised as graphs and transitions are families of (partial) graph

Research partially supported by the Italian MIUR projects PRIN 2017FTXR7S "IT-MaTTerS" and 20228KXFN2 "STENDHAL" and by the University of Pisa project PRA_2022_99 "FM4HD".

M. Fernández and C. M. Poskitt (Eds.): ICGT 2023, LNCS 13961, pp. 22–42, 2023.
https://doi.org/10.1007/978-3-031-36709-0_2

morphisms. These logics combine temporal and spatial reasoning and allow to express the possible transformation of a graph topology over time.

Quantified Temporal Logics. Under usual temporal logics, such as LTL and CTL [13], the states of the model are taken as atomic, with propositions holding for entire states: on the other hand, one of the defining characteristics of graph logics is that they permit to reason and to express properties on the individual elements of the graph. Despite their undecidability [15,26], quantified temporal logics have been advocated in this setting due to their expressiveness and the possibility for quantification to range over the elements in the states of the model.

The semantical models of these logics require some ingenuity, though. Consider a simple model with two states s_0, s_1, two transitions $s_0 \rightarrow s_1$ and $s_1 \rightarrow s_0$, and an item i that appears only in s_0. Is the item i being destroyed and recreated, or is it just an identifier being reused multiple times? This issue is denoted in the literature as the *trans-world identity problem* [2,25]. A solution consists in fixing a single set of universal items, which gives identity to each individual appearing in the states of the model. Since each item i belongs to this universal domain, it is exactly the same individual after every temporal evolution in s_1. However, this means that transitions basically behave as injections among the items of the states, and this view is conceptually difficult to reconcile with the simple model sketched above where we describe the destruction and recreation of a given item. Similarly, the possibility of cloning items is then ruled out, since it is impossible to accomodate it with the idea of evolution steps as injections.

Counterpart Semantics. A solution to this problem was proposed by Lewis [31] with the *counterpart paradigm*: instead of a universal set of items, each state identifies a local set of elements, and (possibly partial) morphisms connect them by carrying elements from one state to the other. This allows us to speak formally about entities that are destroyed, duplicated, (re)created, or merged, and to adequately deal with the identity problem of individuals between graphs.

In [16], a counterpart-based semantics is used to introduce a set-theoretical semantics of a μ-calculus with second-order quantifiers. This modal logic provides a formalism that enriches states with algebras and transitions with partial homomorphisms, subsuming the case of graph logics. These models are generalised to a categorical setting in [17] by means of relational presheaves, building on the ideas presented in [20,21]. The models are represented with categories and (families of) relational presheaves, which give a categorical representation for the states-as-algebras approach with partial homomorphisms. The temporal advancement of a system is captured by equipping categories with the notion of *one-step* arrows for a model, and the categorical framework is then used to introduce a second-order linear temporal logic QLTL.

Classical Semantics and Positive Normal Form. We start by introducing the notion of counterpart models in Sect. 2, and present the (admittedly rather straightforward) syntax of our temporal logic QLTL as well as its counterpart-based semantics in Sect. 3, using a standard set-theoretic perspective, with satisfiability given inductively as a logical predicate. Unlike [16,17] where the models use *partial functions*, we generalise to *relations*, thus modelling the duplication of elements by allowing it to have multiple counterparts. In Sect. 4 we present some results on the positive normal forms, where the models may use either partial morphisms or relations, and we highlight their differences. Positive normal forms (i.e., where negation is defined only for atomic formulae) are a standard tool of temporal logics, since they simplify its theoretical treatment as well as facilitating model checking algorithms [5,27]. The use of relations instead of (possibly partial) functions weakens the expressiveness of such normal forms, and requires the introduction of additional operators for the logics. However, the duplication of individuals is a central feature of graph transformation formalisms such as Sequi-Pushout [7], and thus worthy of investigation.

Temporal Logics in Agda. An additional contribution of our work is a computer-assisted formalisation using the dependently typed proof assistant Agda [37] of the models, semantics, and positive normal forms of the logic presented in this paper. We introduce the main aspects of the mechanisation in Sect. 5, which can be adapted for counterpart-based models whose worlds are algebras on any multi-sorted signature, even if for the sake of presentation in this paper we restrict our attention to graph signatures. A formal presentation of a temporal logic in a proof assistant has several advantages: it solidifies the correctness and coherence of the mathematical ideas presented in the work, as they can be independently inspected and verified concretely by means of a software tool; moreover, the mechanisation effectively provides a playground in which the mechanisms and validity of these logics can be expressed, tested, and experimented with.

To the best of our knowledge, few formalisations of temporal logics have been provided with a proof assistant, and none of these comes equipped with a counterpart-based semantics. This work constitutes a step towards the machine-verified use of temporal logics by embedding in an interactive proof assistant a quantified extension of LTL that can reason on individual elements of states.

2   Counterpart Models

This section introduces our models for system evolution. We consider the instantiation of counterpart models to the case where each world is not associated to a mere (structureless) set of individuals, but to a directed graph with its evolution in time being represented by suitable relations preserving the graph structure.

Definition 2.1. *A **(directed) graph** is a 4-tuple $G := \langle N, E, s, t \rangle$ such that N is a set of nodes, E is a set of edges, and $s, t : E \to N$ are two functions assigning a source node $s(e) \in N$ and a target node $t(e) \in N$ to each edge $e \in E$, respectively. The set of all directed graphs is denoted as* Graphs.

Definition 2.2. *A **graph (relational) morphism** between two graphs $G = \langle N, E, s, t \rangle$ and $G' = \langle N', E', s', t' \rangle$ is a pair $R := \langle R_N, R_E \rangle$ such that $R_N \subseteq N \times N'$ and $R_E \subseteq E \times E'$ are relations between nodes and edges of the graphs such that $e_1 R_E e_2$ implies $s(e_1) R_N s'(e_2)$ and $t(e_1) R_N t'(e_2)$. Given graphs G, G', the set of graph morphisms is denoted* GraphRel$(G, G') \subseteq \mathscr{P}((N \times N') \times (E \times E'))$.

Definition 2.3. *A **counterpart model** is a triple* $\mathfrak{M} := \langle W, D, \mathcal{C} \rangle$ *such that*

- *W is a non-empty set of elements, called* worlds,
- *$D : W \to$ Graphs is a function assigning a directed graph to each world,*
- *$\mathcal{C} : W \times W \to \mathscr{P}(\mathsf{GraphRel}(D(\omega), D(\omega')))$ is a function assigning to every pair $\langle \omega, \omega' \rangle$ a set of graph morphisms $\mathcal{C}\langle \omega, \omega' \rangle \subseteq \mathsf{GraphRel}(D(\omega), D(\omega'))$, where every $C \in \mathcal{C}\langle \omega, \omega' \rangle$ is a graph morphism between the graphs associated to the two worlds. We refer to these as **atomic counterpart relations**.*

Given two worlds ω and ω', the set $\mathcal{C}\langle \omega, \omega' \rangle$ is the collection of atomic transitions from ω to ω', defining the possible ways we can access worlds with a *one-step transition* in the system. When the set $\mathcal{C}\langle \omega, \omega' \rangle$ is empty, there are no atomic transitions from ω to ω'. Each atomic relation $C \in \mathcal{C}\langle \omega, \omega' \rangle$ connects the nodes and edges between two worlds ω and ω', intuitively identifying them as the same component after one time evolution of the model. For example, if we consider two nodes $n \in D(\omega)_N$ and $n' \in D(\omega')_N$ and a relation $C \in \mathcal{C}\langle \omega, \omega' \rangle$, if $\langle n, n' \rangle \in C_N$ then n' represents a future development of the node n via C.

Definition 2.4. *A node $s' \in D(\omega')_N$ **is the counterpart of** $s \in D(\omega)_N$ through a counterpart relation C whenever $\langle s, s' \rangle \in C_N$, and similarly for edges.*

Example 2.1. (Counterpart model). We give an example of a counterpart model by indicating the set of worlds $\{\omega_0, \omega_1, \omega_2\}$ and the cardinality of the sets of relations $\mathcal{C}\langle \omega, \omega' \rangle$ in Fig. 1; the graph structures associated to each world and the graph morphisms connecting them are shown in Fig. 2. There are two counterpart relations $C_1, C_2 \in \mathcal{C}\langle \omega_1, \omega_2 \rangle$ between ω_1 and ω_2, and we use blue dashed and green dotted lines to distinguish C_1 and C_2, respectively.

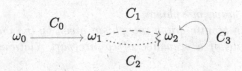

Fig. 1. Graphical representation of the worlds and accessibility relations of a model.

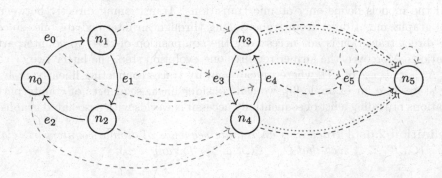

Fig. 2. Graphical representation of a model.

The use of *relations* as transitions allows us to model the removal of edges and nodes of a graph, by having no counterpart for them in the next state. For example, if there is no edge $e' \in D(\omega')_E$ such that $\langle e, e' \rangle \in C_E$, we conclude that the edge e has been removed by C. Similarly, the duplication of a node is represented by connecting it with two instances of the counterpart relation, e.g. by having elements $n'_1, n'_2 \in D(\omega')_N$ such that $\langle n, n'_1 \rangle \in C_N$ and $\langle n, n'_2 \rangle \in C_N$.

The simple counterpart model in Example 2.1 displays the effects of merging and deletion. The first counterpart relation C_0 merges the nodes n_0 and n_2 of ω_0, yet this act does not generate a cycle: in fact, e_2 is deleted, since it is not connected to a counterpart in ω_1. Similarly, both counterpart relations C_1 and C_2 merge the nodes n_3 and n_4 of ω_1, while they differ in which edge they remove when transitioning from ω_1 to ω_2, by deleting either the edge e_3 or e_4, respectively. Note that in both cases the nodes n_3 and n_4 need to be preserved, albeit possibly merged, in order to ensure that the relations considered are actually (relational) morphisms of graphs.

2.1   Counterpart Relations and Traces

We assume hereafter a fixed counterpart model $\mathfrak{M} := \langle W, D, \mathcal{C} \rangle$ and we formally introduce the idea behind counterpart relations. We indicate composition of graph morphisms in diagrammatic order: as an example, given $C \in$ GraphRel(G_1, G_2) and $C' \in$ GraphRel(G_2, G_3), the composite graph morphism is denoted with $C; C' \in$ GraphRel(G_1, G_3) and is such that $(C; C')_N = \{(a, c) \mid \exists b \in G_{2N}. \langle a, b \rangle \in C_N \wedge \langle b, c \rangle \in C'_N\}$, and similarly for edges.

Definition 2.5. *A graph morphism* $C \in$ GraphRel$(D(\omega), D(\omega'))$ *is a **counterpart relation** of the model if one of the following cases holds*

- *C is the identity graph morphism;*
- *$C \in \mathcal{C}\langle \omega, \omega' \rangle$ is an atomic graph morphism given by the model $\mathfrak{M}$;*
- *C is the composite of a sequence of counterpart relations $C_0; \cdots; C_n$ with $C_i \in \mathcal{C}\langle \omega_i, \omega_{i+1} \rangle$.*

Note that the composition $C; C' \in$ GraphRel$(D(\omega_1), D(\omega_3))$ of two atomic counterpart relations $C \in \mathcal{C}\langle \omega_1, \omega_2 \rangle$ and $C' \in \mathcal{C}\langle \omega_2, \omega_3 \rangle$ might not be atomic, and the models define only atomic transitions. Transitioning directly between two graphs might differ from transitioning through an intermediate one, since the direct transition is not necessarily the composition of the two counterpart relations. Moreover, the former requires one evolution step, the latter two.

As is the case of LTL where we can identify traces connecting linearly evolving states, see for example [3], we can consider linear sequences of counterpart relations providing a list of sequentially accessible worlds with associated graphs.

Definition 2.6. *A **trace** σ is an infinite sequence of atomic counterpart relations $(C_0, C_1, \dots)$ such that $C_i \in \mathcal{C}\langle \omega_i, \omega_{i+1} \rangle$ for any $i \geq 0$.*

In other words, a trace identifies a path in the graph induced by the counterpart model, with worlds as nodes and atomic counterpart relations as edges. Given a trace $\sigma = (C_0, C_1, \dots)$, we use $\sigma_i := (C_i, C_{i+1}, \dots)$ to denote the trace obtained by excluding the first i counterpart relations. We use $\omega_0, \omega_1, \dots$ and ω_i to indicate the worlds of the trace σ whenever it is clear from the context. Similarly, we denote with $C_{\leq i}$ the composite relation $C_0; \dots ; C_{i-1}$ from the first world ω_0 up to the i-th world ω_i through the relations given by the trace σ. Should $i = 0$, the relation $C_{\leq 0}$ is the identity graph morphism on ω_0.

3   Quantified Linear Temporal Logic

We present the syntax and semantics of our (first-order) quantified linear temporal logic QLTL by adopting a standard set-theoretic presentation.

3.1   Syntax and Semantics of QLTL

Since free variables may appear inside formulae, we recall the usual presentation of context and terms-in-context. For the sake of simplicity we consider algebras whose terms represent directed graphs, as per Definition 2.1. More precisely, the signature has two sorts E and N and two functions symbols s and t, obviously representing the source and target functions on edges, respectively. We assume a set of sorted variables $\mathcal{X} = \mathcal{X}_N \uplus \mathcal{X}_E$ and define a typed context Γ as a finite subset of such variables, using the notation $[\Gamma]\, n : N$ to indicate that the term n has type N and it is constructed in a typed context Γ, that is, its free variables are contained among those occurring in Γ (and similarly for $[\Gamma]\, e : E$).

In order to give a simpler presentation for the semantics of temporal logics, it is customary to exclude the elementary constructs that can be expressed in terms of other operators. We thus present QLTL with a minimal set of standard operators and derive conjunction and universal quantification by using negation.

Definition 3.1. (QLTL). *Let Γ be a typed context on the set of variables $\mathcal{X}$. The set $\mathcal{F}_\Gamma^{QLTL}$ of QLTL formulae is generated by the following rules*

$$\psi := \mathsf{true} \mid e_1 =_E e_2 \mid n_1 =_N n_2$$

$$\phi := \psi \mid \neg\phi \mid \phi \vee \phi \mid \exists_N x.\phi \mid \exists_E x.\phi \mid \mathsf{O}\phi \mid \phi\mathsf{U}\phi \mid \phi\mathsf{W}\phi,$$

where $[\Gamma]\, e_i : E$ and $[\Gamma]\, n_i : N$ for $i = 1, 2$.

The above definition actually provides formulae-in-context: we use the notation $[\Gamma]\phi$ to indicate that a formula ϕ belongs to $\mathcal{F}_\Gamma^{QLTL}$ for the typed context Γ. Clearly, saying that $[\Gamma]\phi$ is the same as stating that $\mathrm{fv}(\phi) \subseteq \Gamma$, i.e. the free variables of ϕ are contained among those occurring in Γ. For both terms and formulae we omit the bracketed context whenever it is unnecessary to specify it.

The letter ψ denotes the set of *atomic formulae*, built out of two sorted equivalence predicates. Given two edge terms $e_1, e_2 : E$, the formula $e_1 =_E e_2$ indicates that the two edges coincide in the graph associated to the current world, and similarly for two node terms $n_1, n_2 : N$ and the formula $n_1 =_N n_2$.

The existential operators $\exists_N x.\phi$ and $\exists_E x.\phi$ can be used to express the existence of a node (edge, respectively) in the current graph satisfying a certain property ϕ, where the variable x is allowed to appear as a free variable of ϕ.

The *next* operator $\bigcirc\phi$ expresses the fact that a certain property ϕ has to be true at the next state. The *until* operator $\phi_1 U \phi_2$ indicates that the property ϕ_1 has to hold at least until the property ϕ_2 becomes true, which must hold at the present or future time. Finally, the *weak until* operator $\phi_1 W \phi_2$ is similar to the $\phi_1 U \phi_2$ operator, but allows for counterparts to exist indefinitely without ever reaching a point where ϕ_2 holds, provided that ϕ_1 also keeps holding indefinitely.

The dual operators are syntactically expressed by false $:= \neg$true, $\phi_1 \wedge \phi_2 := \neg(\neg\phi_1 \vee \neg\phi_2)$, and $\forall_N x.\phi := \neg\exists_N x.\neg\phi$ and similarly for edges. Note that, differently from classical LTL, the until and the weak until operators are not self-dual: this fact will be discussed and made explicit in Remark 3.3.

3.2   Satisfiability

To present the notion of satisfiability of a formula with respect to a counterpart model we introduce the definition of *assignment* for a context in a world.

Definition 3.2. (Assignment). *An **assignment** in the world $\omega \in W$ for the typed context Γ is a pair of functions $\mu := \langle \mu_N, \mu_E \rangle$ such that $\mu_N : \Gamma_N \to D(\omega)_N$ and $\mu_E : \Gamma_E \to D(\omega)_E$. We use the notation $\mathcal{A}_\omega^\Gamma$ to indicate the set of assignments μ defined in ω for the typed context Γ.*

Moreover, we denote by $\mu[x \mapsto_\tau n] \in \mathcal{A}_\omega^{\Gamma,(x:\tau)}$ the assignment obtained by extending the domain of μ with $n \in D(\omega)_\tau$ at the variable $x \notin \Gamma$, omitting the type $\tau \in \{N, E\}$ whenever clear from context. We indicate with $\Gamma, (x : \tau)$ the context Γ extended with an additional variable x with sort τ.

An assignment μ for a typed context Γ provides the interpretation of terms-in-context whose context is (contained in) Γ: it allows for evaluating the free variables of the term, thus defining its semantics (with respect to that assignment). This will be used in Definition 3.5 to provide a meaning for the equalities $n_1 =_N n_2$ and $e_1 =_E e_2$ for nodes and edges, respectively.

Definition 3.3. (Assignment on terms). *Given an assignment $\mu \in \mathcal{A}_\omega^\Gamma$, we indicate with $\mu^* := \langle \mu_N^*, \mu_E^* \rangle$ the interpretation of μ on a term-in-context $[\Gamma]\, n : N$ or $[\Gamma]\, e : E$, given inductively by the rules: $\mu_E^*(x) := \mu_E(x)$, $\mu_N^*(y) := \mu_N(y)$, $\mu_N^*(s(e)) := D(\omega)_s(\mu_E^*(e))$, and $\mu_N^*(t(e)) := D(\omega)_t(\mu_E^*(e))$.*

We now look at how to transport assignments over counterpart relations. The intuition is that we have to connect the nodes and the edges in the image of two assignments when there is a counterpart relation among the worlds, and the items of the underlying graphs are related point-wise.

Definition 3.4. (Counterpart relations on assignments). *Given a counterpart relation $C \in \mathsf{GraphRel}(D(\omega_1), D(\omega_2))$ and two assignments $\mu_1 \in \mathcal{A}_{\omega_1}^\Gamma$ and $\mu_2 \in \mathcal{A}_{\omega_2}^\Gamma$ on the context Γ, we say that the assignments μ_1 and μ_2*

are **counterpart related** if $\langle \mu_{1N}(x), \mu_{2N}(x) \rangle \in C_N$ for any $x \in \Gamma_N$ and $\langle \mu_{1E}(x), \mu_{2E}(x) \rangle \in C_E$ for any $x \in \Gamma_E$. We indicate this with the notation $\langle \mu_1, \mu_2 \rangle \in C$.

We can now introduce the notion of satisfiability of a QLTL formula with respect to a trace σ and an assignment μ.

Definition 3.5. (QLTL satisfiability). *Given a QLTL formula-in-context* $[\Gamma]\phi$, *a trace* $\sigma = (C_0, C_1, \dots)$, *and an assignment* $\mu \in \mathcal{A}_{\omega_0}^{\Gamma}$ *in the first world of* σ, *we inductively define the* satisfiability relation *as follows*

- $\sigma, \mu \vDash \mathsf{true}$;
- $\sigma, \mu \vDash e_1 =_E e_2$ *if* $\mu_E^*(e_1) = \mu_E^*(e_2)$;
- $\sigma, \mu \vDash n_1 =_N n_2$ *if* $\mu_N^*(n_1) = \mu_N^*(n_2)$;
- $\sigma, \mu \vDash \neg \phi$ *if* $\sigma, \mu \nvDash \phi$;
- $\sigma, \mu \vDash \phi_1 \vee \phi_2$ *if* $\sigma, \mu \vDash \phi_1$ *or* $\sigma, \mu \vDash \phi_2$;
- $\sigma, \mu \vDash \exists_N x.\phi$ *if there is a node* $n \in D(\omega_0)_N$ *such that* $\sigma, \mu[x \mapsto n] \vDash \phi$;
- $\sigma, \mu \vDash \exists_E x.\phi$ *if there is an edge* $e \in D(\omega_0)_E$ *such that* $\sigma, \mu[x \mapsto e] \vDash \phi$;
- $\sigma, \mu \vDash \mathsf{O}\phi$ *if there is* $\mu_1 \in \mathcal{A}_{\omega_1}^{\Gamma}$ *such that* $\langle \mu, \mu_1 \rangle \in C_0$ *and* $\sigma_1, \mu_1 \vDash \phi$;
- $\sigma, \mu \vDash \phi_1 \mathsf{U}\phi_2$ *if there is an* $\bar{n} \geq 0$ *such that*
  1. *for any* $i < \bar{n}$, *there is* $\mu_i \in \mathcal{A}_{\omega_i}^{\Gamma}$ *such that* $\langle \mu, \mu_i \rangle \in C_{\leq i}$ *and* $\sigma_i, \mu_i \vDash \phi_1$;
  2. *there is* $\mu_{\bar{n}} \in \mathcal{A}_{\omega_{\bar{n}}}^{\Gamma}$ *such that* $\langle \mu, \mu_{\bar{n}} \rangle \in C_{\leq \bar{n}}$ *and* $\sigma_{\bar{n}}, \mu_{\bar{n}} \vDash \phi_2$;
- $\sigma, \mu \vDash \phi_1 \mathsf{W}\phi_2$ *if one of the following holds*
  - *the same conditions for* $\phi_1 \mathsf{U}\phi_2$ *apply; or*
  - *for any* i *there is* $\mu_i \in \mathcal{A}_{\omega_i}^{\Gamma}$ *such that* $\langle \mu, \mu_i \rangle \in C_{\leq i}$ *and* $\sigma_i, \mu_i \vDash \phi_1$.

3.3 Examples

We provide some examples of satisfiability for QLTL formulae on the running example in Fig. 2 to illustrate how our counterpart semantics works in practice. Take for example the trace given by $\sigma := (C_0, C_1, C_3, C_3, \dots)$, thus considering the case where e_4 is the only edge preserved when transitioning from ω_1.

Example 3.1. (Allocation and deallocation). As anticipated in Sect. 2, one of the main advantages of a counterpart semantics is the possibility to reason about existence, deallocation, duplication, and merging of elements in the system and its evolution. Consider for example the following shorthand formulae

$$\mathbf{present}_\tau(x) := \exists_\tau y.x =_\tau y$$
$$\mathbf{nextPreserved}_\tau(x) := \mathbf{present}_\tau(x) \wedge \mathsf{O}\,\mathbf{present}_\tau(x)$$
$$\mathbf{nextDealloc}_\tau(x) := \mathbf{present}_\tau(x) \wedge \neg\mathsf{O}\,\mathbf{present}_\tau(x)$$

The formula $\mathbf{present}_\tau(x)$ captures the existence of a node or an edge at the current moment. We can combine this predicate with the *next* operator to talk about elements that are present in the current world and that will still be present at the next step, which we condense with the $\mathbf{nextPreserved}_\tau(x)$ formula. We can similarly refer to elements that are now present but that will be deallocated at the next step by considering the formula $\mathbf{nextDealloc}_\tau(x)$. Indeed we have

$$\sigma_0, \{x : \mathsf{E} \mapsto e_0 \} \vDash \mathbf{nextPreserved}_E(x); \; \sigma_1, \{x : \mathsf{N} \mapsto n_3\} \nvDash \mathbf{nextDealloc}_N(x)$$
$$\sigma_0, \{x : \mathsf{N} \mapsto n_1\} \vDash \mathbf{nextPreserved}_N(x); \; \sigma_1, \{x : \mathsf{E} \mapsto e_3 \} \vDash \mathbf{nextDealloc}_E(x)$$
$$\sigma_0, \{x : \mathsf{E} \mapsto e_2 \} \nvDash \mathbf{nextPreserved}_E(x); \; \sigma_1, \{x : \mathsf{E} \mapsto e_4 \} \nvDash \mathbf{nextDealloc}_E(x)$$

Example 3.2. (Graph structure). Moreover, our syntax allows us to define formulae that exploit the algebraic structure of our graphs, and combine them with the temporal operators to state properties about how the graph evolves in time. We illustrate this by providing the following formulae

$$\mathbf{loop}(e) := s(e) =_N t(e)$$
$$\mathbf{hasLoop}(n) := \exists_E e.s(e) =_N n \wedge \mathbf{loop}(e)$$
$$\mathbf{composable}(x,y) := t(x) =_N s(y)$$
$$\mathbf{haveComposition}(x,y) := \mathbf{composable}(x,y) \wedge \exists_E e.(s(e) =_N s(x) \wedge t(e) =_N t(y))$$
$$\mathbf{adjacent}(x,y) := \exists_E e.((s(e) =_N x \wedge t(e) =_N y) \vee (t(e) =_N x \wedge s(e) =_N y))$$

which capture, respectively, the following scenarios: we can check whether a given edge of the graph is a loop with $\mathbf{loop}(x)$, or verify with $\mathbf{hasLoop}(x)$ that the only node having a loop is n_5; alternatively, we can express this fact by stating that the loop belongs to the entire world using $\mathbf{hasLoop}$

$$\sigma_0, \{x \mapsto e_0\} \nvDash \mathbf{loop}(x); \; \sigma_0, \{x \mapsto n_0\} \nvDash \mathbf{hasLoop}(x); \; \sigma_0, \{\} \nvDash \exists_N x.\mathbf{hasLoop}(x)$$
$$\sigma_1, \{x \mapsto e_3\} \nvDash \mathbf{loop}(x); \; \sigma_1, \{x \mapsto n_3\} \nvDash \mathbf{hasLoop}(x); \; \sigma_1, \{\} \nvDash \exists_N x.\mathbf{hasLoop}(x)$$
$$\sigma_2, \{x \mapsto e_5\} \vDash \mathbf{loop}(x); \; \sigma_2, \{x \mapsto n_5\} \vDash \mathbf{hasLoop}(x); \; \sigma_2, \{\} \vDash \exists_N x.\mathbf{hasLoop}(x)$$

Since $\exists_N x.\mathbf{hasLoop}(x)$ is a closed formula, the empty assignment is the only one the formula can be valued on. Thus, we obtain the classical notion of a formula simply providing the binary information of being true or false, with no choice of individuals needed to satisfy it. Finally, we can express some properties about the existence of intermediate nodes and composability of edges in the graph

$$\sigma_0, \{x \mapsto n_0, y \mapsto n_1\} \vDash \mathbf{adjacent}(x,y); \quad \sigma_0, \{x \mapsto e_0, y \mapsto e_1\} \vDash \mathbf{composable}(x,y)$$
$$\sigma_1, \{x \mapsto n_3, y \mapsto n_4\} \vDash \mathbf{adjacent}(x,y); \quad \sigma_0, \{x \mapsto e_0, y \mapsto e_1\} \vDash \mathsf{O}\mathbf{composable}(x,y)$$
$$\sigma_1, \{x \mapsto n_3, y \mapsto n_4\} \vDash \mathsf{O}\mathbf{adjacent}(x,y); \; \sigma_1, \{x \mapsto e_3, y \mapsto e_4\} \nvDash \mathsf{O}\mathbf{composable}(x,y)$$

$$\sigma_0, \{x \mapsto e_0, y \mapsto e_1\} \nvDash \mathbf{haveComposition}(x,y)$$
$$\sigma_0, \{x \mapsto e_0, y \mapsto e_1\} \nvDash \mathsf{O}\mathbf{haveComposition}(x,y)$$
$$\sigma_2, \{x \mapsto e_5, y \mapsto e_5\} \vDash \mathsf{O}\mathbf{haveComposition}(x,y)$$

Remark 3.1. (Eventually and always operators). As in LTL, we can define the additional *eventually* $\Diamond\phi$ and *always* $\Box\phi$ operators as $\Diamond\phi := \mathsf{true}\mathsf{U}\phi$ and $\Box\phi := \phi\mathsf{W}\mathsf{false}$, respectively. Alternatively, their semantics can be presented directly as

- $\sigma, \mu \vDash \Diamond\phi$ if there are $i \geq 0$ and $\mu_i \in \mathcal{A}_{\omega_i}^{\Gamma}$ s.t. $\langle \mu, \mu_i \rangle \in C_{\leq i}$ and $\sigma_i, \mu_i \vDash \phi$.
- $\sigma, \mu \vDash \Box\phi$ if for any $i \geq 0$ there is $\mu_i \in \mathcal{A}_{\omega_i}^{\Gamma}$ s.t. $\langle \mu, \mu_i \rangle \in C_{\leq i}$ and $\sigma_i, \mu_i \vDash \phi$.

Example 3.3. (Temporal evolution). We can use these operators to express the evolution of the graph after an unspecified amount of steps

$$\mathbf{willMerge}_\tau(x,y) := x \neq_\tau y \wedge \Diamond(x =_\tau y)$$
$$\mathbf{alwaysPreserved}_\tau(x) := \Box\mathbf{present}_\tau(x)$$
$$\mathbf{willBecomeLoop}(e) := \neg\mathbf{loop}(e) \wedge \Diamond\mathbf{loop}(e)$$

In the example in Fig. 1 for the same trace $\sigma = (C_0, C_1, C_3, \dots)$ we have

$$\sigma_0, \{\} \vDash \exists n. \exists m. \mathbf{willMerge}_N(n, m), \qquad \sigma_0, \{\} \nvDash \forall e. \Diamond \mathbf{loop}(e),$$
$$\sigma_0, \{\} \nvDash \forall e. \mathbf{alwaysPreserved}_E(e), \qquad \sigma_0, \{\} \vDash \exists e. \Diamond \Box \mathbf{loop}(e),$$
$$\sigma_0, \{\} \vDash \exists e. \mathbf{willBecomeLoop}(e), \qquad \sigma_0, \{\} \vDash \exists x. \exists y. \neg \Diamond \mathbf{composable}(x, y),$$
$$\sigma_0, \{\} \vDash (\exists e. s(e) \neq t(e)) \mathsf{U}(\exists x. \forall y. x = y), \sigma_0, \{\} \vDash (\exists e. s(e) = t(e)) \mathsf{W} \neg (\exists x. \mathbf{loop}(e)).$$

Remark 3.2. (Quantifier elision for unbound variables). A relevant difference with standard quantified logics is that in QLTL we cannot elide quantifications where the variable introduced does not appear in the subformula. Assuming $\equiv$ to denote semantical equivalence and taking any ϕ with $x \notin \mathrm{fv}(\phi)$, we have that in general $\exists x.\phi \not\equiv \phi$ and, similarly, $\forall x.\phi \not\equiv \phi$. More precisely, the above equivalences hold whenever ϕ does not contain any temporal operator and the current world $D(\omega)$ being considered is not empty.

Consider a world ω with a single node $D(\omega)_N = \{s\}$, no edges, and a single looping counterpart relation $\mathcal{C}\langle \omega, \omega \rangle = \{C\}$ where $C = \emptyset$ is the empty counterpart relation. The trace is given by $\sigma = (C, C, \dots)$. By taking the empty assignment $\{\}$ and the closed formula $\phi = \mathsf{O}(\mathsf{true})$, one can easily check that $\sigma, \{\} \vDash \mathsf{O}(\mathsf{true})$, but $\sigma, \{\} \nvDash \exists_N x. \mathsf{O}(\mathsf{true})$. The reason is that, once an assignment is extended with some element, stepping from one world to the next one requires every individual of the assignment to be preserved and have a counterpart in the next world. Alternatively, we could have restricted assignments in the semantics so that counterparts are required only for the free variables occurring in the formula. For example, the definition for the *next* $\mathsf{O}\phi$ operator would become

$$- \ \sigma, \mu \vDash \mathsf{O}\phi \text{ if there is } \mu_1 \in A_{\omega_1}^{\mathrm{fv}(\phi)} \text{ such that } \langle \mu_{|\mathrm{fv}(\phi)}, \mu_1 \rangle \in C_0 \text{ and } \sigma_1, \mu_1 \vDash \phi.$$

For ease of presentation in this work and with respect to our Agda implementation, we consider the case where all elements in the context have a counterpart.

Remark 3.3. (Until and weak until are incompatible). In standard LTL, the *until* $\phi_1 \mathsf{U} \phi_2$ and *weak until* $\phi_1 \mathsf{W} \phi_2$ operators have the same expressivity, and can be defined in terms of each other by the equivalences $\phi_1 \mathsf{U} \phi_2 \equiv_{\mathsf{LTL}} \neg(\neg\phi_2 \mathsf{W}(\neg\phi_1 \wedge \neg\phi_2)), \phi_1 \mathsf{W} \phi_2 \equiv_{\mathsf{LTL}} \neg(\neg\phi_2 \mathsf{U}(\neg\phi_1 \wedge \neg\phi_2))$. However, this is not the case in QLTL. Similarly, it might at first seem reasonable to define the standard *always* operator in QLTL with $\Box\phi := \neg\Diamond\neg\phi$. However, this definition does not align with the semantics provided in Remark 3.1. This characteristic of QLTL is again due to the fact that we are in the setting of (possibly deallocating) relations, and we formally explain and present an intuition for this when we introduce the PNF semantics in Sect. 4. The LTL equivalences can be obtained again by restricting to models whose counterpart relations are total functions: this allows us to consider a unique trace of counterparts that are always defined, which brings our models back to a standard LTL-like notion of trace.

4   Positive Normal Form for QLTL

Positive normal forms are a standard presentation of temporal logics: they can be used to simplify constructions and algorithms on both the theoretical and implemention side [5,27]. Their use is crucial for semantics based on fixpoints, such as in [16], while still preserving the expressiveness of the original presentation. As we remark in Sect. 5, providing a negation-free semantics for our logic also ensures that it can be more easily manipulated in a proof assistant where definitions and proofs are *constructive*. Moreover, the positive normal form conversion serves as a concrete procedure that can be used interactively to automatically convert formulae into their positive normal form, which is proven in the proof assistant to be correct. In this section we present an explicit semantics for the positive normal form of QLTL, which we denote as PNF.

4.1   Semantics of PNF

As observed in Remark 3.3, to present the positive normal form we need additional operators to adequately capture the negation of temporal operators. Thus, we introduce a new flavour of the next operator, called *next-forall* $\mathsf{A}\phi$. Similarly, we have to introduce a dual for *until* $\phi_1\mathsf{U}\phi_2$ and *weak until* $\phi_1\mathsf{W}\phi_2$, which we indicate as the *then* $\phi_1\mathsf{T}\phi_2$ and *until-forall* $\phi_1\mathsf{F}\phi_2$ operators, respectively.

Definition 4.1. (QLTL in PNF). *Let Γ be a typed context on the set of variables $\mathcal{X}$. The set $\mathcal{F}_\Gamma^{PNF}$ of formulae of QLTL in **positive normal form** is generated by the following rules*

$$\psi := \mathsf{true} \mid e_1 =_E e_2 \mid n_1 =_N n_2$$

$$\phi := \psi \mid \neg\psi \mid \phi \vee \phi \mid \phi \wedge \phi \mid \exists_\tau x.\phi \mid \forall_\tau x.\phi \mid \mathsf{O}\phi \mid \mathsf{A}\phi \mid \phi\mathsf{U}\phi \mid \phi\mathsf{F}\phi \mid \phi\mathsf{W}\phi \mid \phi\mathsf{T}\phi,$$

where $[\Gamma]\ e_i : E$ and $[\Gamma]\ n_i : N$ for $i = 1, 2$ and $\tau \in \{N, E\}$.

The intuition for the *next-forall* $\mathsf{A}\phi$ operator is that it allows us to capture the case where a counterpart of an individual does not exist at the next step: if any counterpart exists, it is required to satisfy the formula ϕ.

Similarly to the *until* $\phi_1\mathsf{U}\phi_2$ operator, the *until-forall* $\phi_1\mathsf{F}\phi_2$ operator allows us to take a sequence of graphs where ϕ_1 is satisfied for some steps until ϕ_2 holds. The crucial observation is that *every* intermediate counterpart satisfying ϕ_1 and the conclusive counterparts must satisfy ϕ_2. Such counterparts are not required to exist, and indeed any trace consisting of all empty counterpart relations always satisfies both $\phi_1\mathsf{F}\phi_2$ and $\phi_1\mathsf{T}\phi_2$. Similarly to the *weak until* $\phi_1\mathsf{W}\phi_2$ operator, the *then* $\phi_1\mathsf{T}\phi_2$ operator corresponds to a *weak until-forall* and can be validated by a trace where all counterparts satisfy ϕ_1 without ever satisfying ϕ_2.

We now provide a satisfiability relation for PNF formulae by specifying the semantics just for the additional operators, omitting the ones that do not change.

Definition 4.2. (QLTL in PNF satisfiability). *Given a QLTL formula-in-context $[\Gamma]\phi$ in positive normal form, a trace $\sigma = (C_0, C_1, \dots)$, and an assignment $\mu \in \mathcal{A}_{\omega_0}^\Gamma$ in the first world of σ, we inductively define the satisfiability relation with respect to the additional operators as follows*

- $\sigma, \mu \vDash \neg\psi$ *if* $\sigma, \mu \nvDash \psi$;
- $\sigma, \mu \vDash \phi_1 \wedge \phi_2$ *if* $\sigma, \mu \vDash \phi_1$ *and* $\sigma, \mu \vDash \phi_2$;

– $\sigma, \mu \models \forall_\tau x.\phi$ if for any $s \in D(\omega_0)_\tau$ we have that $\sigma, \mu[x \mapsto s] \models \phi$;
– $\sigma, \mu \models \mathsf{A}\phi$ if for any $\mu_1 \in \mathcal{A}_{\omega_1}^\Gamma$ such that $\langle \mu, \mu_1 \rangle \in C_0$ we have that $\sigma_1, \mu_1 \models \phi$;
– $\sigma, \mu \models \phi_1 \mathsf{F} \phi_2$ if there is an $\bar{n} \geq 0$ such that
    1. for any $i < \bar{n}$ and $\mu_i \in \mathcal{A}_{\omega_i}^\Gamma$ such that $\langle \mu, \mu_i \rangle \in C_{\leq i}$ we have $\sigma_i, \mu_i \models \phi_1$;
    2. for any $\mu_{\bar{n}} \in \mathcal{A}_{\omega_{\bar{n}}}^\Gamma$ such that $\langle \mu, \mu_{\bar{n}} \rangle \in C_{\leq \bar{n}}$ we have $\sigma_{\bar{n}}, \mu_{\bar{n}} \models \phi_2$;
– $\sigma, \mu \models \phi_1 \mathsf{T} \phi_2$ if one of the following holds
   • the same conditions for $\phi_1 \mathsf{F} \phi_2$ apply; or
   • for any i and $\mu_i \in \mathcal{A}_{\omega_i}^\Gamma$ such that $\langle \mu, \mu_i \rangle \in C_{\leq i}$ we have $\sigma_i, \mu_i \models \phi_1$.

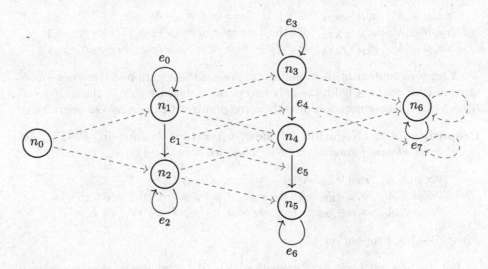

Fig. 3. Graphical representation of a counterpart model.

Example 4.1. We illustrate with the example in Fig. 3 the possibility for a counterpart relation to duplicate both edges and nodes of a graph, as well as providing some concrete cases for the new operators defined in Definition 4.2. For example, we have that $\sigma_0, \{x \mapsto n_0\} \models \mathsf{A}(\mathbf{hasLoop}(x))$, but $\sigma_1, \{x \mapsto n_1\} \not\models \mathsf{A}(\mathbf{hasLoop}(x))$ since n_4 is a counterpart of n_1 but does not have a loop. Moreover, we have that $\sigma_1, \{x \mapsto e_1\} \models \mathbf{hasLoop}(s(x))\mathsf{U}(\mathbf{loop}(x))$, but $\sigma_1, \{x \mapsto e_1\} \not\models \mathbf{hasLoop}(s(x))\mathsf{F}(\mathbf{loop}(x))$ because we also require for e_5 to have a loop at its source since it is a counterpart of e_1. Notice that $\sigma_2, \{x \mapsto e_5\} \models \mathsf{A}(\mathbf{loop}(x))$ since there is no counterpart at the next step, and indeed we similarly have that $\sigma_2, \{x \mapsto n_5\} \models \mathbf{hasLoop}(x)\mathsf{F}(\mathsf{false})$. Finally, we have that $\sigma_2, \{x \mapsto e_4\} \models \mathbf{hasLoop}(s(x))\mathsf{T}(\neg\mathbf{loop}(x))$ because the intermediate condition always holds.

4.2 Negation of QLTL and PNF

The crucial observation that validates the PNF presented in Sect. 4 is that the negation of *next* $O\phi$, *until* $\phi_1 U \phi_2$, and *weak until* $\phi_1 W \phi_2$ formulae can now be expressed inside the logic. We indicate with $\vDash_{QLTL}$ and $\vDash_{PNF}$ the satisfiability relations for formulae in standard QLTL and QLTL in PNF, respectively.

Proposition 4.1. (Negation is expressible in PNF).
(💾 Relational.Negation)
Let ψ, ψ_1, ψ_2 be atomic formulae in PNF. Then we have

$$\forall \sigma, \mu \in \mathcal{A}_{\omega_0}^{\Gamma} . \ \sigma, \mu \vDash_{QLTL} \neg O(\psi) \quad \iff \quad \sigma, \mu \vDash_{PNF} A(\neg \psi)$$
$$\forall \sigma, \mu \in \mathcal{A}_{\omega_0}^{\Gamma} . \ \sigma, \mu \vDash_{QLTL} \neg(\psi_1 U \psi_2) \iff \sigma, \mu \vDash_{PNF} (\neg \psi_2) T(\neg \psi_1 \wedge \neg \psi_2)$$
$$\forall \sigma, \mu \in \mathcal{A}_{\omega_0}^{\Gamma} . \ \sigma, \mu \vDash_{QLTL} \neg(\psi_1 W \psi_2) \iff \sigma, \mu \vDash_{PNF} (\neg \psi_2) F(\neg \psi_1 \wedge \neg \psi_2)$$

A converse statement that similarly expresses the negation of these new operators in PNF does not hold: the only exception is the easy case of the *next-forall* $A\phi$ operator, whose negation directly corresponds with the *next* $O\phi$ operator.

Proposition 4.2. (Negation of new operators is *not* in PNF). *Let ψ, ψ_1, ψ_2 be atomic formulae in PNF. Then we have*

$$\forall \sigma, \mu \in \mathcal{A}_{\omega_0}^{\Gamma} . \ \sigma, \mu \nvDash_{PNF} A(\psi) \quad \iff \quad \sigma, \mu \vDash_{PNF} O(\neg \psi)$$
$$\forall \sigma, \mu \in \mathcal{A}_{\omega_0}^{\Gamma} . \ \sigma, \mu \nvDash_{PNF} \psi_1 T \psi_2 \quad \nLeftrightarrow \quad \sigma, \mu \vDash_{PNF} (\neg \psi_2) U(\neg \psi_1 \wedge \neg \psi_2)$$
$$\forall \sigma, \mu \in \mathcal{A}_{\omega_0}^{\Gamma} . \ \sigma, \mu \nvDash_{PNF} \psi_1 F \psi_2 \quad \nLeftrightarrow \quad \sigma, \mu \vDash_{PNF} (\neg \psi_2) W(\neg \psi_1 \wedge \neg \psi_2)$$

Proof. See [18, Proposition 4.2].

Let us now go back to Proposition 4.1. We exploit the correspondence between operators given there to define a translation $\overline{\cdot} : \mathcal{F}^{QLTL} \rightarrow \mathcal{F}^{PNF}$ from the QLTL syntax presented in Definition 3.5 to the current one in PNF. What is noteworthy is that such translation preserves the equivalence of formulae.

Theorem 4.3. (PNF equivalence). *(💾 Relational.Conversion) Let $\overline{\cdot}$: $\mathcal{F}^{QLTL} \rightarrow \mathcal{F}^{PNF}$ be the syntactical translation that replaces negated temporal operators with their equivalent ones in PNF by pushing negation down to atomic formulae. For any QLTL formula $[\Gamma]\phi \in \mathcal{F}^{QLTL}$ we have*

$$\forall \sigma, \mu \in \mathcal{A}_{\omega_0}^{\Gamma} . \ \sigma, \mu \vDash_{QLTL} \phi \iff \sigma, \mu \vDash_{PNF} \overline{\phi}$$

Contrary to what happens in LTL, the usual expansion laws where each operator is defined in terms of itself do not hold in QLTL for the case of counterpart relations, as shown by the following result.

Proposition 4.4. (Expansion laws do *not* hold in QLTL). *We have the following inequalities in PNF*

$$\phi_1 U \phi_2 \not\equiv \phi_2 \vee (\phi_1 \wedge O(\phi_1 U \phi_2)) \qquad \phi_1 F \phi_2 \not\equiv \phi_2 \vee (\phi_1 \wedge A(\phi_1 F \phi_2))$$
$$\phi_1 W \phi_2 \not\equiv \phi_2 \vee (\phi_1 \wedge O(\phi_1 W \phi_2)) \qquad \phi_1 T \phi_2 \not\equiv \phi_2 \vee (\phi_1 \wedge A(\phi_1 T \phi_2))$$

Proof. See [18, Proposition 4.4].

Remark 4.1. (Functional counterpart relations). The previous results can be reframed in the case in which each counterpart relation is a *partial function*, following the definition of counterpart models given in [16,17]. It turns out that under the assumption of partial functions we recover all the equivalences stated in Proposition 4.2 ($\circlearrowleft$ Functional.Negation) as well as the expansion laws of Proposition 4.4 ($\circlearrowleft$ Functional.ExpansionLaws). The latter can be used to provide a presentation of the temporal operators as *least fixpoint* ($\phi_1 U \phi_2$, $\phi_1 F \phi_2$) and *greatest fixpoints* ($\phi_1 W \phi_2$, $\phi_1 T \phi_2$) of a suitably defined operator; this definition based on fixpoints would coincide with the semantics of the operators given in Definition 4.2 only in the case of partial (or total) functions.

5   Agda Formalisation

This section presents an overview of an additional contribution of this work: a complete formalisation of the semantics of QLTL and its PNF using the dependently typed programming language and proof assistant Agda [37]. We provide a brief exposition and usage of our development in [18], showing how the temporal evolution of the running example in Fig. 2 can be concretely modelled in Agda. The complete formalisation of the logic along with the PNF results is available at https://github.com/iwilare/algebraic-temporal-logics.

5.1   Formalisation Aspects

Our formalisation work consists in the mechanisation of all the aspects presented in the paper: we start by defining the notion of counterpart relations and traces of relational morphisms as models of the logic, and provide a representation for (well-typed and well-scoped) syntax for formulae of QLTL and PNF along with their satisfiability semantics. Then, we provide a conversion function from QLTL to PNF along with proofs of correctness and completeness of the procedure; finally, using the defined framework, we prove among other equivalences the relevant expansion laws introduced in Sect. 4.2 for the functional setting.

For the sake of presentation in this paper we restricted our attention to graph signatures, and a concrete example in Agda of our library instantiated on the signature of directed graphs is available in [18]. However, we remark that our implementation is general enough to model algebras over any generic multi-sorted signature. In particular, this means that, by specifying a suitable signature, the class of models (and formulae) considered by the logic can be extended to the case of any graphical formalism that admits an algebraic representation on a multi-sorted signature, such as trees, hypergraphs, and so on.

Moreover, given the constructive interpretation of the formalisation, proving the correctness and completeness of PNF with respect to QLTL also yields a concrete procedure that can convert formulae into their positive normal form version. We describe now how the main components provided by our formalisation can be employed by the user to interact with the proof assistant.

Signature Definition. Using the definitions given in our formalisation, the user can write their own algebraic signature that will be used to represent the system of interest as algebras on the signature. For example, by defining the signature of graphs Gr the user can reason on the temporal evolution of graphs, using (relational) graph homomorphisms as counterpart relations between worlds.

Formula Construction. After having provided the signature of interest, the user can construct formulae using the full expressiveness of QLTL and can reason on equality of terms constructed according to the signature. This allows the user to express properties that combine both logical quantifiers as well as exploiting the specific structure of the system, possibly composing and reusing previously defined formulae. The infrastructure provided by the formalisation is such that the formulae constructed by the user are inherently checked to be well-scoped and well-typed with respect to the sorts of the signature, e.g. edges and nodes in the case of graphs. The user can freely use negation in formulae, and can (optionally) use the procedures we defined to automatically convert formulae to their PNF, which as we have seen in Sect. 4 can be particularly counterintuitive in the counterpart setting with respect to standard temporal logics.

Model Definition. The models of the system at various time instances can be constructed by the user, following again the signature provided. Then, the user specifies a series of symbolic worlds and indicates the possible transitions that can be taken by defining a relation on the worlds. Then, an algebra of the signature must be assigned to each world, and the connection of worlds is translated into a morphism between the algebras which the user provides. The transitions of the models are checked by Agda to preserve the algebraic structure of the worlds considered, thus corresponding to the notion of graph morphisms; this step is relatively straightforward as the automation available in Agda helps with proving the structure-preservation of the maps. Traces between worlds are defined using a coinductive definition of traces using sized types [11], thus allowing for infinite (repeating) traces to be modelled and defined by the user.

Validation of Formulae in the Model. Using the library the user can prove that a specific model satisfies a given formula; our formalisation automatically simplifies the goal that must be proven to verify the formula, and the user is guided by the proof assistant by automatically constructing the skeleton of the proof term.

5.2   Intuitionistic Proof Assistant

In our setting, some crucial usability issues need to be mentioned. Agda is a proof assistant based on the *intuitionistic* interpretation of mathematics [24]. This means that some useful logical principles often used in the setting of temporal logics are *not provable* in the system, such as the law of excluded middle or the De Morgan laws to switch connectives and quantifiers whenever negation appears in subformulae. Thus, without assuming these logical principles, the embedding of our temporal logic QLTL would actually be restricted to the *intuitionistic* fragment; in practice, this is not particularly problematic since classical reasoning can simply be assumed as axiom, and allows the equivalences mentioned to be recovered. This, however, would be undesirable from the user's perspective, as they would have to explicitly use these classical axioms in their proofs. In order to tackle these usability aspects and the treatment of negation in the intuitionistic setting, we take the following approach: the formulae of the logic are expressed in Agda using a full *positive normal form* similar to the one presented in Sect. 4, which we have proven correct in Agda in Theorem 4.3 by postulating classical principles. This effectively shifts the burden of dealing (classically) with negation from the user to the implementer, while also giving them complete accessibility over the extended set of correct quantifiers. Moreover, the correctness proof of the conversion to PNF constitutes both a theoretical guarantee that no expressive power is either gained or lost in the presentation of the logic, as well as a concrete algorithm that the user can execute to convert their formulae into equivalent ones in PNF, for which validity can be easier to prove.

5.3 Automation

Embedding a temporal logic in a proof assistant allows the user to exploit the *assistant* aspect of the tool, for example, by aiding the user in showing (or even prove automatically) that a certain formula in a model is satisfied or not.

In Agda, this automation aspect is limited, especially if compared to proof assistants where automation and the use of tactics is a core aspect of the software environment, such as Coq [6], Lean [34], and Isabelle [36]. The Agda synthesizer Agsy [32] is the main helper tool in Agda implementing a form of automated proof search. Unfortunately, Agsy only provides general-purpose searching procedures and its theorem proving capabilities are nowhere near those of specialised model checking algorithms. Still, the goal-oriented interactivity available in Agda is an invaluable tool in proving theorems step-by-step and *manually* verify formulae in our setting, and the assisted introduction of constructors allows the user to quickly generate the proof structure needed to validate temporal formulae.

6 Related Works

Up to the early 2010s,s, there has been a series of papers devoted to some variants of quantified logics for expressing properties of graphs and of graph evolutions. Our models are inspired by the counterpart-based logics explored in the context of a μ-calculus with fixpoints in [16], and we refer there for an overview of and a comparison with the by-then current proposals, such as the well-known [1], all favouring an approach based on universal domains. Among the follow-ups of the works mentioned there, there is [19] and the related [44,47], which further explore one of the relevant tools developed in the graph community, GROOVE. To some extent, the present paper and its companion [17], which introduces the categorical semantics of second-order QLTL, are summarising a previous thread of research concerning counterpart models, including its implementation. And in fact, the categorical semantics for counterpart models appears of interest in itself in the literature on modal logics, as witnessed by the works surveyed in [17].

Concerning the formalisation of temporal logics in (constructive) proof assistants, the topic has a long history, see e.g. [8,45,48]. A practical application and comparison with modern model checkers is in [14], and a fully verified LTL model checker is implemented in the Isabelle theorem prover. In [38], a verified proof-search program is formalised in Agda for standard CTL, together with a toolbox to implement well-typed proof-searching procedures; a similar embedding of constructive LTL in Agda is provided in [29] for the verification of functional reactive programs. Our proof-of-concept implementation of QLTL in Agda witnesses the possibility to move towards the formalisation of quantified temporal logics for proof assistants, an issue sparsely tackled in the literature.

Concerning graph computation models (GCMs), we find in the literature several formalisms that use graph-specific definitions where syntactical statements on nodes, edges, sources of edges, targets of edges, and equalities are first-class citizens in the logic to express properties on the system under analysis. The last decade has seen a series of papers advocating quantified temporal logics as a formalism for the specification of GCMs properties. We offer a short review of some of the most recent proposals appeared in the literature, focussing on the dichotomy between the universal domains and the counterpart-based approaches.

Graph Programs/Flow Graphs. The use of monadic-second order logics to prove properties of graph-based programming languages has been advocated in [39,46], where the emphasis is placed on distilling post-conditions formulae from a graph transformation rule and a precondition formula. A more abstract meta-model for run-time verification is proposed in [4,33], where a control flow graph can be instantiated to concrete models and the properties are given by first-order formulae. Despite the differences, in both cases the resulting analysis is akin to the adoption of a universal domain approach.

Metric Logics, I. The use of traces and first-order specifications is a key ingredient of runtime verification. A relevant proposal is the use of metric first-order temporal logic (MFOTL) [40,41], investigated with respect to the expressiveness of suitable fragments in [28] or to duality results akin to our PNF in [35]. These logics allows to reason on the individual components of states, using (arbitrary) sets of relations as models, which allows for different kinds of graphs to be encoded. The core difference with our line of work is that, contrary to standard models of MFOTL, we allow for variable domains in the temporal structure and for nodes to be created and destroyed using counterpart relations.

Metric Logics, II. A graph-oriented approach to MFOTL is given by Metric Temporal Graph Logic (MTGL) [23,43], which allows to model properties on the structures and the attributes of the state and has been used in the context of formal testing [42]. Here traces are pairs of injective spans representing a rule, and are equivalent to our partial graph morphisms. The syntax is tailored over such rules, so that ϕ_G refers to a formula over a given graph G, and a one step $\exists(f, \phi_H)$ is indexed over a mono $f : G \rightarrow H$, roughly representing the partial morphism induced by a rule. Thus, besides our use of relations, identity and preservation/deletion of elements seem to be left implicit, and the exploration of the connection with counterpart-based QLTL is among our future endeavours.

7   Conclusions and Future Works

We have seen how a set-theoretic semantics for a first-order linear-time temporal logic QLTL can be presented in the counterpart setting. We saw how its syntax and semantics can be naturally used in an algebraic setting to express properties of direct graphs and their evolution in time, and how the notions and models presented in the previous sections can be formalised and practically experimented with in a proof assistant based on dependent type theory such as Agda. We have investigated some results on the positive normal forms of this logic in the case of relations and partial functions, and argued for their usefulness both in practice and in the case of constructive proof assistants.

We identify a variety of possible expansions for our work.

Second-Order. Our theoretical presentation and formalisation work focuses on the first-order aspects of QLTL. The semantics in [16,17] allows also for the quantification over sets of elements. This is impractical in Agda due to the typical formalisation of subsets as predicates, which would be cumbersome to present in concrete examples, e.g. when expressing universal quantification and extensional equality over subsets of elements. A possible extension could be to investigate practical encodings and possible automation techniques to introduce second-order quantification for counterpart-based temporal logics.

CTL and Other Logics. The quantified temporal logics presented here focus on providing a restricted yet sufficiently powerful set of operators and structures. These logics could be extended to alternative constructs and models, such as those of CTL [13]. Extending our logic to more complex models seems a straightforward task, which might however cause a combinatorial explosion in the temporal operators required to obtained a positive normal form.

Automation and Solvers. We highlighted how the proofs required to validate temporal formulae need to be provided manually by the user. Considerable amount of effort has been spent in interfacing proof assistants with external solvers and checkers to both reuse existing work and algorithms and to provide more efficient alternatives to the automation given by proof assistants. The traditional way of employing proof automation is through the use of *internal* and *external* solvers: the first technique uses the reflection capabilities of Agda to allow a (verified) solver and proof-searching procedure to be written in Agda itself, in the spirit of [14,30,38]. The second mechanism consists in writing bindings to external programs, such as model checkers or SMT and SAT solvers, so that proving the formula or providing a counterexample is offloaded to a more efficient and specialised program. A possible extension of this work would be the implementation of either of these mechanisms to the setting of counterpart semantics.

Finite Traces. A current trend in artificial intelligence is the study of temporal formulas over *finite* traces [22], due to applications in planning and reinforcement learning. Our models seem to be well-suited to tackle such a development, since each finite trace can be thought of as an infinite one terminating with a cycle in an empty graph, thus inheriting all the issues we highlighted about positive normal forms for our logic.

References

1. Baldan, P., Corradini, A., König, B., Lluch Lafuente, A.: A temporal graph logic for verification of graph transformation systems. In: Fiadeiro, J.L., Schobbens, P.-Y. (eds.) WADT 2006. LNCS, vol. 4409, pp. 1–20. Springer, Heidelberg (2007). https://doi.org/10.1007/978-3-540-71998-4_1
2. Belardinelli, F.: Quantified Modal Logic and the Ontology of Physical Objects. Ph.D. work, Scuola Normale Superiore of Pisa (2004–2005)
3. Blackburn, P., van Benthem, J., Wolter, F. (eds.): Handbook of Modal Logic, vol. 3. North Holland (2007)
4. Búr, M., Marussy, K., Meyer, B.H., Varró, D.: Worst-case execution time calculation for query-based monitors by witness generation. ACM Trans. Embed. Comput. Syst. **20**(6), 1–36 (2021)
5. Bustan, D., Flaisher, A., Grumberg, O., Kupferman, O., Vardi, M.Y.: Regular vacuity. In: Borrione, D., Paul, W. (eds.) CHARME 2005. LNCS, vol. 3725, pp. 191–206. Springer, Heidelberg (2005). https://doi.org/10.1007/11560548_16
6. Coq Development Team: The Coq Proof Assistant Reference Manual (2016)
7. Corradini, A., Heindel, T., Hermann, F., König, B.: Sesqui-pushout rewriting. In: Corradini, A., Ehrig, H., Montanari, U., Ribeiro, L., Rozenberg, G. (eds.) ICGT 2006. LNCS, vol. 4178, pp. 30–45. Springer (2006)
8. Coupet-Grimal, S.: An axiomatization of linear temporal logic in the calculus of inductive constructions. J. Logic Comput. **13**(6), 801–813 (2003)
9. Courcelle, B.: The monadic second-order logic of graphs. I. Recognizable sets of finite graphs. Inform. Comput. **85**(1), 12–75 (1990)
10. Courcelle, B.: The monadic second-order logic of graphs. XII. Planar graphs and planar maps. Theor. Comput. Sci. **237**(1), 1–32 (2000)
11. Danielsson, N.A.: Up-to techniques using sized types. In: POPL 2018, pp. 43:1–43:28. ACM (2018)
12. Dawar, A., Gardner, P., Ghelli, G.: Expressiveness and complexity of graph logic. Inf. Comput. **205**(3), 263–310 (2007)
13. Emerson, E.A.: Temporal and modal logic. In: van Leeuwen, J. (ed.) Handbook of Theoretical Computer Science, Volume B: Formal Models and Semantics, pp. 995–1072. Elsevier and MIT Press (1990)
14. Esparza, J., Lammich, P., Neumann, R., Nipkow, T., Schimpf, A., Smaus, J.-G.: A fully verified executable LTL model checker. In: Sharygina, N., Veith, H. (eds.) CAV 2013. LNCS, vol. 8044, pp. 463–478. Springer, Heidelberg (2013). https://doi.org/10.1007/978-3-642-39799-8_31
15. Franconi, E., Toman, D.: Fixpoint extensions of temporal description logics. In: Calcanese, D., De Giacomo, G., Franconi, E. (eds.) DL 2003. CEUR Workshop Proceedings, vol. 81 (2003)
16. Gadducci, F., Lluch-Lafuente, A., Vandin, A.: Counterpart semantics for a second-order μ-calculus. Fundamenta Informaticae **118**(1–2), 177–205 (2012)

17. Gadducci, F., Trotta, D.: A presheaf semantics for quantified temporal logics. CoRR abs/2111.03855 (2021)
18. Gadducci, F., Laretto, A., Trotta, D.: Specification and verification of a linear-time temporal logic for graph transformation. CoRR abs/2305.03832 (2023)
19. Ghamarian, A.H., de Mol, M., Rensink, A., Zambon, E., Zimakova, M.: Modelling and analysis using GROOVE. Int. J. Softw. Tools Technol. Trans. **14**(1), 15–40 (2012)
20. Ghilardi, S., Meloni, G.: Modal and tense predicate logic: models in presheaves and categorical conceptualization. In: Borceux, F. (ed.) Categorical Algebra and its Applications. LNM, vol. 1348, pp. 130–142. Springer (1988)
21. Ghilardi, S., Meloni, G.: Relational and partial variable sets and basic predicate logic. J. Symbol. Logic **61**(3), 843–872 (1996)
22. Giacomo, G.D., Vardi, M.Y.: Synthesis for LTL and LDL on finite traces. In: Yang, Q., Wooldridge, M.J. (eds.) IJCAI 2015, pp. 1558–1564. AAAI Press (2015)
23. Giese, H., Maximova, M., Sakizloglou, L., Schneider, S.: Metric temporal graph logic over typed attributed graphs. In: Hähnle, R., van der Aalst, W. (eds.) FASE 2019. LNCS, vol. 11424, pp. 282–298. Springer, Cham (2019). https://doi.org/10.1007/978-3-030-16722-6_16
24. Girard, J., Lafont, Y., Taylor, P.: Proofs and Types, Cambridge Tracts in Theoretical Computer Science, vol. 7. Cambridge University Press (1989)
25. Hazen, A.: Counterpart-theoretic semantics for modal logic. J. Philos. **76**(6), 319–338 (1979)
26. Hodkinson, I., Wolter, F., Zakharyaschev, M.: Monodic fragments of first-order temporal logics: 2000–2001 A.D. In: Nieuwenhuis, R., Voronkov, A. (eds.) LPAR 2001. LNCS (LNAI), vol. 2250, pp. 1–23. Springer, Heidelberg (2001). https://doi.org/10.1007/3-540-45653-8_1
27. Huang, S., Cleaveland, R.: A tableau construction for finite linear-time temporal logic. J. Logic Algebr. Meth. Program. **125**, 100743 (2022)
28. Hublet, F., Basin, D., Krstić, S.: Real-time policy enforcement with metric first-order temporal logic. In: Atluri, V., Di Pietro, R., Jensen, C.D., Meng, W. (eds.) ESORICS 2022. LNCS, vol. 13555, pp. 211–232. Springer, Cham (2022). https://doi.org/10.1007/978-3-031-17146-8_11
29. Jeffrey, A.: LTL types FRP: linear-time temporal logic propositions as types, proofs as functional reactive programs. In: Claessen, K., Swamy, N. (eds.) PLPV 2012, pp. 49–60. ACM (2012)
30. Kokke, P., Swierstra, W.: Auto in Agda. In: Hinze, R., Voigtländer, J. (eds.) MPC 2015. LNCS, vol. 9129, pp. 276–301. Springer, Cham (2015). https://doi.org/10.1007/978-3-319-19797-5_14
31. Lewis, D.K.: Counterpart theory and quantified modal logic. J. Philos. **65**(5), 113–126 (1968)
32. Lindblad, F., Benke, M.: A tool for automated theorem proving in Agda. In: Filliâtre, J.-C., Paulin-Mohring, C., Werner, B. (eds.) TYPES 2004. LNCS, vol. 3839, pp. 154–169. Springer, Heidelberg (2006). https://doi.org/10.1007/11617990_10
33. Marussy, K., Semeráth, O., Babikian, A.A., Varró, D.: A specification language for consistent model generation based on partial models. J. Object Technol. **19**(3), 1–22 (2020)
34. Moura, L., Ullrich, S.: The lean 4 theorem prover and programming language. In: Platzer, A., Sutcliffe, G. (eds.) CADE 2021. LNCS (LNAI), vol. 12699, pp. 625–635. Springer, Cham (2021). https://doi.org/10.1007/978-3-030-79876-5_37

35. Huerta, Y., Munive, J.J.: Relaxing safety for metric first-order temporal logic via dynamic free variables. In: Dang, T., Stolz, V. (eds.) RV 2022. LNCS, vol. 13498, pp. 45–66. Springer, Cham (2022). https://doi.org/10.1007/978-3-031-17196-3_3

36. Nipkow, T., Paulson, L.C., Wenzel, M.: Isabelle/HOL - A Proof Assistant for Higher-Order Logic. LNCS, vol. 2283, pp. 67–104. Springer, Heidelberg (2002). https://doi.org/10.1007/3-540-45949-9_5

37. Norell, U.: Dependently typed programming in Agda. In: Kennedy, A., Ahmed, A. (eds.) TLDI 2009, pp. 1–2. ACM (2009)

38. O'Connor, L.: Applications of applicative proof search. In: Chapman, J., Swierstra, W. (eds.) TyDe@ICFP 2016, pp. 43–55. ACM (2016)

39. Poskitt, C.M., Plump, D.: Monadic second-order incorrectness logic for GP 2. J. Logic Algebr. Meth. Program. **130**, 100825 (2023)

40. Schneider, J., Basin, D., Krstić, S., Traytel, D.: A formally verified monitor for metric first-order temporal logic. In: Finkbeiner, B., Mariani, L. (eds.) RV 2019. LNCS, vol. 11757, pp. 310–328. Springer, Cham (2019). https://doi.org/10.1007/978-3-030-32079-9_18

41. Schneider, J., Traytel, D.: Formalization of a monitoring algorithm for metric first-order temporal logic. Archive of Formal Proofs (2019)

42. Schneider, S., Maximova, M., Sakizloglou, L., Giese, H.: Formal testing of timed graph transformation systems using metric temporal graph logic. Int. J. Softw. Tools Technol. Transf. **23**(3), 411–488 (2021). https://doi.org/10.1007/s10009-020-00585-w

43. Schneider, S., Sakizloglou, L., Maximova, M., Giese, H.: Optimistic and pessimistic on-the-fly analysis for metric temporal graph logic. In: Gadducci, F., Kehrer, T. (eds.) ICGT 2020. LNCS, vol. 12150, pp. 276–294. Springer, Cham (2020). https://doi.org/10.1007/978-3-030-51372-6_16

44. Smid, W., Rensink, A.: Class diagram restructuring with GROOVE. In: Gorp, P.V., Rose, L.M., Krause, C. (eds.) TTC 2013. EPTCS, vol. 135, pp. 83–87 (2013)

45. Sprenger, C.: A verified model checker for the modal μ-calculus in Coq. In: Steffen, B. (ed.) TACAS 1998. LNCS, vol. 1384, pp. 167–183. Springer, Heidelberg (1998). https://doi.org/10.1007/BFb0054171

46. Wulandari, G.S., Plump, D.: Verifying graph programs with monadic second-order logic. In: Gadducci, F., Kehrer, T. (eds.) ICGT 2021. LNCS, vol. 12741, pp. 240–261. Springer, Cham (2021). https://doi.org/10.1007/978-3-030-78946-6_13

47. Zambon, E., Rensink, A.: Recipes for coffee: Compositional construction of JAVA control flow graphs in GROOVE. In: Müller, P., Schaefer, I. (eds.) Principled Software Development. LNCS, pp. 305–323. Springer, Cham (2018). https://doi.org/10.1007/978-3-319-98047-8_19

48. Zanarini, D., Luna, C., Sierra, L.: Alternating-time temporal logic in the calculus of (Co)inductive constructions. In: Gheyi, R., Naumann, D. (eds.) SBMF 2012. LNCS, vol. 7498, pp. 210–225. Springer, Heidelberg (2012). https://doi.org/10.1007/978-3-642-33296-8_16

Finding the Right Way to Rome: Effect-Oriented Graph Transformation

Jens Kosiol[1]([✉]) [iD], Daniel Strüber[2,3] [iD], Gabriele Taentzer[1] [iD],
and Steffen Zschaler[4] [iD]

[1] Philipps-Universität Marburg, Marburg, Germany
{kosiolje,taentzer}@mathematik.uni-marburg.de
[2] Chalmers | University of Gothenburg, Gothenburg, Sweden
danstru@chalmers.se
[3] Radboud University, Nijmegen, The Netherlands
[4] King's College London, London, UK
szschaler@acm.org

Abstract. Many applications of graph transformation require rules that change a graph without introducing new consistency violations. When designing such rules, it is natural to think about the desired outcome state, i.e., the desired *effect*, rather than the specific steps required to achieve it; these steps may vary depending on the specific rule-application context. Existing graph-transformation approaches either require a separate rule to be written for every possible application context or lack the ability to constrain the maximal change that a rule will create. We introduce *effect-oriented graph transformation*, shifting the semantics of a rule from specifying actions to representing the desired effect. A single effect-oriented rule can encode a large number of *induced* classic rules. Which of the *potential* actions is executed depends on the application context; ultimately, all ways lead to Rome. If a graph element to be deleted (created) by a potential action is already absent (present), this action need not be performed because the desired outcome is already present. We formally define effect-oriented graph transformation, show how matches can be computed without explicitly enumerating all induced classic rules, and report on a prototypical implementation of effect-oriented graph transformation in Henshin.

Keywords: Graph transformation · Double-pushout approach · Consistency-preserving transformations

1 Introduction

Applications of graph transformation such as model synchronisation [13,14,20] or search-based optimisation [5,19] require graph-transformation rules that combine a change to the graph with repair [28] operations to ensure transformations are consistency sustaining or even improving [21]. For any given graph constraint, there are typically many different ways in which it can be violated, requiring

M. Fernández and C. M. Poskitt (Eds.): ICGT 2023, LNCS 13961, pp. 43–63, 2023.
https://doi.org/10.1007/978-3-031-36709-0_3

slightly different specific changes to repair the violation. As a result, it is often easier to think about the desired *effect* of a repairing graph transformation rule rather than the specific transformations required. We would like to be able to reach a certain state of the graph—defined in terms of the presence or absence of particular graph elements (the *effect*)—even if, in different situations, a different set of specific changes is required to achieve this.

Existing approaches to graph transformation make it difficult to precisely capture the effect of a rule without explicitly specifying the specific set of changes required. For example, the *double-pushout approach (DPO)* to graph transformation [10, 11] has gained acceptance as an underlying formal semantics for *graph* and *model transformation rules* in practice as a simple and intuitive approach: A transformation rule simply specifies which graph elements are to be deleted and created when it is applied; that is, a rule prescribes exactly all the *actions* to be performed. For graph repair, this effectively forces one to specify every way in which a constraint can be violated and the specific changes to apply in this case, so that the right rule can be applied depending on context.

On the other end of the spectrum, the *double-pullback approach* [18] is much more flexible. Here, rules only specify minimal changes. However, there is no way of operationally constraining the maximal possible change.

There currently exists no approach to graph transformation that allows the *effect* of rules to be specified concisely and precisely without specifying every action that needs to be taken. In this paper, we introduce the notion of *effect-oriented graph transformations*. In this approach, graph-transformation rules encode a, potentially large, number of *induced rules*. This is achieved by differentiating basic actions that have to be performed by any transformation consistent with the rule and *potential* actions that only have to be performed if they are required to achieve the intended rule *effect*. Depending on the application context, a different set of actions will be executed—all ways lead to Rome. We provide an algorithm for selecting the right set of actions depending on context, without having to explicitly enumerate all possibilities—we efficiently find the right way to Rome.

Thus, the paper makes the following contributions:

1. We define the new notion of effect-oriented graph transformation rules and discuss different notions of consistent matches for these;
2. We provide an algorithm for constructing a complete match and a transformation given a partial match for an effect-oriented transformation rule. The algorithm is efficient in the sense that it avoids computing and matching all induced rules explicitly;
3. We report on a prototypical implementation of effect-oriented transformations in Henshin; and
4. We compare our approach to existing approaches to graph transformation showing that it does indeed provide new expressivity.

The rest of this paper is structured as follows. First, we introduce a running example (Sect. 2) and briefly recall basic preliminaries (Sect. 3). Section 4 introduces effect-oriented rules and transformations and several notions of construct-

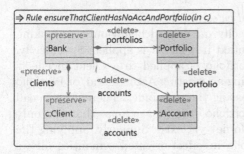 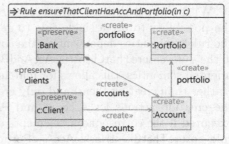

Fig. 1. Example rules, shown in the integrated visual syntax of Henshin [1, 29]. The LHS of a rule consists of all red and grey elements (additionally annotated with delete or preserve), the RHS of all green (additionally annotated as create) and grey elements. (Color figure online)

ing matches. Section 5 explains in more detail one algorithm for constructing matches for effect-oriented rules and reports on a prototype implementation in Henshin. In Sect. 6, we discuss how existing applications can benefit from effect-oriented rules and transformations, and compare our new approach to graph transformation with other approaches that could be used to achieve these goals. We conclude in Sect. 7. We provide some additional results and explanations and proofs of our formal statements in an extended version of this paper [22].

2 Running Example

We use the well-known banking example [23] and adapt it slightly to illustrate our newly introduced concept of effect-oriented transformations. Assume that the context of this example is specified in a meta-model formalised as a type graph (not shown) for the banking domain in which a Bank has Clients, Accounts and Portfolios. A Client may have Accounts which may be associated with a Portfolio.

Imagine a scenario where it is to be ensured that a Client has an Account with a Portfolio. To realise this condition in a rule-based manner so that no unnecessary elements are created, at least three rules (and a programme to coordinate their application) are required: A rule that checks whether a Client already has an Account with a Portfolio, a rule that adds a Portfolio to an existing Account of the Client, and a rule that creates all the required structure; this last rule is shown as the rule ensureThatClientHasAccAndPortfolio in Fig. 1.

An analogous problem exists if the Accounts and Portfolios of a Client are to be removed. The rule ensureThatClientHasNoAccAndPortfolio in Fig. 1 is the rule that deletes the entire structure, and additional rules are needed to delete Accounts that are not associated with Portfolios. In general, the number of rules needed and the complexity of their coordination depend on the size of the structure to occur together and hence, to be created (deleted).

With our new notion of *effect-oriented rules* and *transformations*, we provide
the possibility to use a single rule to specify *all* desired behaviours by making
the rule's semantics dependent on the context in which it is applied. Specifically,
if the rule ensureThatClientHasAccAndPortfolio is applied to a Client in effect-
oriented semantics, this allows for matching an existing Account and/or Portfolio
(rather than creating them) and only creating the remainder. Therefore, we call
the creation actions for Account and Portfolio *potential actions* that are only
executed if the corresponding elements do not yet exist. Similarly, applying the
rule ensureThatClientHasNoAccAndPortfolio in effect-oriented semantics allows
deleting nothing (if the matched Client has no Account and no Portfolio) or
only an Account (that does not have a Portfolio). So here the deletion actions
are potential actions that are only executed if the corresponding elements are
present. We will allow for some of the actions of an effect-oriented rule to be
mandatory. Note that in a Henshin rule, we would need to provide additional
annotation to differentiate mandatory from potential actions. We will define
different strategies for this kind of matching that may be appropriate for different
application scenarios.

3   Preliminaries

In this section, we briefly recall basic preliminaries. Throughout our paper, we
work with *typed graphs* and leave the treatment of attribution and type inheritance to future work. For brevity, we omit the definitions of *nested graph conditions* and their *shift* along morphisms [12,16]. We also omit basic notions
from category theory; in particular, we omit standard facts about adhesive categories [11,24] (of which typed graphs are an example). While these are needed
in our proofs, the core ideas in this work can be understood without their knowledge.

Definition 1 (Graph. Graph morphism). *A graph $G = (V_G, E_G, src_G,$
$tar_G)$ consists of a set of nodes (or vertices) V_G, a set of edges E_G, and source
and target functions $src_G, tar_G \colon E_G \to V_G$ that assign a source and a target
node to each edge.*

*A graph morphism $f = (f_V, f_E)$ from a graph G to a graph H is a pair of
functions $f_V \colon V_G \to V_H$ and $f_E \colon E_G \to E_H$ that both commute with the source
and target functions, i.e., such that $src_H \circ f_E = f_V \circ src_G$ and $tar_H \circ f_E =
f_V \circ tar_G$. A graph morphism is injective/surjective/bijective if both f_V and f_E
are. We denote injective morphisms via a hooked arrow, i.e., as $f \colon G \hookrightarrow H$.*

Typing helps equip graphs with meaning; a *type graph* provides the available
types for elements and morphisms assign the elements of *typed graphs* to those.

Definition 2 (Type graph. Typed graph). *Given a fixed graph TG (the
type graph), a typed graph $G = (G, type_G)$ (over TG) consists of a graph G
and a morphism $type_G \colon G \to TG$. A typed morphism $f \colon G \to H$ between typed
graphs G and H (typed over the same type graph TG) is a graph morphism that
satisfies $type_H \circ f = type_G$.*

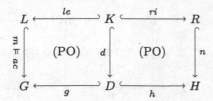

Fig. 2. A rule-based transformation in the Double-Pushout approach

Throughout this paper, we assume all graphs to be typed over a given type graph and all morphisms to be typed morphisms. However, for (notational) simplicity, we let this typing be implicit and just speak of graphs and morphisms. Moreover, all considered graphs are finite.

Definition 3 (Rules and transformations). *A rule $\rho = (p, ac)$ consists of a plain rule p and an* application condition *ac. The plain rule is a span of injective morphisms of typed graphs $p = (L \xleftarrow{le} K \xrightarrow{ri} R)$; its graphs are called* left-hand side (LHS), interface, *and* right-hand side (RHS), *respectively. The application condition ac is a nested condition [16] over L.*

Given a rule $\rho = (L \xleftarrow{le} K \xrightarrow{ri} R, ac)$ and an injective morphism $m \colon L \hookrightarrow G$, a (direct) transformation $G \Longrightarrow_{\rho,m} H$ from G to H (in the Double-Pushout approach) is given by the diagram in Fig. 2 where both squares are pushouts and m satisfies the application condition ac, denoted as $m \models ac$. If such a transformation exists, the morphism m is called a match *and rule ρ is applicable at match m; in this case, n is called the* comatch *of the transformation. An injective morphism $m \colon L \hookrightarrow G$ with $m \models ac$ from the LHS of a rule to some graph G is called a* pre-match.

For a rule to be applicable at a pre-match m, there must exist a pushout complement for $m \circ le$; in categories of graph-like structures, an elementary characterisation can be given in terms of the *dangling condition* [11, Fact 3.11]: A rule is applicable at a pre-match m if and only if m does not map a node to be deleted in L to a node in G with an incident edge that is not also to be deleted.

Application conditions can be 'shifted' along morphisms in a way that preserves their semantics [12, Lemma 3.11]. We presuppose this operation in our definition of *subrules* without repeating it. Our notion of a subrule is a simplification of the concept of *kernel* and *multi-rules* [15].

Definition 4 (Subrule). *Given a rule $\rho = (L \xleftarrow{le} K \xrightarrow{ri} R, ac)$, a subrule of ρ is a rule $\rho' = (L' \xleftarrow{le'} K' \xrightarrow{ri'} R', ac')$ together with a subrule embedding $\iota \colon \rho' \hookrightarrow \rho$ where $\iota = (\iota_L, \iota_K, \iota_R)$ and $\iota_X \colon X' \hookrightarrow X$ is an injective morphism for $X \in \{L, K, R\}$ such that both squares in Fig. 3 are pullbacks and $ac \equiv \mathrm{Shift}(\iota_L, ac')$.*

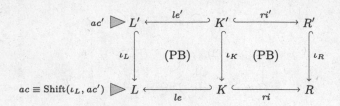

Fig. 3. Subrule ρ' of a rule ρ

4   Effect-Oriented Rules and Transformations

The intuition behind effect-oriented semantics is that a rule prescribes the state that should prevail after its application, not the actions to be performed. In this section, we develop this approach. We introduce *effect-oriented rules* as a compact way to represent a whole set of *induced rules*. All induced rules share a common *base rule* as subrule (prescribing actions to be definitively performed) but implement different choices of the *potential actions* allowed by the effect-oriented rule. In a second step, we develop a semantics for effect-oriented rules; it depends on the larger context of effect-oriented transformations which of the induced rules is actually applied. Here, we implement the idea that potential deletions of an effect-oriented rule are to be performed if a suitable element exists but can otherwise be skipped. In contrast, a potential creation is only to be performed if there is not yet a suitable element. This maximises the number of deletions to be made while minimising the number of creations. We propose two ways in which this 'maximality' and 'minimality' can be formally defined.

4.1   Effect-Oriented Rules as Representations of Rule Sets

In an *effect-oriented rule*, a *maximal rule* extends a *base rule* by *potential actions*. Here, and in all of the following, we assume that the left and right morphisms of rules and morphisms between rules (such as subrule embeddings) are actually inclusions. This does not lose generality (as the desired situation can always be achieved via renaming of elements) but significantly eases the presentation.

Definition 5 (Effect-oriented rule). *An* effect-oriented rule $\rho_e = (\rho_b, \rho_m, \iota)$ *is a rule* $\rho_m = (L_m \xleftarrow{le_m} K_m \xrightarrow{ri_m} R_m, ac_m)$, *called* maximal rule, *together with a subrule* $\rho_b = (L_b \xleftarrow{le_b} K_b \xrightarrow{ri_b} R_b, ac_b)$, *called* base rule, *and a subrule embedding* $\iota \colon \rho_b \hookrightarrow \rho_m$ *such that* $K_b = K_m$ *(and* ι_K *is an identity).*

The potential deletions *of the maximal rule* ρ_m *are the elements of* $(L_m \setminus K_m) \setminus L_b = L_m \setminus L_b$; *analogously, its* potential creations *are the elements of* $(R_m \setminus K_m) \setminus R_b = R_m \setminus R_b$. *Here, and in the following, '\' denotes the componentwise difference on the sets of nodes and edges.*

While requiring $K_b = K_m$ restricts the expressiveness of effect-oriented rules, it suffices for our purposes and allows for simpler definitions of their matching.

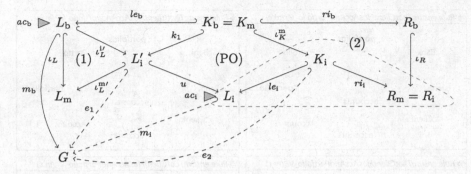

Fig. 4. Construction of an induced rule ρ_i (indicated via the red, dashed border) and of its match

If, during matching, potential actions would compete with potential interface elements for elements to which they can be mapped, developing notions of maximality of matches becomes more involved.

Example 1. We consider the rules from Fig. 1 as the maximal rules of effect-oriented rules. In each case, there are different possibilities as to which subrule of the rule to choose as the base rule. Our convention that the interfaces of the base and maximal rules of an effect-oriented rule coincide specifies that in each case the base rule contains at least the interface that is to be preserved. This minimal choice renders all deletions and creations potential.

Specifically, for the rule ensureThatClientHasAccAndPortfolio, one can assume that all elements to be created belong only to the maximal rule and represent potential creations. A possible alternative is to consider as the base rule the rule that creates Account (together with its incoming edges), making the creation of a Portfolio and its incoming edges potential. Further combinations are possible.

An effect-oriented rule ρ_e represents a set of *induced rules*. The induced rules are constructed by extending the base rule of ρ_e with potential deletions and creations from the maximal rule. However, we require every induced rule to have the same RHS as the maximal rule. Potential creations are omitted in an induced rule by also incorporating them into the interface. This ensures that the state that is represented by the RHS of the maximal rule holds after applying an induced rule, even if not all potential creations are performed.

Definition 6 (Induced rules). *Given an effect-oriented rule $\rho_e = (\rho_b, \rho_m, \iota)$, every rule $\rho_i = (L_i \xleftarrow{le_i} K_i \xrightarrow{ri_i} R_i, ac_i)$ is one of its induced rules if it is constructed in the following way (see (the upper part of) Fig. 4):*

1. *There is a factorisation (1) $\iota_L = \iota_L^{m'} \circ \iota_L^{i'}$ of $\iota_L \colon L_b \hookrightarrow L_m$ into two inclusions $\iota_L^{m'}$ and $\iota_L^{i'}$.*
2. *There is a factorisation (2) $\iota_R \circ ri_b = ri_m = ri_i \circ \iota_K^m$ of $ri_m \colon K_m \hookrightarrow R_m$ into two inclusions ri_i and ι_K^m such that the square (2) is a pullback.*

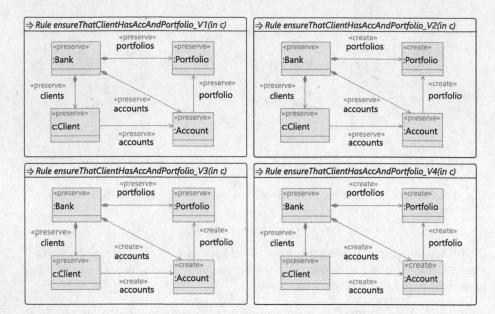

Fig. 5. Four induced rules arising from rule ensureThatClientHasAccAndPortfolio

3. (L_i, u, le_i) *are computed as pushout of the pair of morphisms* (k_1, ι_K^m), *where* $k_1 := \iota_L^{i\prime} \circ le_b$; *in that, we choose* L_i *such that* u *and* le_i *become inclusions (employing renaming if necessary).*

4. *The application condition* ac_i *is computed as* $\mathrm{Shift}(\iota_L^i, ac_b)$, *where* $\iota_L^i := u \circ \iota_L^{i\prime}$.

   The size of an induced rule ρ_i *is defined as* $|\rho_i| := |L_i' \setminus L_b| + |K_i \setminus K_b|$.

Example 2. Consider again rule ensureThatClientHasAccAndPortfolio as a maximal rule and its preserved elements as the base rule. This effect-oriented rule has 26 induced rules, namely all rules that stereotype some of the «create»-elements of ensureThatClientHasAccAndPortfolio as «preserve». Figure 5 shows a selection of those, namely induced rules that reduce undesired reuse of elements. These are the rules where, *together with a node that is to be preserved (instead of being created), all adjacent edges that lead to preserved elements are also preserved.* We call this the *weak connectivity condition* and discuss and formalise it in [22]. An example for an induced rule that is not depicted is the rule that creates a new portfolio-edge between existing Accounts and Portfolios.

Next, consider the effect-oriented rule where the maximal rule ensureThatClientHasAccAndPortfolio is combined with the base rule that already creates an Account with its two incoming edges. Here, the Account cannot become a context in any induced rule because its creation is already required by the base rule. (This is ensured by the factorisation (2) in Fig. 4 being a pullback.) The induced rules of ensureThatClientHasNoAccAndPortfolio are obtained in a similar way.

Our first result states that an induced rule actually contains the base rule of its effect-oriented rule as a subrule. In particular, this ensures that, if an induced rule is applied, all actions specified by the base rule are performed.

Proposition 1 (Base rule as subrule of induced rule). *If ρ_i is an induced rule of the effect-oriented rule $\rho_e = (\rho_b, \rho_m, \iota)$, then ρ_b is a subrule of ρ_i via the embedding $(\iota_L^i, \iota_K^i, \iota_R^i)$, where $\iota_L^i := u \circ \iota_L^{i\prime}$, $\iota_K^i := \iota_K^m$ and $\iota_R^i := \iota_R$ (compare Fig. 4).*

Next, we show that an effect-oriented rule indeed compactly represents a potentially large set of rules. The exact number of induced rules depends on how the edges of the maximal rule are connected.

Proposition 2 (Number of induced rules). *For an effect-oriented rule $\rho_e = (\rho_b, \rho_m, \iota)$, the number n of its induced subrules (up to isomorphism) satisfies:*

$$2^{|V_{L_m} \setminus V_{L_b}| + |V_{R_m} \setminus V_{R_b}|} \le n \le 2^{|V_{L_m} \setminus V_{L_b}| + |E_{L_m} \setminus E_{L_b}| + |V_{R_m} \setminus V_{R_b}| + |E_{R_m} \setminus E_{R_b}|} \ .$$

While our definition of induced rules is intentionally liberal, in many application cases it may be sensible to limit the kind of considered induced rules to avoid undesired reuse (e.g., connecting an Account to an already existing Portfolio of another Client). In [22], we provide a definition that enables that.

4.2   Matching Effect-Oriented Rules

In this section, we develop different ways to match effect-oriented rules. *Effect-oriented transformations* get their semantics from 'classical' Double-Pushout transformations using the induced rules of the applied effect-oriented rule. We will develop different ways in which an existing context determines which induced rule of an effect-oriented rule should be applied at which match.

We assume a pre-match for the base rule of an effect-oriented rule to be given and try to to extend this pre-match to a match for an appropriate induced rule. The notion of *compatibility* captures this extension relationship. In [22], we present two technical lemmas that further characterise compatible matches.

Definition 7 (Compatibility). *Given an effect-oriented rule $\rho_e = (\rho_b, \rho_m, \iota)$, a pre-match $m_b : L_b \hookrightarrow G$ for its base rule and a match $m_i : L_i \hookrightarrow G$ for one of its induced rules ρ_i are compatible if $m_i \circ \iota_L^i = m_b$, where $\iota_L^i = u \circ \iota_L^{i\prime} : L_b \hookrightarrow L_i$ stems from the subrule embedding of ρ_b into ρ_i (compare Fig. 4 and Proposition 1).*

An induced rule ρ_i can be matched compatibly to m_b if it has a match m_i such that m_b and m_i are compatible.

Given an effect-oriented rule and a pre-match for its base rule, there can be many different induced rules for which there is a compatible match. The following definition introduces different strategies for selecting such a rule and match, so that the corresponding applications form transformations that are complete in terms of deletion and creation actions to achieve the intended effect, which is why they are called effect-oriented transformations. Their common core is that in any effect-oriented transformation, a pre-match of a base rule is extended by potential creations and deletions from the maximal rule such that no further extension is possible. Intuitively, this ensures that all possible potential deletions but only necessary potential creations are performed (in a sense we make formally precise in Theorem 1). A stricter notion is to maximise the number of reused elements.

Definition 8 (Local completeness. Maximality. Effect-oriented transformation). *Given a pre-match $m_b \colon L_b \to G$ for the base rule ρ_b of an effect-oriented rule $\rho_e = (\rho_b, \rho_m, \iota)$, and a match m_i for one of its induced rules ρ_i that is compatible with m_b, ρ_i and m_i are locally complete w.r.t. m_b if (see Fig. 4):*

1. *Local completeness of additional deletions: Any further factorisation $\iota_L = \iota_L^{m''} \circ \iota_L^{iv}$ into inclusions (with domain resp. co-domain L_i'') such that there exists a non-bijective inclusion $j \colon L_i' \hookrightarrow L_i''$ with $j \circ \iota_L^{iv} = \iota_L^{iv'}$ and $\iota_L^{m''} \circ j = \iota_L^{m'}$ meets one of the following two criteria.*
   - *Not matchable: There is no injective morphism $e_1' \colon L_i'' \hookrightarrow G$ with $e_1' \circ \iota_L^{iv'} = m_b$.*
   - *Not applicable: Such an e_1' exists, but the morphism $m_i' \colon L_i'' \to G$ which it induces together with the* right *extension match e_2 of m_i (where L_i'' is the LHS of the induced rule that corresponds to this further factorisation) is not injective.*

2. *Local completeness of additional creations: Any further factorisation $ri_m = ri_i' \circ \iota_K^{m'}$ into inclusions (with domain resp. co-domain K_i') such that there is a non-bijective inclusion $j \colon K_i \hookrightarrow K_i'$ with $j \circ \iota_K^m = \iota_K^{m'}$ and $ri_i' \circ j = ri_i$ meets one of the following two criteria.*
   - *Not matchable: There is no injective morphism $e_2' \colon K_i' \hookrightarrow G$ with $e_2' \circ \iota_K^{m'} = m_b \circ le_b$.*
   - *Not applicable: Such an e_2' exists, but the morphism $m_i' \colon L_i'' \to G$ which it induces together with the* left *extension match e_1 of m_i (where L_i'' is the LHS of the induced rule that corresponds to this further factorisation) is not injective.*

An effect-oriented transformation $t \colon G \Longrightarrow H$ via ρ_e is a double-pushout transformation $t \colon G \Longrightarrow_{\rho_i, m_i} H$, where ρ_i is an induced rule of ρ_e and ρ_i is locally complete w.r.t. $m_b := m_i \circ \iota_L^i$, the induced pre-match for ρ_b. The semantics of an effect-oriented rule is the collection of all of its effect-oriented transformations.

A transformation $t \colon G \Longrightarrow_{\rho_i, m_i} H$ via an induced rule ρ_i of a given effect-oriented rule ρ_e is globally maximal (w.r.t. G) if for any other transformation $t' \colon G \Longrightarrow_{\rho_i', m_i'} H'$ via an induced rule ρ_i' of ρ_e, it holds that $|\rho_i| \geq |\rho_i'|$. Such a transformation t is locally maximal if for any other transformation $t' \colon G \Longrightarrow_{\rho_i', m_i'} H'$ via an induced rule ρ_i' of ρ_e where the induced pre-matches m_b and m_b' for the base rule coincide, it holds that $|\rho_i| \geq |\rho_i'|$. In all of these situations, we also call the match m_i and the rule ρ_i locally complete or locally/globally maximal.

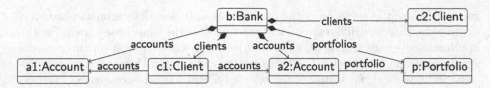

Fig. 6. Tiny example instance for the banking domain

Example 3. To illustrate the different kinds of matching for effect-oriented rules, we again consider ensureThatClientHasAccAndPortfolio as a maximal rule whose «preserve»-elements form the base rule and apply it according to different semantics to the example instance depicted in Fig. 6. First, we consider the base match that maps the Client-node of the rule to Client c1 in the instance. Extending this base match in a *locally complete* fashion requires one to reuse the existing Portfolio and one of the existing Accounts. Choosing Account a2 leads to induced rule ensureThatClientHasAccAndPortfolio_V1 (Fig. 5) because there already exists an edge to Portfolio p. In contrast, choosing Account a1 leads to a transformation that creates a portfolio-edge from a1 to p (where the underlying induced rule is not depicted in Fig. 5). Both transformations are locally complete; in particular, locally complete matching is not deterministic. If, for semantic reasons, one wants to avoid transformations like the second one and only allows the induced rules that are depicted in Fig. 5, applying ensureThatClientHasAccAndPortfolio_V2 at Account a1 becomes locally complete.

The unique match for ensureThatClientHasAccAndPortfolio_V1 is the only locally maximal match compatible with the chosen base match in our example and is also globally maximal. To see a locally maximal match that is not globally maximal, we consider the base match that maps to Client c2 instead of c1. Here, the locally maximal match reuses a2, p and the portfolio-edge between them and creates the missing edges from c2 to a2 and p. Choosing Account a1 instead of a2 does not provide a locally maximal match as one cannot reuse a portfolio-edge then (reducing the size of the induced rule by 1). The globally maximal match remains unchanged as, by definition, it does not depend on a given base match. It is that evident that in a larger example, every Client with an Account with Portfolio constitutes a globally maximal match. Thus, also globally maximal matching is non-deterministic. In fact, one can even construct examples where globally maximal matches for different induced rules (of equal size) exist.

Unlike potential creations, potential deletions may require backtracking to find a match. To see this, consider the rule ensureThatClientHasNoAccAndPortfolio as a maximal rule whose «preserve»-elements form the base rule. Assuming that an Account can be connected to multiple Clients, a locally complete prematch for an induced rule is not automatically a match for it. One has to look for an Account that is only connected to the matched Client.

The above example shows that none of the defined notions of transformation is deterministic. The situation is similar to Double-Pushout transformations in general, where the selection of the match is usually non-deterministic; however, in our case, there are different possible outcomes for the same base match. For the applications we are aiming at, such as rule-based search, graph repair or model synchronisation (see Sect. 6.1), this is not a problem. In these, it is often sufficient to know that, for instance, the selected Client has an Account and Portfolio after applying the rule ensureThatClientHasAccAndPortfolio but not necessarily important which ones.

It is easy to see that every globally maximal transformation via an induced rule is also locally maximal, and that every locally maximal transformation is locally complete. The definition of an effect-oriented transformation thus captures the weakest case and also covers locally and globally maximal transformations.

Proposition 3 (Relations between different kinds of effect-oriented transformations). *Given an effect-oriented rule $\rho_e = (\rho_b, \rho_m, \iota)$ and a graph G, every transformation $t: G \Longrightarrow_{\rho_i, m_i} H$ via an induced rule ρ_i of ρ_e that is globally maximal is also locally maximal. Every locally maximal transformation is also locally complete for its induced pre-match m_b for the base rule ρ_b.*

Intuitively, the maximal rule of an effect-oriented rule specifies a selection of potential actions. The induced rules result from the different possible combinations of potential actions. The next theorem clarifies which effects can ultimately occur *after* an effect-oriented transformation: If an element remains for which a matching potential deletion was specified by the effect-oriented rule, one of two alternatives took place: Either, the potential deletion was performed but on a different element (*alternative action*)—if there is more than one way to match an element potentially to be deleted. Or, that element was matched to by a potential creation (*alternative creation*). The latter can happen when a rule specifies potential creations and deletions for elements of the same type (at comparable positions). Similarly, if x denotes a performed potential creation but there had been an element y to which x could have been matched, y was used by another potential creation or deletion (*alternative action*). In particular, Theorem 1 also shows that effect-oriented rules can specify alternative actions. Their application can be non-deterministic, where one of several possible actions is chosen at random.

Theorem 1 (Characterising effect-oriented transformations). *Let $\rho_e = (\rho_b, \rho_m, \iota)$ be an effect-oriented rule and $t: G \Longrightarrow_{\rho_i, m_i} H$ an effect-oriented transformation via one of its induced rules ρ_i (compare Fig. 4 for the following).*

Let $x \in L_m \setminus L_b$ be an element that represents a potential deletion of ρ_e and let K_b^+ be the extension of K_b with x (if defined as graph) and $\iota^+: K_b \hookrightarrow K_b^+$ the corresponding inclusion. If there exists an injective morphism $m^+: K_b^+ \hookrightarrow H$ with $m^+ \circ \iota^+ = n_i \circ \iota_R \circ ri_b$, where n_i is the comatch of t, then either

1. (Alternative action): *the element x belongs to L_i; in particular, an element of the same type as x (and in comparable position) was deleted from G by t; or*
2. (Alternative creation): *the element $m^+(x)$ of H has a pre-image from R_i under n_i, i.e., it was first created by t or matched by a potential creation.*

Similarly, let $x \in R_i \setminus (K_i \cup R_b)$ represent one of the potential creations of ρ_e that have been performed by t. Let K_b^+ be the extension of K_b with x (if defined as graph) and $\iota^+ : K_b \hookrightarrow K_b^+$ the corresponding inclusion. Then either

1. *(Alternative action): for every injective morphism $m_b^+ : K_b^+ \hookrightarrow G$ with $m_b^+ \circ \iota^+ = m_i \circ le_i \circ \iota_K^m$, the element $m_b^+(x) \in G$ has a pre-image from L_i under m_i (i.e., it is already mapped to by another potential action); or*
2. *(Non-existence of match): no injective morphism $m_b^+ : K_b^+ \hookrightarrow G$ with $m_b^+ \circ \iota^+ = m_i \circ le_i \circ \iota_K^m$ exists.*

5 Locally Complete Matches—Algorithm and Implementation

In this section, we present an algorithm for the computation of a locally complete match (and a corresponding induced rule) from an effect-oriented rule and a pre-match for its base rule. Starting with such a pre-match is well-suited for practical applications such as model repair and rule-based search, where a match of the base rule is often already fixed and needs to be complemented by (some of) the actions of the maximal rule. Note that this pre-match is common to all matches of all induced rules. Searching for the pre-match once and extending it contextually is generally much more efficient than searching for the matches of the induced rules from scratch. Moreover, we simultaneously compute a locally complete match and its corresponding induced rule and thus avoid first computing all induced rules (of which there can be many, cf. Proposition 2) and trying to match them in order of their size. The correctness of our algorithm is shown in Theorem 2 below. Note that we have focussed on the correctness of the algorithm and, apart from the basic efficiency consideration above, further optimisations for efficiency (incorporating ideas from [3]) are reserved for future work. We provide a short comment on our use of backtracking in [22]. In Sect. 5.2 we report on a prototype implementation of effect-oriented transformations in Henshin using this algorithm.

5.1 An Algorithm for Computing Locally Complete Matches

We consider the problem of finding a locally complete match from a given pre-match. So-called *rooted rules*, i.e., rules where a partial match is fixed (or at least can be determined in constant time), have been an important part of the development of rule-based algorithms for graphs that run in linear time [3,6]. Note that we are not looking for an induced rule and match that lead to a maximal transformation but only to a non-extensible, i.e., locally complete one. This has the effect that the dangling-edge condition remains the only possible source of backtracking in the matching process.

In Algorithm 1, we outline a function that extends a pre-match for a base rule of an effect-oriented rule to a compatible, locally complete match for a corresponding induced rule. The input to our algorithm is an effect-oriented rule

$\rho_e = (\rho_b, \rho_m, \iota)$ (the parameter *rule*), a graph G (the parameter *graph*) and a pre-match $m_b \colon L_b \hookrightarrow G$ for ρ_b (the parameter *currentMappings*). It returns a match m_i for an induced rule ρ_i of ρ_e such that m_i and m_b are compatible and m_i is locally complete; it returns null if and only if no such compatible, locally complete match exists. We outline the matching of nodes and consequently consider *currentMappings* to be a list of node mappings; from this, one can infer the matching of edges. The computed match also represents the corresponding induced rule. We provide the details for these conventions in [22].

The search for a match starts with initialising *unboundNodes* with the potential actions of the given effect-oriented rule (line 5). Then the function *findExtension* recursively tries to match those, extending the pre-match (line 6). Function *findExtension* works as follows. For the unbound node n at the currently considered *position*, all available candidates, i.e., all nodes x in the graph G to which n can be mapped, are collected with the function *findExtensionCandidates* (line 10). A candidate x must satisfy the following properties: (i) no other node may already map to x, i.e., x does not yet occur in *currentMappings* (injectivity condition) and (ii) the types of n and x must coincide (consistency condition). If candidates exist, for each candidate x, the algorithm tries to map n to x until a solution is found. To do this, the set of current mappings is extended by the pair (n, x) (line 13). If n was the last node to be matched (line 14) and the morphism defined by *currentMappings* satisfies the dangling-edge condition (line 15), the result is returned as solution (line 16). If the dangling-edge condition is violated, the selected candidate is removed and the next one is tried (line 17). If n was not the last node to be mapped (line 18), the function findExtension is called for the extended list of current mappings and the next unmatched node from *unboundNodes* (line 19). If this leads to a valid solution, this solution is returned (lines 20–21). Otherwise, the pair (n, x) is removed from *currentMappings* (line 23), and the next candidate is tried. If candidates exist but none of them lead to a valid solution, null is returned (line 33). If no candidate exists (line 25), either the current mapping or null is returned as the solution (if n was the last node to assign; lines 26–29) or *findExtension* is called for the next *position* (line 31), i.e., the currently considered node n is omitted from the mapping (and, hence, from the induced rule).

Algorithm 1. Computation of a locally complete match

```
1  input: effect−oriented rule (ρb, ρm, ι), graph G, and a pre−match mb
2  output: locally complete match mi compatible with mb
3
4  function findLocallyCompleteMatch(rule, graph, currentMappings)
5     unboundNodes = V_Lm \ V_Lb ⊔ V_Rm \ V_Rb;
6     return findExtension(currentMappings, graph, unboundNodes, 0);
7
8  function findExtension(currentMappings, graph, unboundNodes, position)
9     n = unboundNodes.get(position);
10    candidates = findExtensionCandidates(currentMappings, graph, n);
11    if (!candidates.isEmpty())
12       for each x in candidates
```

```
13        currentMappings.put(n,x);
14        if (position == unboundNodes.size() − 1) //last node to be matched
15          if (danglingEdgeCheck(graph, currentMappings))
16            return currentMappings;
17          else currentMappings.remove(n,x);
18        else //map next unbound node
19          nextSolution = findExtension(currentMappings, graph, unboundNodes,
                  position + 1);
20          if (nextSolution != null)
21            return nextSolution;
22          else //try next candidate
23              currentMappings.remove(n,x);
24      end for //no suitable candidate found
25    else //there is no candidate for the current node
26      if (position == unboundNodes.size() − 1) //last node to be matched
27        if (danglingEdgeCheck(graph, currentMappings))
28          return currentMappings;
29        else return null;
30      else //map next unbound node
31        return findExtension(currentMappings, graph, unboundNodes, position+1);
32    //no candidate led to a valid mapping
33    return null;
```

Theorem 2 (Correctness of Algorithm 1). *Given an effect-oriented rule* $\rho_e = (\rho_b, \rho_m, \iota)$, *a graph* G, *and a pre-match* $m_b \colon L_b \hookrightarrow G$, *Algorithm 1 terminates and computes an induced rule* ρ_i *with a match* m_i *such that* m_i *and* m_b *are compatible and* ρ_i *is locally complete w.r.t.* m_b. *In particular, Algorithm 1 returns* null *if and only if no induced rule of* ρ_e *can be matched compatibly with* m_b *and returns* ρ_b *as induced rule with match* m_b *if and only if* m_b *is locally complete and a match.*

5.2   Implementation

In this section, we present a prototypical implementation of effect-oriented transformations in Henshin [1,29], a model transformation language based on graph transformation. From the user perspective, the implementation includes two major classes for applying a given effect-oriented rule $\rho_e = (\rho_b, \rho_m, \iota)$ to a host graph G: the class `LocallyCompleteMatchFinder` for finding a locally complete match, and the class `EffectOrientedRuleApplication` to apply a rule $\rho_e = (\rho_b, \rho_m, \iota)$ at such a match. We assume that $\rho_e = (\rho_b, \rho_m, \iota)$ is provided as simple Henshin rule representing the maximal rule ρ_m, where the base rule ρ_b is implicitly represented by the preserved part of the rule. The host graph G is provided, as usual in Henshin, in the form of a model instance for a given meta-model (representing the type graph).

The implementation follows the algorithm presented in Sect. 5.1. Our main design goal was to reuse the existing interpreter core of Henshin, with its functionalities for matching and rule applications, as much as possible. In particular, in `LocallyCompleteMatchFinder`, we derive the base rule ρ_b by creating a copy of ρ_m with creations and deletions removed, and feeding it into the interpreter core to obtain a pre-match m_b on G. For cases where a pre-match m_b can be found, we provide an implementation of Algorithm 1 that produces a partial match $\hat{m}_m$ incorporating the mappings of m_b and additional mappings for elements to be deleted and elements not to be created. In order to treat elements of different actions consistently, we perform these steps on an intermediate rule ρ_{gr}, called the *grayed* rule, in which creations are converted to preserve actions. In `EffectOrientedRuleApplication`, we first derive the induced rule ρ_i from $\hat{m}_m$, such that $\hat{m}_m$ is a complete match for ρ_i. For the actual rule application, we feed ρ_i together with $\hat{m}_m$ into the Henshin interpreter core using a classical rule application.

We have tested the implementation using our running example. For this purpose, we specified all rules and an example graph. Our implementation behaved completely as expected. The source code of the implementation and the example are available online at https://github.com/dstrueber/effect-oriented-gt.

6   Related Work

In this section, we describe how existing practical applications could benefit from the use of effect-oriented transformations (Sect. 6.1) and relate effect-oriented graph transformations to other graph transformation approaches (Sect. 6.2).

6.1   Benefiting from Effect-Oriented Graph Transformation

There are several application cases in the literature where graph transformation has been used to achieve certain states. In the following, we recall graph repair, where a consistent graph is to be achieved, and model synchronisation, where consistent model relations are to be achieved after one of the models has been changed. A slightly different case is service matching, where a specified service should be best covered by descriptions of existing services.

In their rule-based approach to *graph repair* [28], Sandmann and Habel repair (sub-)conditions of the form $\exists(a\colon B \hookrightarrow C)$ using the (potentially large) set of rules that, for every graph B' between B and C, contains the rule that creates C from B'. Negative application conditions (NACs) ensure that the rule with the largest possible B' as LHS is selected during repair. With effect-oriented graph transformation, the entire set of rules derived by them can be represented by a single effect-oriented rule that has the identity on B as base rule and $B \hookrightarrow C$ as maximal rule. Moreover, we do not need to use NACs, since locally complete matching achieves the desired effect.

In the context of model synchronisation, a very similar situation occurs in [26]. There, Orejas et al. define consistency between pairs of models via *patterns*. For synchronisation, a whole set of rules is derived from a single pattern to account for the different ways in which consistency might be restored (i.e., to create the missing elements). Again, NACs are used to control the application of the rules. As above, we can represent the whole set of rules as a single effect-oriented rule.

Fritsche, Kosiol, et al. extend *TGG-based model synchronisation processes* to achieve higher incrementality using special repair rules [13,14,20]. Elements to be deleted according to these rules may be deleted for other reasons during the synchronisation process, destroying the matches needed for the repair rules. In [14], this problem is avoided by only considering edits where this cannot happen. In [13], Fritsche approaches this problem pragmatically by omitting such deletions on-the-fly—if an element is already missing that needs be deleted to restore consistency, the consistency has already been restored locally and the deletion can simply be skipped. More formally, Kosiol in [20] presents a set of subrules of a repair rule, where the maximal one matchable can always be chosen to perform the propagation. This whole set of rules can also be elegantly represented by a single effect-oriented rule.

In [2], Arifulina addresses the heterogeneity of *service specifications and descriptions*. She develops a method for matching service specifications by finding a maximal partial match for a rule that specifies a service. Apart from the fact that Arifulina allows the partial match to also omit context elements, the problem can also be formulated as finding a globally maximal match (in our sense) of the service-specifying rule in the LHS of available service description rules.

6.2   Relations to Other Graph Transformation Approaches

There are several approaches to graph transformation that take the variability of the transformation context into account. We recall each of them briefly and discuss the commonalities and differences to effect-oriented graph transformation.

Other Semantics for Applying Single Transformation Rules. Graph transformation approaches such as the single pushout approach [25], the sesquipushout approach [9], AGREE [7], PBPO [8], and PBPO$^+$ rewriting [27] are more expressive than DPO rewriting, as they allow some kind of copying or merging of elements or (implicitly specified) side effects. Rules are defined as (extended) spans in all these approaches. Therefore, they also specify sets of actions that must be executed in order to apply a rule. AGREE, PBPO, and PBPO$^+$ would enable one to specify what we call potential deletions; however, in these approaches their specification is far more involved. None of the mentioned approaches supports specifying potential creations that can be omitted depending on the currently considered application context.

In the *Double-Pullback approach*, 'a rule specifies only what at least has to happen on a system's state, but it allows to observe additional effects which may be caused by the environment' [18, p. 85]. In effect-oriented graph transformation, the base rule also specifies what has to happen as a minimum. But the additional effects are not completely arbitrary, as the maximal rule restricts the additional actions. Moreover, these additional actions are to be executed only if the desired state that they specify does not yet exist. This suggests that double-pullback transformations are a more general concept than effect-oriented transformations with locally complete matching.

Effect-oriented Transformations via Multiple Transformation Rules. In the following, we discuss how graph transformation concepts that apply several rules in a controlled way can be used to emulate effect-oriented transformations.

Graph programs [6,17] usually provide control constructs for rule applications such as sequential application, conditional and optional applications, and application loops. To emulate effect-oriented graph transformations, the base rule would be applied first and only once. For each induced rule, we would calculate the remainder rule, which is the difference to the base rule, i.e., it specifies all actions of the induced rule that are not specified in the base rule. To choose the right remainder rule, we would need a set of additional rules that check which actions still need to be executed in the given instance graph. Depending on these checks, the appropriate remainder rule is selected and applied.

Amalgamated transformations [4,12] are useful when graph transformation with universally quantified actions are required. They provide a formal basis for *interaction schemes* where a *kernel rule* is applied exactly once and additional *multi-rules*, extending the kernel rule, are applied as often as possible at matches extending the one of the kernel rule. To emulate the behaviour of an effect-oriented transformation by an interaction scheme, the basic idea is to generate the set of multi-rules as all possible induced rules. Application conditions could be used to control that the 'correct' induced rule is applied.

A compact representation of a rule with several variants is given by *variability-based (VB) rules* [31]. A VB rule represents a set of rules with a common core. Elements that only occur in a subset of the rules are annotated with so-called *presence conditions*. A VB rule could compactly represent the set of induced rules of an effect-oriented rule, albeit in a more complicated way, by explicitly defining a list of features and using them to annotate variable parts. An execution semantics of VB rules has been defined for single graphs [31] and sets of variants of graphs [30]. However, for VB rules, the concept of driving the instance selection by the availability of a match with certain properties has not been developed.

7   Conclusion

Effect-oriented graph transformation supports the modelling of systems in a more declarative way than the graph transformation approaches in the literature. The specification of basic actions is accompanied by the specification of desired states to be achieved. Dependent on the host graph, the application of a base rule is extended to the application of an induced rule that performs exactly the actions required to achieve the desired state. We have discussed that effect-oriented transformations are well suited to specify graph repair and model synchronisation strategies, since change actions can be accompanied by actions that restore consistency within a graph or between multiple (model) graphs. We have outlined how existing approaches to graph transformation can be used to emulate effect-oriented transformation but lead to accidental complexity that effect-oriented transformations can avoid.

In the future, we are especially interested in constructing effect-oriented rules that induce consistency-sustaining and -improving transformations [21]. Examining the computational complexity of different approaches for their matching, developing efficient algorithms for the computation of their matches, elaborating conflict and dependency analysis for effect-oriented rules, and combining effect-orientation with multi-amalgamation are further topics of theoretical and practical interest.

Acknowledgements. We thank the anonymous reviewers for their constructive feedback. Parts of the research for this paper have been performed while J. K. was a Visiting Research Associate at King's College London. This work has been partially supported by the Deutsche Forschungsgemeinschaft (DFG), grant TA 294/19-1.

References

1. Arendt, T., Biermann, E., Jurack, S., Krause, C., Taentzer, G.: Henshin: advanced concepts and tools for in-place EMF model transformations. In: Petriu, D.C., Rouquette, N., Haugen, Ø. (eds.) Model Driven Engineering Languages and Systems – 13th International Conference, MODELS 2010, Oslo, Norway, October 3–8, 2010, Proceedings, Part I. Lecture Notes in Computer Science, vol. 6394, pp. 121–135. Springer (2010). https://doi.org/10.1007/978-3-642-16145-2_9
2. Arifulina, S.: Solving heterogeneity for a successful service market. Ph.D. thesis, University of Paderborn, Germany (2017). https://doi.org/10.17619/UNIPB/1-13
3. Bak, C., Plump, D.: Rooted graph programs. Electron. Commun. Eur. Assoc. Softw. Sci. Technol. **54** (2012). https://doi.org/10.14279/tuj.eceasst.54.780
4. Boehm, P., Fonio, H.-R., Habel, A.: Amalgamation of graph transformations with applications to synchronization. In: Ehrig, H., Floyd, C., Nivat, M., Thatcher, J. (eds.) CAAP 1985. LNCS, vol. 185, pp. 267–283. Springer, Heidelberg (1985). https://doi.org/10.1007/3-540-15198-2_17
5. Burdusel, A., Zschaler, S., John, S.: Automatic generation of atomic multiplicity-preserving search operators for search-based model engineering. Softw. Syst. Model. **20**(6), 1857–1887 (2021). https://doi.org/10.1007/s10270-021-00914-w
6. Campbell, G., Courtehoute, B., Plump, D.: Fast rule-based graph programs. Sci. Comput. Program. **214**, 102727 (2022). https://doi.org/10.1016/j.scico.2021.102727
7. Corradini, A., Duval, D., Echahed, R., Prost, F., Ribeiro, L.: AGREE – algebraic graph rewriting with controlled embedding. In: Parisi-Presicce, F., Westfechtel, B. (eds.) ICGT 2015. LNCS, vol. 9151, pp. 35–51. Springer, Cham (2015). https://doi.org/10.1007/978-3-319-21145-9_3
8. Corradini, A., Duval, D., Echahed, R., Prost, F., Ribeiro, L.: The PBPO graph transformation approach. J. Log. Algebr. Meth. Program. **103**, 213–231 (2019). https://doi.org/10.1016/j.jlamp.2018.12.003

9. Corradini, A., Heindel, T., Hermann, F., König, B.: Sesqui-pushout rewriting. In: Corradini, A., Ehrig, H., Montanari, U., Ribeiro, L., Rozenberg, G. (eds.) Graph Transformations, Third International Conference, ICGT 2006, Natal, Rio Grande do Norte, Brazil, September 17–23, 2006, Proceedings. Lecture Notes in Computer Science, vol. 4178, pp. 30–45. Springer (2006). https://doi.org/10.1007/11841883_4

10. Corradini, A., Montanari, U., Rossi, F., Ehrig, H., Heckel, R., Löwe, M.: Algebraic approaches to graph transformation - part I: basic concepts and double pushout approach. In: Rozenberg, G. (ed.) Handbook of Graph Grammars and Computing by Graph Transformations, Volume 1: Foundations, pp. 163–246. World Scientific (1997)

11. Ehrig, H., Ehrig, K., Prange, U., Taentzer, G.: Fundamentals of Algebraic Graph Transformation. Monographs in Theoretical Computer Science. An EATCS Series, Springer (2006). https://doi.org/10.1007/3-540-31188-2

12. Ehrig, H., Golas, U., Habel, A., Lambers, L., Orejas, F.: $\mathcal{M}$-adhesive transformation systems with nested application conditions. part 1: parallelism, concurrency and amalgamation. Math. Struct. Comput. Sci. **24**(4) (2014). https://doi.org/10.1017/S0960129512000357

13. Fritsche, L.: Local Consistency Restoration Methods for Triple Graph Grammars. Ph.D. thesis, Technical University of Darmstadt, Germany (2022). http://tuprints.ulb.tu-darmstadt.de/21443/

14. Fritsche, L., Kosiol, J., Schürr, A., Taentzer, G.: Avoiding unnecessary information loss: correct and efficient model synchronization based on triple graph grammars. Int. J. Softw. Tools Technol. Transf. **23**(3), 335–368 (2021). https://doi.org/10.1007/s10009-020-00588-7

15. Golas, U., Habel, A., Ehrig, H.: Multi-amalgamation of rules with application conditions in $\mathcal{M}$-adhesive categories. Math. Struct. Comput. Sci. **24**(4) (2014). https://doi.org/10.1017/S0960129512000345

16. Habel, A., Pennemann, K.: Correctness of high-level transformation systems relative to nested conditions. Math. Struct. Comput. Sci. **19**(2), 245–296 (2009). https://doi.org/10.1017/S0960129508007202

17. Habel, A., Plump, D.: Computational completeness of programming languages based on graph transformation. In: Honsell, F., Miculan, M. (eds.) Foundations of Software Science and Computation Structures, 4th International Conference, FOSSACS 2001 Held as Part of the Joint European Conferences on Theory and Practice of Software, ETAPS 2001 Genova, Italy, April 2–6, 2001, Proceedings. LNCS, vol. 2030, pp. 230–245. Springer (2001). https://doi.org/10.1007/3-540-45315-6_15

18. Heckel, R., Ehrig, H., Wolter, U., Corradini, A.: Double-pullback transitions and coalgebraic loose semantics for graph transformation systems. Appl. Categorical Struct. **9**(1), 83–110 (2001). https://doi.org/10.1023/A:1008734426504

19. Horcas, J.M., Strüber, D., Burdusel, A., Martinez, J., Zschaler, S.: We're not gonna break it! consistency-preserving operators for efficient product line configuration. IEEE Trans. Softw. Eng. (2022). https://doi.org/10.1109/TSE.2022.3171404

20. Kosiol, J.: Formal Foundations for Information-Preserving Model Synchronization Processes Based on Triple Graph Grammars. Ph.D. thesis, University of Marburg, Germany (2022). https://archiv.ub.uni-marburg.de/diss/z2022/0224

21. Kosiol, J., Strüber, D., Taentzer, G., Zschaler, S.: Sustaining and improving graduated graph consistency: a static analysis of graph transformations. Sci. Comput. Program. **214**, 102729 (2021)

22. Kosiol, J., Strüber, D., Taentzer, G., Zschaler, S.: Finding the right way to rome: effect-oriented graph transformation (2023). https://arxiv.org/abs/2305.03432
23. Krause, C.: Bank accounts example. Online (2023). https://wiki.eclipse.org/Henshin/Examples/Bank_Accounts
24. Lack, S., Sobociński, P.: Adhesive and quasiadhesive categories. RAIRO Theor. Inform. Appl. **39**(3), 511–545 (2005). https://doi.org/10.1051/ita:2005028
25. Löwe, M.: Algebraic approach to single-pushout graph transformation. Theor. Comput. Sci. **109**(1&2), 181–224 (1993). https://doi.org/10.1016/0304-3975(93)90068-5
26. Orejas, F., Guerra, E., de Lara, J., Ehrig, H.: Correctness, completeness and termination of pattern-based model-to-model transformation. In: Kurz, A., Lenisa, M., Tarlecki, A. (eds.) Algebra and Coalgebra in Computer Science, Third International Conference, CALCO 2009, Udine, Italy, 7–10 September 2009. Proceedings. LNCS, vol. 5728, pp. 383–397. Springer (2009). https://doi.org/10.1007/978-3-642-03741-2_26
27. Overbeek, R., Endrullis, J., Rosset, A.: Graph rewriting and relabeling with pbpo$^+$. In: Gadducci, F., Kehrer, T. (eds.) Graph Transformation – 14th International Conference, ICGT 2021, Held as Part of STAF 2021, Virtual Event, June 24–25, 2021, Proceedings. LNCS, vol. 12741, pp. 60–80. Springer (2021). https://doi.org/10.1007/978-3-030-78946-6_4
28. Sandmann, C., Habel, A.: Rule-based graph repair. In: Echahed, R., Plump, D. (eds.) Proceedings Tenth International Workshop on Graph Computation Models, GCM@STAF 2019, Eindhoven, The Netherlands, 17th July 2019. EPTCS, vol. 309, pp. 87–104 (2019). https://doi.org/10.4204/EPTCS.309.5
29. Strüber, D., Born, K., Gill, K.D., Groner, R., Kehrer, T., Ohrndorf, M., Tichy, M.: Henshin: A usability-focused framework for EMF model transformation development. In: ICGT 2017: International Conference on Graph Transformation, pp. 196–208. Springer, Cham (2017). https://doi.org/10.1007/978-3-319-61470-0_12
30. Strüber, D., Peldszus, S., Jürjens, J.: Taming multi-variability of software product line transformations. In: FASE, pp. 337–355 (2018). https://doi.org/10.1007/978-3-319-89363-1_19
31. Strüber, D., Rubin, J., Arendt, T., Chechik, M., Taentzer, G., Plöger, J.: Variability-based model transformation: formal foundation and application. Formal Aspects Comput. **30**(1), 133–162 (2017). https://doi.org/10.1007/s00165-017-0441-3

Moving a Derivation Along a Derivation Preserves the Spine

Hans-Jörg Kreowski, Sabine Kuske, Aaron Lye, and Aljoscha Windhorst[✉]

University of Bremen, Department of Computer Science,
P.O.Box 33 04 40, 28334 Bremen, Germany
{kreo,kuske,lye,windhorst}@uni-bremen.de

Abstract. In this paper, we investigate the relationship between two elementary operations on graph-transformational derivations: moving a derivation along a derivation based on parallel and sequential independence on one hand and restriction of a derivation by clipping off vertices and edges that are never matched by a rule application throughout the derivation on the other hand. As main result, it is shown that moving a derivation preserves its spine being the minimal restriction.

1 Introduction

A major part of the study of graph transformation concerns the fundamental properties of derivations. In this paper, we contribute to this study by relating two operations on derivations both based on long-known concepts: *moving* and *restriction*.

The first operation, introduced in Sect. 3, moves derivations along derivations. We consider two variants: *forward moving* and *backward moving*. It is well-known and often considered that two applications of rules to the same graph can be applied one after the other with the same result if they are parallel independent. As an operation, this strong confluence may be called *conflux*. Analogously, two successive rule applications can be interchanged if they are sequentially independent, called *interchange* as an operation. The situation is depicted in Fig. 1.

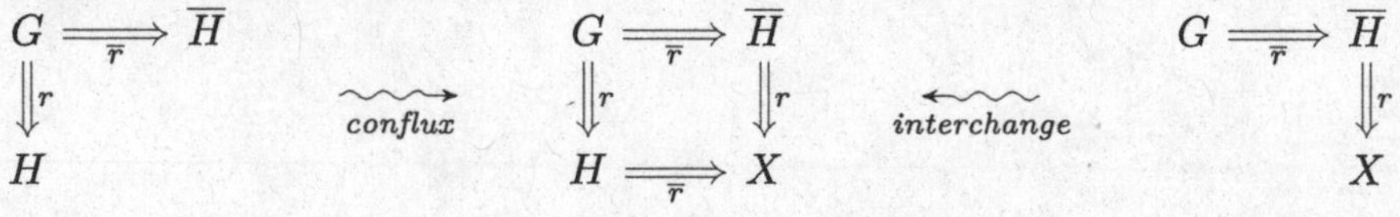

$$G \xRightarrow{\overline{r}} \overline{H} \qquad\qquad G \xRightarrow{\overline{r}} \overline{H} \qquad\qquad G \xRightarrow{\overline{r}} \overline{H}$$

$$\Big\Vert r \qquad \rightsquigarrow \atop conflux \qquad \Big\Vert r \quad\quad \Big\Vert r \qquad \leftsquigarrow \atop interchange \qquad\qquad \Big\Vert r$$

$$H \qquad\qquad\qquad H \xRightarrow{\overline{r}} X \qquad\qquad\qquad\qquad X$$

Fig. 1. *conflux* and *interchange*

The confluence property of parallel independent rule applications was announced by Rosen in [1] and proved by Ehrig and Rosen in [2]; the interchangeability of sequentially independent rule applications was introduced and studied by Ehrig and Kreowski in [3] (see, e.g., Ehrig, Corradini et al. and Ehrig et al. [4–6] for comprehensive surveys). Both results are also known as local Church-Rosser theorems.

It is not difficult to show that the operations can be generalized to derivations. If P and $\overline{P}$ are sets of rules such that each two applications of a P-rule and a $\overline{P}$-rule to the same graph are parallel independent, then derivations $d = (G \overset{*}{\underset{P}{\Longrightarrow}} H)$ and $\overline{d} = (G \overset{*}{\underset{\overline{P}}{\Longrightarrow}} \overline{H})$ induce derivations $d' = (\overline{H} \overset{*}{\underset{P}{\Longrightarrow}} X)$ and $\overline{d'} = (H \overset{*}{\underset{\overline{P}}{\Longrightarrow}} X)$ for some graph X by iterating *conflux* as long as possible. Analogously, two successive derivations $\overline{d} = (G \overset{*}{\underset{\overline{P}}{\Longrightarrow}} \overline{H})$ and $d' = (\overline{H} \overset{*}{\underset{P}{\Longrightarrow}} X)$ induce derivations $d = (G \overset{*}{\underset{P}{\Longrightarrow}} H)$ and $\overline{d'} = (H \overset{*}{\underset{\overline{P}}{\Longrightarrow}} X)$ for some graph H provided that each successive application of a $\overline{P}$-rule and a P-rule are sequentially independent. The situation is depicted in Fig. 2.

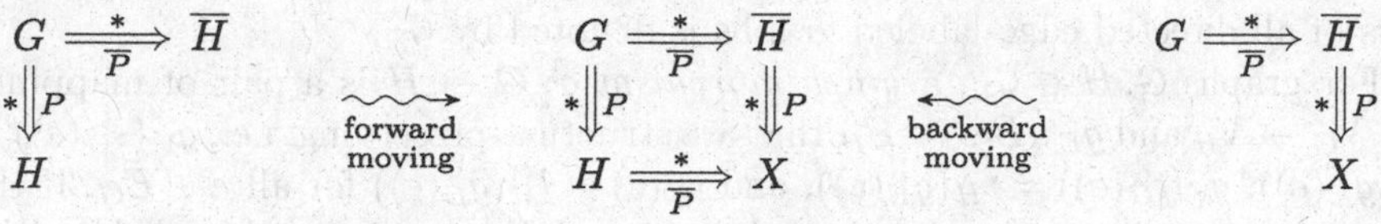

Fig. 2. Forward and backward moving

We interpret these two operations on derivations as moving d forward along $\overline{d}$ (and symmetrically $\overline{d}$ along d) on one hand and as moving d' backward along $\overline{d}$.

The second operation, recalled in Sect. 4, restricts a derivation to a subgraph of its start graph provided that the *accessed part* of the derivation is covered by the subgraph. The accessed part consists of all vertices and edges of the start graph that are accessed by some of the matching morphisms of the left-hand sides of applied rules within the derivation. The *restriction* clips off the complement of the subgraph being an invariant part of all graphs throughout the derivation. As the accessed part is a subgraph itself, the restriction to it, which we call *spine*, is minimal and consists only of vertices and edges essential for the derivation. If $\hat{d}$ is a restriction of a derivation d, then d is an extension of $\hat{d}$, and $\hat{d}$ is embedded into a larger context. Extension and restriction are investigated in some detail by Ehrig, Kreowski, and Plump [7–10].

Finally, we state and prove the main result of this paper in Sect. 5: Forward moving and backward moving preserve the spine of the moved derivation. This investigation is motivated by our recently proposed graph-transformational approach for proving the correctness of reductions between NP-problems in [11]. Such proofs require certain derivations of one kind to be constructed from certain derivations of another kind where, in addition, the initial graphs of both are connected by a derivation of a third kind. We have provided a toolbox of operations on derivations that allow such constructions where one essential operation is moving. Moreover, in some sample correctness proofs the phenomenon of the main result could be observed and used favorably. Our running example stems from this context.

2   Preliminaries

In this section, we recall the basic notions and notations of graphs and rule-based graph transformation as far as needed in this paper.

Graphs. Let Σ be a set of labels with $* \in \Sigma$. A (directed edge-labeled) *graph* over Σ is a system $G = (V, E, s, t, l)$ where V is a finite set of *vertices*, E is a finite set of *edges*, $s, t \colon E \to V$ and $l \colon E \to \Sigma$ are mappings assigning a *source*, a *target* and a *label* to every edge $e \in E$. An edge e with $s(e) = t(e)$ is called a *loop*. An edge with label $*$ is called an *unlabeled edge*. In drawings, the label $*$ is omitted. *Undirected edges* are pairs of edges between the same vertices in opposite directions. The components V, E, s, t, and l of G are also denoted by V_G, E_G, s_G, t_G, and l_G, respectively. The empty graph is denoted by $\emptyset$. The class of all directed edge-labeled graphs is denoted by $\mathcal{G}_\Sigma$.

For graphs $G, H \in \mathcal{G}_\Sigma$, a *graph morphism* $g \colon G \to H$ is a pair of mappings $g_V \colon V_G \to V_H$ and $g_E \colon E_G \to E_H$ that are structure-preserving, i.e., $g_V(s_G(e)) = s_H(g_E(e))$, $g_V(t_G(e)) = t_H(g_E(e))$, and $l_G(e) = l_H(g_E(e))$ for all $e \in E_G$. If the mappings g_V and g_E are bijective, then G and H are *isomorphic*, denoted by $G \cong H$. If they are inclusions, then G is called a *subgraph* of H, denoted by $G \subseteq H$. For a graph morphism $g \colon G \to H$, the image of G in H is called a *match* of G in H, i.e., the match of G with respect to the morphism g is the subgraph $g(G) \subseteq H$. It may be noted that a match is always associated with a particular graph morphism.

To simplify the handling with the two mappings g_V and g_E of a graph morphism $g \colon G \to H$, we may write $g(x)$ for $x \in V_G \cup E_G$ with $g(x) = g_V(x)$ for $x \in V_G$ and $g(x) = g_E(x)$ for $x \in E_G$.

Rules and Rule Application. A *rule* $r = (L \supseteq K \subseteq R)$ consists of three graphs $L, K, R \in \mathcal{G}_\Sigma$ such that K is a subgraph of L and R. The components L, K, and R are called *left-hand side*, *gluing graph*, and *right-hand side*, respectively.

The application of r to a graph G consists of the following three steps. (1) Choose a match $g(L)$ of L in G. (2) Remove the vertices of $g_V(V_L) \setminus g_V(V_K)$ and the edges of $g_E(E_L) \setminus g_E(E_K)$ yielding Z, i.e., $Z = G - (g(L) - g(K))$. (3) Add $V_R \setminus V_K$ disjointly to V_Z and $E_R \setminus E_K$ disjointly to E_Z yielding (up to isomorphism) the graph $H = Z + (R - K)$ where all edges keep their sources, targets and labels except for $s_H(e) = g_V(s_R(e))$ for $e \in E_R \setminus E_K$ with $s_R(e) \in V_K$ and $t_H(e) = g_V(t_R(e))$ for $e \in E_R \setminus E_K$ with $t_R(e) \in V_K$.

The construction is subject to the *dangling condition*, i.e., an edge the source or target of which is in $g(L) - g(K)$ is also in $g(L) - g(K)$, ensuring that Z becomes a subgraph of G so that H becomes a graph automatically. Moreover, we require the *identification condition*, i.e., if different items of L are mapped to the same item in $g(L)$, then they are items of K.

The construction produces a right matching morphism $h \colon R \to H$ given by $h(x) = g(x)$ for $x \in K$ and $h(x) = in_{R-K}(x)$ for $x \in R - K$ where $in_{R-K} \colon R - K \to H = Z + (R - K)$ is the injection associated to the disjoint union. The right matching morphism h satisfies obviously the dangling and identification conditions with respect to the inverted rule $r^{-1} = (R \supseteq K \subseteq L)$ such that r^{-1} can be applied to H yielding G with intermediate graph Z.

The application of r to G w.r.t. g is called *direct derivation* and is denoted by $G \underset{r}{\Longrightarrow} H$ (where g is kept implicit). A *derivation* from G to H is a sequence of direct derivations $G_0 \underset{r_1}{\Longrightarrow} G_1 \underset{r_2}{\Longrightarrow} \cdots \underset{r_n}{\Longrightarrow} G_n$ with $G_0 = G$, $G_n \cong H$ and $n \geq 0$. If $r_1, \cdots, r_n \in P$, then the derivation is also denoted by $G \underset{P}{\overset{n}{\Longrightarrow}} H$. If the length of the derivation does not matter, we write $G \underset{P}{\overset{*}{\Longrightarrow}} H$.

The construction of the resulting graph of a rule application involves a disjoint union so that it is only unique up to isomorphism. This is no problem with respect to the considered graphs in most cases as they are usually assumed to be abstract, meaning that the names of vertices and edges are not significant. For the handling of graphs, this has two consequences. On one hand, one can rename vertices and edges without any harm. And on the other hand, one can deal with concrete graphs as representatives of abstract graphs whenever it is convenient. For example, if one considers a rule application, then one can assume without loss of generality that the intermediate graph is a subgraph of the resulting graph. We make use of these facts in the following in such a way that certain graphs can be considered as equal rather than isomorphic and inclusions can be used rather than injections. In this way, various arguments become easier.

It may be noted that the chosen notion of rule application fits into the DPO framework as introduced in [12] (see, e.g., [5] for a comprehensive survey).

Independence. Let $r = (L \supseteq K \subseteq R)$ and $\overline{r} = (\overline{L} \supseteq \overline{K} \subseteq \overline{R})$ be two rules. Two direct derivations $G \underset{r}{\Longrightarrow} H$ and $G \underset{\overline{r}}{\Longrightarrow} \overline{H}$ with matches $g(L)$ and $\overline{g}(\overline{L})$ are *parallel independent* if $g(L) \cap \overline{g}(\overline{L}) \subseteq g(K) \cap \overline{g}(\overline{K})$. Successive direct derivations $G \underset{\overline{r}}{\Longrightarrow} \overline{H} \underset{r}{\Longrightarrow} X$ with the right match $\overline{h}(\overline{R})$ and the (left) match $g'(L)$ are *sequentially independent* if $\overline{h}(\overline{R}) \cap g'(L) \subseteq \overline{h}(\overline{K}) \cap g'(K)$. It is well-known that parallel independence induces the direct derivations $H \underset{\overline{r}}{\Longrightarrow} X$ and $\overline{H} \underset{r}{\Longrightarrow} X$ with matches $\overline{g}'(\overline{L}) = \overline{g}(\overline{L})$ and $g'(L) = g(L)$ respectively and that sequential independence induces the derivation $G \underset{r}{\Longrightarrow} H \underset{\overline{r}}{\Longrightarrow} X$ with matches $g(L) = g'(L)$ and $\overline{g}'(\overline{L}) = \overline{g}(\overline{L})$.

The two constructions are called *conflux* and *interchange* respectively (cf. Figure 1 in the Introduction).

Restriction of a Rule Application. Let $G \underset{r}{\Longrightarrow} H$ be a rule application with $r = (L \supseteq K \subseteq R)$ based on the matching morphism $g \colon L \to G$. Let Z be the intermediate graph and h the right matching morphism. Let $G' \subseteq G$ with $g(L) \subseteq G'$. Then one can define a graph morphism $g' \colon L \to G'$ by $g'(x) = g(x)$ for $x \in G$ that satisfies the dangling condition and the identification condition inheriting all the properties from g. Therefore, r can be applied to G' yielding the intermediate graph $Z' = G' - (g'(L) - g'(K)) = G' - (g(L) - g(K)) \subseteq G - (g(L) - g(K)) = Z$, the resulting graph $H' = Z' + (R - K)$ and the right matching morphism $h' \colon R \to$

H' defined by $h'(x) = g'(x) = g(x) = h(x)$ for $x \in K$ and $h'(x) = in'_{R-K}(x)$ for $x \in R - K$. As Z and $in_{R-K}(R - K)$ are disjoint in H and $Z' \subseteq Z$, one can choose $in'_{R-K}(x) = in_{R-K}(x) = h(x)$ for $x \in R - K$ such that $H' \subseteq H$ without loss of generality. The constructed rule application $G' \underset{r}{\Longrightarrow} H'$ is called *restriction* of $G \underset{r}{\Longrightarrow} H$ to G' and denoted by $(G \underset{r}{\Longrightarrow} H)|G'$.

Example 1. This first part of our running example provides the rule sets P_{color} and P_{dual} and a sample derivation for each of the two sets. P_{color} contains the following rules for some label a and $i = 1, ..., k$ for some $k \in \mathbb{N}$:

The rules can be used to check the k-colorability of an unlabeled and undirected graph by constructing the following derivations: Apply *add_loop* to each vertex of the start graph, *add_color(i)* once for each $i = 1, \ldots, k$ and *choose_color(i)* for $i = 1, \ldots, k$ as long as possible. A derivation of this kind represents a k-coloring of the start graph if the derived graph does not contain any of the triangles for $i = 1, \ldots, k$. A sample derivation of this kind is $d_{color} =$

The start graph is the complete bipartite graph $K_{3,3}$. The derived graph represents a 2-coloring of $K_{3,3}$.

P_{dual} contains the following rules for some label b:

The rules can be used to construct the dual graph of an unlabeled, simple, loop-free, undirected graph by constructing the following derivation: apply *double_edge* for each edge once, followed by *add_edge* for each pair of vertices that are not connected and finally *remove_pair* as long as possible. A sample derivation of this kind is $d_{dual} =$

where each thick line represents a pair of edges labeled by b and $*$ respectively. The start graph is again the complete bipartite graph $K_{3,3}$ and the derived graph is the dual graph of $K_{3,3}$.

3 Moving Derivations Along Derivations Using Independence

In this section, we introduce two types of moving derivations along derivations: *forward moving* in 3.1 and *backward moving* in 3.2. The forward case gets two derivations with the same start graph, and one derivation is moved to the derived graph of the other derivation (cf. left part of Fig. 2 in the Introduction). The backward case gets the sequential composition of two derivations, and the second derivation is moved to the start graph of the first derivation (cf. right part of Fig. 2). The forward construction is based on parallel independence and iterated application of *conflux* while the backward construction is based on sequential independence and iterated application of *interchange*.

Definition 1. *Let $(P, \overline{P})$ be a pair of sets of rules.*

1. *Then $(P, \overline{P})$ is called* parallel independent *if each two direct derivations of the form $G \underset{r}{\Longrightarrow} H$ and $G \underset{\overline{r}}{\Longrightarrow} \overline{H}$ with $r \in P$ and $\overline{r} \in \overline{P}$ are parallel independent,*
2. *and $(P, \overline{P})$ is called* sequentially independent *if each derivation of the form $G \underset{\overline{r}}{\Longrightarrow} \overline{H} \underset{r}{\Longrightarrow} X$ with $r \in P$ and $\overline{r} \in \overline{P}$ is sequentially independent.*

Example 2. The pair of sets of rules (P_{color}, P_{dual}) is parallel and sequentially independent. A P_{color}-rule application and a P_{dual}-rule application to the same graph would not be parallel independent if one removes something while the other accesses it. This does not happen. The only removing rules are *choose_color* and *remove_pair*. If the rule *choose_color* is applied, then an a-loop is removed, but a-loops are not accessible by P_{dual}-rules. If *remove_pair* is applied, then a b-edge and $*$-edge are removed, but none of these edges can be accessed by a P_{color}-rule. An application of a P_{dual}-rule followed by an application of a P_{color}-rule would not be sequentially independent if the first generates something that the second accesses or if the right match of the first step accesses something that is removed by the second step. This does not happen. The rule *double_edge* adds a b-edge, the rule *add_edge* a $*$-edge, but none of them can be accessed by a P_{color}-rule. And the only removing P_{color}-rule is *choose_color* removing an a-loop, but no right-hand side of a P_{dual}-rule contains such a loop. Summarizing the statement holds.

It may be noticed that this is not the case for the pair (P_{dual}, P_{color}). Clearly it is parallel independent as the order of the rule application is not significant in this case. But if one applies first the rule *choose_color* generating an edge and then *double_edge* to this newly generated edge, then the sequential independence fails.

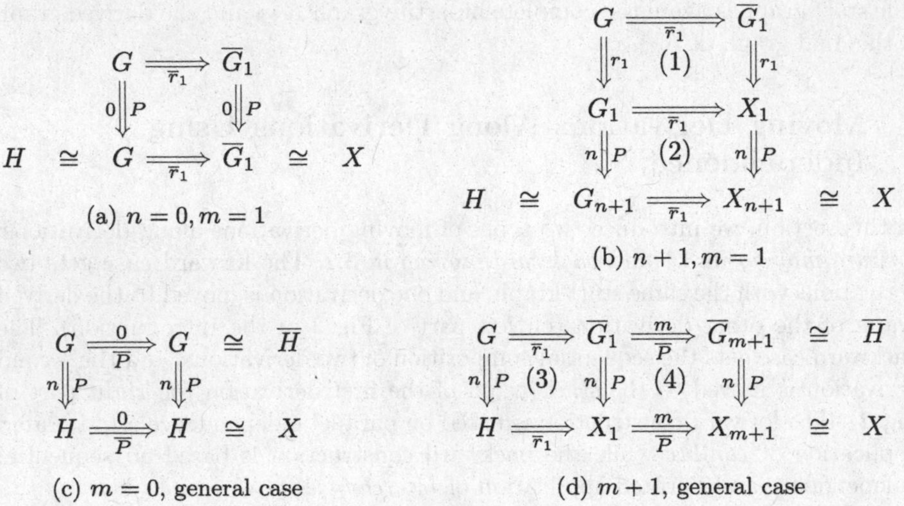

$$
\begin{array}{ccc}
G \xrightarrow{\overline{r}_1} \overline{G}_1 \\
0 \Big\| P \qquad 0 \Big\| P \\
H \;\cong\; G \xrightarrow{\overline{r}_1} \overline{G}_1 \;\cong\; X
\end{array}
$$

(a) $n = 0, m = 1$

$$
\begin{array}{ccc}
G \xrightarrow{\overline{r}_1} \overline{G}_1 \\
r_1 \Big\downarrow \quad (1) \quad \Big\downarrow r_1 \\
G_1 \xrightarrow{\overline{r}_1} X_1 \\
n \Big\| P \quad (2) \quad n \Big\| P \\
H \;\cong\; G_{n+1} \xrightarrow{\overline{r}_1} X_{n+1} \;\cong\; X
\end{array}
$$

(b) $n + 1, m = 1$

$$
\begin{array}{ccc}
G \xrightarrow[\overline{P}]{0} G \;\cong\; \overline{H} \\
n \Big\| P \qquad n \Big\| P \\
H \xrightarrow[\overline{P}]{0} H \;\cong\; X
\end{array}
$$

(c) $m = 0$, general case

$$
\begin{array}{ccc}
G \xrightarrow{\overline{r}_1} \overline{G}_1 \xrightarrow[\overline{P}]{m} \overline{G}_{m+1} \;\cong\; \overline{H} \\
n \Big\| P \;\; (3) \;\; n \Big\| P \;\; (4) \;\; n \Big\| P \\
H \xrightarrow{\overline{r}_1} X_1 \xrightarrow[\overline{P}]{m} X_{m+1} \;\cong\; X
\end{array}
$$

(d) $m + 1$, general case

Fig. 3. Diagrams used in the proof of Proposition 1

3.1 Moving Forward Using Parallel Independence

Two parallel independent rule applications do not interfere with each other. Each one keeps the match of the other one intact so that it can be applied after the other one yielding the same result in both cases (cf. left part of Fig. 1). A possible interpretation of this strong confluence property is that the application of r is moved along the application of $\overline{r}$ and - symmetrically - the application of $\overline{r}$ along the application of r. This carries over to derivations.

Proposition 1. *Let* $(P, \overline{P})$ *be a parallel independent pair of sets of rules. Let* $d = (G \xrightarrow[P]{n} H)$ *and* $\overline{d} = (G \xrightarrow[\overline{P}]{m} \overline{H})$ *be two derivations. Then the iteration of* conflux *as long as possible yields two derivations of the form* $H \xrightarrow[\overline{P}]{m} X$ *and* $\overline{H} \xrightarrow[P]{n} X$ *for some graph* X.

Proof. Formally, the construction is done inductively on the lengths of derivations first for $m = 1$ and then in general:

Base $n = 0$ for $m = 1$: Without loss of generality, one can assume that a 0-derivation does not change the start graph so that the direct derivation $G \xrightarrow{\overline{r}_1} \overline{G}_1$ and the 0-derivation $\overline{G}_1 \xrightarrow[P]{0} \overline{G}_1$ fulfill the statement (cf. diagram in Fig. 3a).

Step $n + 1$ for $m = 1$: The derivation d can be decomposed into $G \underset{r_1}{\Longrightarrow} G_1 \underset{P}{\overset{n}{\Longrightarrow}} G_{n+1}$. Then the operation *conflux* can be applied to $G \Longrightarrow G_1$ and $G \underset{\overline{r}_1}{\Longrightarrow} \overline{G}_1$ yielding $G_1 \underset{\overline{r}_1}{\Longrightarrow} X_1$ and $\overline{G}_1 \underset{r_1}{\Longrightarrow} X_1$ as depicted in subdiagram (1) of the diagram in Fig. 3b. Now one can apply the induction hypothesis to $G_1 \underset{\overline{r}_1}{\Longrightarrow} X_1$ and $G_1 \underset{P}{\overset{n}{\Longrightarrow}} G_{n+1}$ yielding $X_1 \underset{P}{\overset{n}{\Longrightarrow}} X_{n+1}$ and $G_{n+1} \underset{\overline{r}_1}{\Longrightarrow} X_{n+1}$ as depicted in subdiagram (2). The composition of the diagrams (1) and (2) completes the construction.

Base $m = 0$ for the general case: Using the same argument as for $n = 0$, one gets the required derivations as the given $G \underset{P}{\overset{n}{\Longrightarrow}} H$ and the 0-derivation $H \underset{\overline{P}}{\overset{0}{\Longrightarrow}} H$ (cf. diagram in Fig. 3c).

Step $m + 1$ for the general case: The derivation $\overline{d}$ can be decomposed into $G \underset{\overline{r}_1}{\Longrightarrow} \overline{G}_1 \underset{\overline{P}}{\overset{m}{\Longrightarrow}} \overline{G}_{m+1}$. The construction for $m = 1$ can be applied to $G \Longrightarrow \overline{G}_1$ and $G \underset{P}{\overset{n}{\Longrightarrow}} H$ yielding $\overline{G}_1 \underset{P}{\overset{n}{\Longrightarrow}} X_1$ and $H \underset{\overline{r}_1}{\Longrightarrow} X_1$ as depicted in subdiagram (3) of the diagram in Fig. 3d. Now one can apply the induction hypothesis to $\overline{G}_1 \underset{\overline{P}}{\overset{m}{\Longrightarrow}} \overline{G}_{m+1}$ and $\overline{G}_1 \underset{P}{\overset{n}{\Longrightarrow}} X_1$ yielding $\overline{G}_{m+1} \underset{P}{\overset{n}{\Longrightarrow}} X_{m+1}$ and $X_1 \underset{\overline{P}}{\overset{m}{\Longrightarrow}} X_{m+1}$ as depicted in subdiagram (4) . The composition of the diagrams (3) and (4) completes the construction.   □

Remark 1. In diagrammatic form, the situation looks as follows:

$$
\begin{array}{ccc}
G \underset{\overline{P}}{\overset{m}{\Longrightarrow}} \overline{H} & & G \underset{\overline{P}}{\overset{m}{\Longrightarrow}} \overline{H} \\
n \Big\Vert P & \rightsquigarrow & n \Big\Vert P \qquad n \Big\Vert P \\
H & & H \underset{\overline{P}}{\overset{m}{\Longrightarrow}} X
\end{array}
$$

The situation of Proposition 1 can be seen as moving d along $\overline{d}$ and $\overline{d}$ along d. The constructed derivation $\overline{H} \underset{P}{\overset{*}{\Longrightarrow}} X$ is called *moved variant* of d and denoted by $move(d, \overline{d})$, and the second constructed derivation $H \underset{\overline{P}}{\overset{*}{\Longrightarrow}} X$ is called moved variant of $\overline{d}$ and denoted by $move(\overline{d}, d)$.

Example 3. As pointed out in Example 2, (P_{color}, P_{dual}) is a parallel independent pair of sets of rules. Therefore, the derivation d_{color} of Example 1 can be moved along the derivation d_{dual} of Example 1 yielding the moved variant $move(d_{color}, d_{dual}) =$

3.2 Moving Backward Using Sequential Independence

Backward moving can be constructed analogously to forward moving if one uses sequential independence and the *interchange*-operation instead of parallel independence and *conflux*. Another possibility is to reduce backward moving to forward moving by means of inverse derivations.

Proposition 2. *Let* $(P, \overline{P})$ *be a sequentially independent pair of sets of rules and* $\overline{d} = (G \underset{\overline{P}}{\overset{m}{\Longrightarrow}} \overline{H})$ *and* $d' = (\overline{H} \underset{P}{\overset{n}{\Longrightarrow}} X)$ *be two derivations. Then the iteration of* interchange *as long as possible yields two derivations of the form* $G \underset{P}{\overset{n}{\Longrightarrow}} H$ *and* $H \underset{\overline{P}}{\overset{m}{\Longrightarrow}} X$ *for some graph* H.

Proof. The sequential independence of $(P, \overline{P})$ implies, obviously, the parallel independence of $(P, (\overline{P})^{-1})$ as right matches of applications of $\overline{P}$-rules become left matches of the corresponding inverse rule applications, and an application of *interchange* becomes an application of *conflux*. Therefore, Proposition 1 can be applied to d' and $(\overline{d})^{-1}$ (deriving $\overline{H}$ into G) yielding $move(d', (\overline{d})^{-1})$ of the form $G \underset{P}{\overset{n}{\Longrightarrow}} H$ and $move((\overline{d})^{-1}, d')$ of the form $X \underset{(\overline{P})^{-1}}{\overset{m}{\Longrightarrow}} H$ such that $move(d', (\overline{d})^{-1})$ and $move((\overline{d})^{-1}, d')^{-1}$ prove the proposition. $\square$

Remark 2. In diagrammatic form, the situation of Proposition 2 looks as follows:

$$
\begin{array}{ccc}
G \underset{\overline{P}}{\overset{m}{\Longrightarrow}} \overline{H} & & G \underset{\overline{P}}{\overset{m}{\Longrightarrow}} \overline{H} \\
 n \Big\| P & \rightsquigarrow & n \Big\| P \qquad n \Big\| P \\
X & & H \underset{\overline{P}}{\overset{m}{\Longrightarrow}} X
\end{array}
$$

This can be seen as moving d' backward along $\overline{d}$. The resulting derivation $move(d', (\overline{d})^{-1})$ is called *backward moved variant* of d' denoted by $evom(d', \overline{d})$.

Example 4. As (P_{color}, P_{dual}) is not only parallel independent, but also sequentially independent, one can move the derivation $move(d_{color}, d_{dual})$ backward along d_{dual} yielding $evom(move(d_{color}, d_{dual}), d_{dual}) = d_{color}$.

4 Accessed Parts, Restrictions and Spines

In this section, we recall another long-known concept: the restriction of a derivation to a subgraph of the start graph. This is always possible if the subgraph contains the accessed part that consists of all vertices and edges of the start graph that are accessed by the matching morphism of some derivation step. As the accessed part is a subgraph itself, the derivation can be restricted to the accessed part, in particular. We call this restriction the spine of the derivation.

4.1 Accessed Parts

We start by defining the accessed part of a derivation.

Definition 2. *Let* $d = (G \underset{P}{\overset{*}{\Rightarrow}} H)$ *be a derivation. Then the* accessed part *of d is the subgraph of G that is defined inductively on the lengths of derivations as follows:*

$$acc(G \underset{P}{\overset{0}{\Rightarrow}} G \cong H) = \emptyset, \ and$$

$$acc(G = G_0 \underset{r_1}{\Longrightarrow} G_1 \underset{P}{\overset{n}{\Rightarrow}} G_{n+1} \cong H) = (acc(G_1 \underset{P}{\overset{n}{\Rightarrow}} G_{n+1}) \cap Z_1) \cup g_1(L_1)$$

where L_1 is the left-hand side of r_1, g_1 is the first matching morphism, and Z_1 is the intermediate graph of the first rule application.

Remark 3. The accessed part of a 0-derivation is the empty graph as nothing is accessed. The inductive construction is meaningful as $acc(G_1 \underset{P}{\overset{n}{\Rightarrow}} G_{n+1})$ is a subgraph of G_1 by induction hypothesis. Without loss of generality, one can assume that Z_1 is also a subgraph of G_1 so that the intersection of both subgraphs is defined being a subgraph of Z_1 and of G_0, therefore. Consequently, the union of this intersection and the match $g_1(L_1)$ provides a subgraph of $G = G_0$.

Example 5. Considering d_{color} in Example 1, $acc(d_{color}) = $ as all vertices of $K_{3,3}$ are accessed by the applications of the rules *add_loop*, *choose_color*(1), and *choose_color*(2), but none of the edges is accessed.

4.2 Restrictions and Spines

It is known that the derivation d can be restricted to every subgraph G' of G with $acc(d) \subseteq G'$ by clipping off $G - G'$ (or their isomorphic counterparts) from all graphs of d keeping all matches invariant.

Definition 3. *Let* $d = (G \underset{P}{\overset{*}{\Rightarrow}} H)$ *be a derivation and $acc(d) \subseteq G' \subseteq G$. The* restriction *is constructed inductively as follows:*

$$((G \underset{P}{\overset{0}{\Rightarrow}} G \cong H)|G') = (G' \underset{P}{\overset{0}{\Rightarrow}} G'), \ and$$

$$((G = G_0 \underset{r_1}{\Longrightarrow} G_1 \underset{P}{\overset{n}{\Rightarrow}} G_{n+1} \cong H)|G') = (G' \underset{r_1}{\Longrightarrow} G_1' \underset{P}{\overset{n}{\Rightarrow}} H') \ where$$

$$(G' \underset{r_1}{\Longrightarrow} G_1') = ((G = G_0 \underset{r_1}{\Longrightarrow} G_1)|G') \ and \ (G_1' \underset{P}{\overset{n}{\Rightarrow}} H') = ((G_1 \underset{P}{\overset{n}{\Rightarrow}} G_{n+1})|G_1').$$

Remark 4. The definition works because of the following reasons. Let $g_1 \colon L_1 \to G_0$ be the matching morphism of $G_0 \underset{r_1}{\Longrightarrow} G_1$. As $g_1(L_1) \subseteq acc(d) \subseteq G'$, the restriction of the first direct derivation to G' is defined (cf. the respective subsection in the preliminaries). And according to the following lemma, G_1' contains $acc(G_1 \underset{P}{\overset{n}{\Rightarrow}} G_{n+1})$ such that the restriction to G_1' is defined using the induction hypothesis.

Lemma 1. *Let $d = (G_0 \underset{r_1}{\Longrightarrow} G_1 \overset{n}{\Longrightarrow} G_{n+1})$ be a derivation, $\hat{d} = (G_1 \overset{n}{\Longrightarrow} G_{n+1})$ be its tail, $G_0' \subseteq G_0$ with $acc(d) \subseteq G_0'$ and $(G_0' \underset{r_1}{\Longrightarrow} G_1') = ((G_0 \underset{r_1}{\Longrightarrow} G_1)|G_0')$ be the restriction of the first rule application. Then $acc(\hat{d}) \subseteq G_1'$.*

Proof. Let $g_1 \colon L_1 \to G_0$ be the matching morphism of $G_0 \underset{r_1}{\Longrightarrow} G_1$ and Z_1 be its intermediate graph. Let $g_1' \colon L_1 \to G_0'$ be the matching morphism of $G_0' \underset{r_1}{\Longrightarrow} G_1'$ and Z_1' be its intermediate graph. Then one gets the following sequence of equalities and inclusions

$$
\begin{aligned}
&acc(\hat{d}) \\
\underset{1}{=}\ &acc(\hat{d}) \cap G_1 \\
\underset{2}{=}\ &acc(\hat{d}) \cap (Z_1 + (R_1 - K_1)) \\
\underset{3}{=}\ &(acc(\hat{d}) \cap Z_1) + (acc(\hat{d}) \cap (R_1 - K_1)) \\
\underset{4}{\subseteq}\ &(acc(\hat{d}) \cap Z_1) + (R_1 - K_1) \\
\underset{5}{=}\ &(acc(\hat{d}) \cap (Z_1 - (g_1(L_1) - g_1(K_1)))) + (R_1 - K_1) \\
\underset{6}{=}\ &(acc(\hat{d}) \cap Z_1) - (g_1(L_1) - g_1(K_1)) + (R_1 - K_1) \\
\underset{7}{\subseteq}\ &(acc(d) - (g_1(L_1) - g_1(K_1))) + (R_1 - K_1) \\
\underset{8}{\subseteq}\ &(G_0' - (g_1(L_1) - g_1(K_1))) + (R_1 - K_1) \\
\underset{9}{=}\ &(G_0' - (g_1'(L_1) - g_1'(K_1))) + (R_1 - K_1) \\
\underset{10}{=}\ &Z_1' + (R_1 - K_1) \\
\underset{11}{=}\ &G_1'
\end{aligned}
$$

using in (1) the inclusion $acc(\hat{d}) \subseteq G_1$ provided by the definition of accessed parts, in (2) and (11) the definition of derived graphs, in (3), (4) and (6) set properties, in (5) the fact $Z_1 = Z_1 - (g_1(L_1) - g_1(K_1))$ provided by the definition of rule applications, in (7) the definition of accessed parts, in (8) the assumption, in (9) the definition of g_1', and in (10) the definition of intermediate graphs. $\square$

Example 6. The derivation d_{color} can be restricted to its accessed part yielding the derivation $d_{color}|acc(d_{color}) =$

It turns out that the set of restrictions of d has some nice properties:

Proposition 3. *1. The restrictions can be partially ordered by $d|G' \leq d|G''$ if $G' \subseteq G''$.*

2. Given two restrictions $d|G'_i$ for $i = 1, 2$, then $d|G'_1 \cap G'_2$ is a restriction, called the intersection.

3. The restriction $d|acc(d)$ is the intersection of all restrictions.

Proof. Given a derivation $d = (G \overset{*}{\Longrightarrow} H)$, then each subgraph G' of G that contains the accessed part $acc(d)$ induces a restriction as constructed above so that there is a 1-to-1 correspondence between the set of those subgraphs and the set of restrictions. It is well-known that the set of subgraphs of a graph that share a common subgraph is partially ordered by inclusion, is closed under intersection and has the common subgraph as minimum being the intersection of all these subgraphs. The 1-to-1 correspondence allows to carry these properties over to the set of restrictions if one chooses the common subgraph as $acc(d)$. □

As the restriction of a derivation to its accessed part plays a prominent role in the next section, we name it.

Definition 4. *The restriction $d|acc(d)$ of a derivation d to its accessed part is called* spine *of d and denoted by $spine(d)$.*

Remark 5. Using the definitions of accessed parts and restrictions, the spine can be characterized as follows:

$$spine(G \overset{0}{\Longrightarrow} G) = \emptyset \overset{0}{\Longrightarrow} \emptyset, \text{ and}$$

$$spine(G_0 \underset{r_1}{\Longrightarrow} G_1 \overset{n}{\Longrightarrow} G_{n+1}) = (G'_0 = acc(G_0 \underset{r_1}{\Longrightarrow} G_1 \overset{n}{\Longrightarrow} G_{n+1}) \underset{r_1}{\Longrightarrow} G'_1 \overset{n}{\Longrightarrow} G'_{n+1})$$

where r_1 is applied to G'_0 using the match $g_1(L_1)$ and $G'_1 \overset{n}{\Longrightarrow} G'_{n+1}$ is the restriction $(G_1 \overset{n}{\Longrightarrow} G_{n+1})|G'_1$.

This construction is meaningful. Due to the definition of accessed parts, $g_1(L_1)$ is a subgraph of the accessed part so that it provides a match of r_1 in G'_0. Due to Lemma 1, the derived graph G'_1 of this rule application contains $acc(G_1 \overset{n}{\Longrightarrow} G_{n+1})$ so that the restriction $(G_1 \overset{n}{\Longrightarrow} G_{n+1})|G'_1$ is defined. This characterization turns out to be quite useful in the next section.

Example 7. As the derivation d_{color} is restricted to its accessed part in Example 6, the derivation given there is the spine of d_{color}. Moreover, it is the spine of $move(d_{color}, d_{dual})$ given in Example 3. This is not a coincidence. In the next section, we show that the spine of a derivation equals the spine of each of its moved variants.

5   Moving Preserves the Spine

As all prerequisites are now available, we can state and prove the main result of this paper: *Moving preserves the spine*. The forward moving is covered in Theorem 1 and the backward case in Corollary 1.

Theorem 1. *Let $(P, \overline{P})$ be a parallel independent pair of sets of rules. Let $d = (G \overset{*}{\underset{P}{\Rightarrow}} H)$ and $\overline{d} = (G \overset{*}{\underset{\overline{P}}{\Rightarrow}} G')$ be two derivations. Then*

$$spine(d) = spine(move(d, \overline{d})).$$

Proof. Following the construction of moving in Proposition 1, the statement is proved, firstly, by induction on the length n of d for length $m = 1$ of $\overline{d}$ and, secondly, by induction on m for arbitrary n.

Base $n = 0$ for $m = 1$: Then $move(d, \overline{d}) = (G' \overset{0}{\underset{P}{\Rightarrow}} G')$ by construction. Therefore, $spine(d) = (\emptyset \overset{0}{\Rightarrow} \emptyset) = spine(move(d, \overline{d}))$.

Step $n + 1$ for $m = 1$: Then d can be decomposed into $G = G_0 \underset{r_1}{\Rightarrow} G_1$ and $tail(d) = (G_1 \overset{n}{\underset{P}{\Rightarrow}} G_{n+1} = H)$, and $d' = move(d, \overline{d})$ can be decomposed into $G' = G'_0 \underset{r_1}{\Rightarrow} G'_1$ and $tail(d') = (G'_1 \overset{n}{\underset{P}{\Rightarrow}} G'_{n+1} = X)$. By construction, *conflux* applied to $G_0 \underset{r_1}{\Rightarrow} G_1$ and $\overline{d}$ yields $\overline{d'} = (G_1 \underset{\overline{P}}{\Rightarrow} G'_1)$ in addition to $G'_0 \underset{r_1}{\Rightarrow} G'_1$. Moreover, $move(tail(d), \overline{d'})) = tail(d')$. As the length of $\overline{d'}$ is 1 and the length of $tail(d)$ is n, the induction hypothesis can be used yielding $spine(tail(d)) = spine(tail(d'))$ and $acc(tail(d)) = acc(tail(d'))$, in particular.

Using this equality and the constructions of the four direct derivations above, one can show now $acc(d) = acc(d')$. Let $g_1 \colon L_1 \to G_0$ be the matching morphism of $G_0 \underset{r_1}{\Rightarrow} G_1$ and Z_1 its intermediate graph, $\overline{g} \colon \overline{L} \to G_0$ be the matching morphism of $\overline{d} = (G_0 \underset{\overline{r}}{\Rightarrow} G'_0)$ with $\overline{r} = (\overline{L} \supseteq \overline{K} \subseteq \overline{R})$ and $\overline{Z}$ its intermediate graph, $g'_1 \colon L_1 \to G'_0$ be the matching morphism of $G'_0 \underset{r_1}{\Rightarrow} G'_1$ and Z'_1 its intermediate graph, and $\overline{g'} \colon L_1 \to G_1$ be the matching morphism of $G_1 \underset{\overline{r}}{\Rightarrow} G'_1$ and $\overline{Z'}$ its intermediate graph. Let $Z = G_0 - ((g_1(L_1) - g_1(K_1)) \cup (\overline{g}(\overline{L}) - \overline{g}(\overline{K})))$, i.e. removing the non-gluing parts of the two matches in G_0. Then Z is also subgraph of Z_1, G_1, $\overline{Z}$, G'_0, Z'_1, G'_1, and $\overline{Z'}$. This implies that $acc(tail(d)) \cap Z = acc(tail(d')) \cap Z$ is subgraph of the eight graphs, in particular, of G_0, $\overline{Z}$, and G'_0. Using the below Lemma 2, one gets $acc(tail(d)) \cap Z = acc(tail(d)) \cap Z_1$ and $acc(tail(d')) \cap Z = acc(tail(d')) \cap Z'_1$. As, moreover, the matches $g_1(L_1)$ and $g'_1(L_1)$ coincide, one gets

$$
\begin{aligned}
acc(d) &= (acc(tail(d)) \cap Z_1) \cup g_1(L_1) \\
&= (acc(tail(d)) \cap Z) \cup g_1(L_1) \\
&= (acc(tail(d')) \cap Z) \cup g'_1(L_1) \\
&= (acc(tail(d')) \cap Z'_1) \cup g'_1(L_1) \\
&= acc(d').
\end{aligned}
$$

It remains to show that the spines of d and d' coincide. As $acc(d) = acc(d')$ contains $g_1(L_1) = g'_1(L_1)$, r_1 can be applied using the matching morphism $\hat{g}_1 \colon L_1 \to acc(d)$ given by $\hat{g}_1(x) = g_1(x)$ for all x in L_1. The derived graph is

$$(acc(d) - (\hat{g}_1(L_1) - \hat{g}_1(K_1))) + in_{R_1 - K_1}(R_1 - K_1)$$
$$= (acc(d) - (g_1(L_1) - g_1(K_1))) + in_{R_1 - K_1}(R_1 - K_1)$$
$$= (acc(d') - (g_1'(L_1) - g_1'(K_1))) + in_{R_1 - K_1}(R_1 - K_1)$$

containing $acc(tail(d)) = acc(tail(d'))$ according to Lemma 1. This is the first direct derivation of the spines of d and d' followed by the restrictions of $tail(d)$ and $tail(d')$ to the derived graph. These restrictions coincide because they share $spine(tail(d)) = spine(tail(d'))$ which are the restriction to $acc(tail(d)) = acc(tail(d'))$, and the additional items of $R_1 - K_1$ are equally renamed by $in_{R_1 - K_1}$ and $in'_{R_1 - K_1}$ and kept invariant throughout the derivation as they are not in the accessed parts. This completes the first induction.

The second induction is simpler.

Base $m = 0$ for arbitrary n: Then $move(d, \overline{d}) = d$ by construction. Therefore, $spine(d) = spine(move(d, \overline{d}))$.

Step $m + 1$ for arbitrary n: Then $\overline{d}$ can be decomposed into $head(\overline{d}) = (G = \overline{G}_0 \underset{\overline{P}}{\Longrightarrow} \overline{G}_1)$ and $tail(\overline{d}) = (\overline{G}_1 \underset{\overline{P}}{\overset{m}{\Longrightarrow}} \overline{G}_{m+1})$. By construction of moving, one gets $move(d, head(\overline{d}))$ and $move(move(d, head(\overline{d})), tail(\overline{d})) = move(d, \overline{d})$. As $head(\overline{d})$ has length 1 and $tail(\overline{d})$ length m, the already proved part of the theorem and the induction hypothesis are applicable such that one gets

$$spine(d) = spine(move(d, head(\overline{d})))$$
$$= spine(move(move(d, head(\overline{d})), tail(\overline{d})))$$
$$= spine(move(d, \overline{d})).$$

This completes the proof. □

Lemma 2. *Let* $(P, \overline{P})$ *be a pair of parallel independent sets of rules,* $d = (G_0 \underset{P}{\overset{n}{\Longrightarrow}} G_n)$ *be a derivation, and* $\overline{d} = (G_0 \underset{\overline{r}}{\Longrightarrow} G_0')$ *be a direct derivation for some* $\overline{r} = (\overline{L} \supseteq \overline{K} \subseteq \overline{R}) \in \overline{P}$ *with the left match* $\overline{g}(\overline{L})$ *and the right match* $\overline{h}(\overline{R})$. *Then the following holds:*

(1) $acc(d) \cap (\overline{g}(\overline{L}) - \overline{g}(\overline{K})) = \emptyset$, *and*
(2) $acc(move(d, \overline{d})) \cap (\overline{h}(\overline{R}) - \overline{h}(\overline{K})) = \emptyset$.

Proof. The statement is proved by induction on the length n of d.

Base: The statement holds for $n = 0$ obviously as $acc(d) = \emptyset = acc(move(d, \overline{d}))$.

Step: If d has length $n + 1$, then it decomposes into $head(d) = (G_0 \underset{r_1}{\Longrightarrow} G_1)$ for some $r_1 = (L_1 \supseteq K_1 \subseteq R_1) \in P$ with the left match $g_1(L_1)$ and the intermediate graph Z_1 and $tail(d) = G_1 \underset{P}{\overset{n}{\Longrightarrow}} G_{n+1}$. The parallel independence of $head(d)$ and $\overline{d}$ provides the property

$$(+): (g_1(L_1)) \cap (\overline{g}(\overline{L}) - \overline{g}(\overline{K})) = \emptyset.$$

Consider now the following equalities plus one inclusion using the definition of acc, the distributivity of intersection and union, property $(+)$, an obvious property of intersection, and the induction hypothesis for $tail(d)$ in this order:

$$
\begin{aligned}
& acc(d) \cap (\overline{g}(\overline{L}) - \overline{g}(\overline{K})) \\
= \; & ((acc(tail(d)) \cap Z_1) \cup g_1(L_1)) \cap (\overline{g}(\overline{L}) - \overline{g}(\overline{K})) \\
= \; & ((acc(tail(d)) \cap Z_1) \cap (\overline{g}(\overline{L}) - \overline{g}(\overline{K}))) \cup (g_1(L_1) \cap (\overline{g}(\overline{L}) - \overline{g}(\overline{K}))) \\
= \; & (acc(tail(d)) \cap Z_1) \cap (\overline{g}(\overline{L}) - \overline{g}(\overline{K})) \\
\subseteq \; & acc(tail(d)) \cap (\overline{g}(\overline{L}) - \overline{g}(\overline{K})) \\
= \; & \emptyset.
\end{aligned}
$$

This proves statement (1). Statement (2) follows analogously using the sequential independence of $\overline{d}$ and $move(head(d))$.   □

Corollary 1. *Let $(P, \overline{P})$ be a sequentially independent pair of sets of rules. Let $\overline{d} = (G \underset{\overline{P}}{\overset{*}{\Longrightarrow}} \overline{H})$ and $d' = (\overline{H} \underset{P}{\overset{*}{\Longrightarrow}} X)$ be two derivations. Then*

$$
spine(d') = spine(evom(d', \overline{d})).
$$

Proof. Using Theorem 1 and the definition of backward moving, one gets $spine(d') = spine(move(d', (\overline{d})^{-1})) = spine(evom(d', \overline{d}))$.   □

We would like to point out that our running example is chosen purposefully as a typical instance of a situation we found quite often when proving the correctness of reductions between NP-problems within a graph-transformational framework. This applies - among others - to the well-known reductions of *clique* to *independent set* to *vertex cover*, of *long paths* to *Hamiltonian paths* to *Hamiltonian cycles* to *traveling salesperson*. For each of these reductions, one can find graph-transformational models of the two involved NP-problems and the reduction such that the derivations of the source problem can be moved along an initial part of the reduction derivations and the derivations of the target problem can be moved along the remainder part of the reduction derivations. Then the preservation of the spines is an important part of the correctness proofs.

6 Conclusion

In this paper, we have investigated the relationship between two elementary operations on derivations: moving a derivation along a derivation based on parallel and sequential independence on one hand and restriction of a derivation by clipping off vertices and edges that are never matched by a rule application throughout the derivation on the other hand. As main result, we have shown that moving a derivation preserves its spine being the minimal restriction. This stimulates further research in at least three directions:

- Sequentialization, parallelization, and shift are three further well-known operations on parallel derivations that are closely related to parallel and sequential independence. Therefore, one may wonder how they behave with respect to the spine. Moreover, one may consider the relation between amalgamation and the generalized confluence of concurrent rule applications and spines.
- We have considered graph transformation within the DPO-approach for labeled directed graphs with multiple edges. As the category of these graphs is a typical adhesive category, there seems to be a good chance that our main result can be generalized for adhesive categories.
- In [11], we have exploited the moving of derivations along derivations and the preservation of spines to prove the correctness of reductions between NP-problems. As such reductions are special kinds of model transformations, it may be of interest to explore whether moving and its properties can be used for proving correctness of other model transformations.

References

1. Rosen, B.K.: A church-rosser theorem for graph grammars (announcement). ACM SIGACT News **7**(3), 26–31 (1975)
2. Ehrig, H., Rosen, B.K.: Commutativity of independent transformations on complex objects. In: IBM Research Report RC 6251, Thomas J. Watson Research Center, Yorktown Height, New York, USA (1976)
3. Ehrig, H., Kreowski, H.-J.: Parallelism of manipulations in multidimensional information structures. In: Mazurkiewicz, A. (ed.) MFCS 1976. LNCS, vol. 45, pp. 284–293. Springer, Heidelberg (1976). https://doi.org/10.1007/3-540-07854-1_188
4. Ehrig, H.: Introduction to the algebraic theory of graph grammars (a survey). In: Claus, V., Ehrig, H., Rozenberg, G. (eds.) Graph Grammars 1978. LNCS, vol. 73, pp. 1–69. Springer, Heidelberg (1979). https://doi.org/10.1007/BFb0025714
5. Corradini, A., Montanari, U., Rossi, F., Ehrig, H., Heckel, R., Löwe, M.: Algebraic approaches to graph transformation part I: basic concepts and double pushout approach. In: Rozenberg, G., (ed.) Handbook of Graph Grammars and Computing by Graph Transformation, vol. 1. Foundations, pp. 163–245. World Scientific, Singapore (1997)
6. Ehrig, H., Ehrig, K., Prange, U., Taentzer, G.: Fundamentals of Algebraic Graph Transformation. Monographs in Theoretical Computer Science. An EATCS Series. Springer (2006)
7. Ehrig, H.: Embedding theorems in the algebraic theory of graph grammars. In: Karpiński, M. (ed.) FCT 1977. LNCS, vol. 56, pp. 245–255. Springer, Heidelberg (1977). https://doi.org/10.1007/3-540-08442-8_91
8. Kreowski, H.-J.: Manipulationen von Graphmanipulationen. PhD thesis, Technische Universität Berlin (1978)
9. Kreowski, H.-J.: A pumping lemma for context-free graph languages. In: Claus, V., Ehrig, H., Rozenberg, G. (eds.) Graph Grammars 1978. LNCS, vol. 73, pp. 270–283. Springer, Heidelberg (1979). https://doi.org/10.1007/BFb0025726
10. Plump, D.: Computing by Graph Rewriting. Habilitation thesis, University of Bremen (1999)
11. Kreowski, H.-J., Kuske, S., Lye, A., Windhorst, A.: A graph-transformational approach for proving the correctness of reductions between NP-problems. In: Heckel,

R., Poskitt, C.M., (eds.) Proceedings of the Thirteenth International Workshop on Graph Computation Models, Nantes, France, 6th July 2022, vol. 374 of Electronic Proceedings in Theoretical Computer Science, pp. 76–93. Open Publishing Association (2022)

12. Ehrig, H., Pfender, M., Schneider, H.-J.: Graph-grammars: an algebraic approach. In: 14th Annual Symposium on Switching and Automata Theory (swat 1973), pp. 167–180. IEEE (1973)

Termination of Graph Transformation Systems Using Weighted Subgraph Counting

Roy Overbeek[✉] and Jörg Endrullis

Vrije Universiteit Amsterdam, Amsterdam, The Netherlands
{r.overbeek,j.endrullis}@vu.nl

Abstract. We introduce a termination method for the algebraic graph transformation framework PBPO$^+$, in which we weigh objects by summing a class of weighted morphisms targeting them. The method is well-defined in rm-adhesive quasitoposes (which include toposes and therefore many graph categories of interest), and is applicable to non-linear rules. The method is also defined for other frameworks, including DPO and SqPO, because we have previously shown that they are naturally encodable into PBPO$^+$ in the quasitopos setting.

Keywords: Graph transformation · Termination · Pullback-Pushout

1 Introduction

Many fields of study related to computation have mature termination theories. See, for example, the corpus for term rewriting systems [34, Chapter 6].

For the study of graph transformation, by contrast, not many termination methods exist, and the ones that do exist are usually defined for rather specific notions of graphs. Although the techniques themselves can be interesting, the latter observation fits somewhat uneasily with the general philosophy of the predominant algebraic graph transformation tradition [12], in which graph transformations are defined and studied in a graph-agnostic manner, by using the language of category theory.

In this paper, we introduce a termination method for PBPO$^+$ [25], a method in the algebraic tradition. We weigh objects G by summing a class of weighted elements (i.e., morphisms of the form $T \to G$), and construct a decreasing measure. Our method enjoys generality across two dimensions:

1. The method is formulated completely in categorical terms, and is well-defined in (locally finite) rm-adhesive quasitoposes. The rm-adhesive quasitoposes include all toposes, and so automatically a large variety of graphs, as well as other structures, such as Heyting algebras [16] and fuzzy presheaves [32].
2. The method is also defined for DPO [13], SqPO [10], AGREE [9] and PBPO [8]. This is because we have recently shown that, in the quasitopos setting, every rule of these formalisms can be straightforwardly encoded as a PBPO$^+$ rule that generates the same rewrite relation [25, Theorem 73].

© The Author(s), under exclusive license to Springer Nature Switzerland AG 2023
M. Fernández and C. M. Poskitt (Eds.): ICGT 2023, LNCS 13961, pp. 81–101, 2023.
https://doi.org/10.1007/978-3-031-36709-0_5

To the best of our knowledge, this is the first termination method applicable in such a broad setting; and the first method that is automatically defined for a variety of well-known algebraic graph transformation frameworks. In addition, the termination method can be applied to non-linear (duplicating) rules.

The paper is structured as follows. We summarize the basic categorical, graph and termination preliminaries (Sect. 2), and we cover the required background on PBPO$^+$ (Sect. 3). Next, we explain and prove our termination method (Sect. 4). After, we amply illustrate our method with a variety of examples (Sect. 5), and then compare our approach to related work (Sect. 6). We close with some concluding remarks and pointers for future work (Sect. 7).

2 Preliminaries

The preliminaries for this paper include basic categorical and graph notions (Sect. 2.1), and a basic understanding of termination (Sect. 2.2).

2.1 Basic Notions

We assume familiarity with basic categorical notions such as (regular) monomorphisms, pullbacks and pushouts [3,26]. We write $\rightarrowtail$ for monos; and $\mathrm{Hom}(\mathbf{C})$, $\mathrm{mono}(\mathbf{C})$, $\mathrm{rm}(\mathbf{C})$ and $\mathrm{iso}(\mathbf{C})$ for the classes of morphisms, monomorphisms, regular monomorphisms and isomorphisms in $\mathbf{C}$, respectively.

Notation 1 (Nonstandard Notation). *Given a class of morphisms* $\mathcal{A}(\mathbf{C})$, *we write* $\mathcal{A}(A, B)$ *to denote the collection of* $\mathcal{A}$-*morphisms from* A *to* B, *leaving* $\mathbf{C}$ *implicit. For sets of objects* S, *we overload* $\mathcal{A}(S, A)$ *to denote* $\bigcup_{X \in S} \mathcal{A}(X, A)$. *If* $\mathcal{A}(\mathbf{C})$ *is a generic class in lemmas, we use* $\rightsquigarrow$ *to denote* $\mathcal{A}$-*morphisms.*

For cospans $A \xrightarrow{f} C \xleftarrow{g} D$, *we write* $\langle f \mid g \rangle$ *to denote the arrow* $B \to D$ *obtained by pulling* f *back along* g.

Definition 2 ($\mathcal{A}$-Local Finiteness). *Let* $\mathcal{A}(\mathbf{C})$ *be a class of morphisms. A category* $\mathbf{C}$ *is* $\mathcal{A}$-*locally finite if* $\mathcal{A}(A, B)$ *is finite for all* $A, B \in \mathrm{Obj}(\mathbf{C})$.

Lemma 3 (Pullback Lemma). *Assume the right square is a pullback and the left square commutes. Then the outer square is a pullback iff the left square is a pullback.* □

$$A \longrightarrow B \longrightarrow C$$
$$\downarrow \qquad \downarrow \mathrm{PB} \downarrow$$
$$D \longrightarrow E \longrightarrow F$$

Definition 4 (Van Kampen Square [18]**).** *A pushout square is said to be* Van Kampen *(VK) if, whenever it lies at the bottom of a commutative cube where the back faces FBAE and FBCG are pullbacks, this implies that the top face is a pushout iff the front faces are pullbacks.*

$$
\begin{array}{ccc}
F & \longrightarrow & G \\
\end{array}
$$

Definition 5 (Rm-Adhesive Category [19]**).** *A category is* rm-adhesive *(a.k.a.* quasiadhesive*) if pushouts along regular monomorphisms exist and are VK.*

Definition 6 (Quasitopos [1,15,35]). *A category* **C** *is a* quasitopos *if it has all finite limits and colimits, it is locally cartesian closed, and it has a regular-subobject classifier.*

Definition 7 (Split Epimorphism). *An epimorphism* $e : A \twoheadrightarrow B$ *is* split *if it has a right-inverse, i.e., if there exists an* $f : B \to A$ *such that* $e \circ f = 1_B$.

For split epimorphisms e, *we let* $e^{\leftharpoonup}$ *denote an arbitrary right-inverse of* e.

Proposition 8 ([1, Prop. 7.59]). *If* e *is a split epi, then* $e^{\leftharpoonup} \in \mathrm{rm}(\mathbf{C})$.   □

Definition 9 (𝒜-Factorization). *An* 𝒜-factorization *of a morphism* $f : A \to C$ *consists of a morphism* $f^{\iota} : A \to B$ *and* 𝒜-morphism $f'' : B \rightsquigarrow C$ *such that* $f = f'' \circ f'$, *and for any other such factorization* $g' : A \to B'$, $g'' : B' \rightsquigarrow C$ *of* f, *there exists a unique* $x : B \to B'$ *making the right diagram commute.*

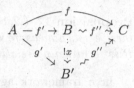

Remark 10. If 𝒜 = mono or 𝒜 = rm then the notion of 𝒜-factorization coincides with the common notion of (regular)-image factorization, so that B is considered to be the image of f. For generality, we intentionally widen the definition to allow for the case where 𝒜 = Hom, in which case in any category, $B = A$, $f' = 1_A$ and $f'' = f$ defines an 𝒜-factorization of f, with $x = g'$ as the unique witness.

Our method is defined fully in categorical terms. For examples, and to guide intuition, we will use the category of edge-labeled multigraphs.

Definition 11 (Graph Notions). *Let a finite label set* $\mathcal{L}$ *be fixed. An (edge-labeled)* (multi)graph G *consists of a set of vertices* V, *a set of edges* E, *source and target functions* $s, t : E \to V$, *and an edge label function* $\ell^E : E \to \mathcal{L}$. *A graph is* unlabeled *if* $\mathcal{L}$ *is a singleton.*

A homomorphism *between graphs* G *and* G' *is a pair of maps* $\phi = (\phi_V : V_G \to V_{G'}, \phi_E : E_G \to E_{G'})$ *satisfying* $(s_{G'}, t_{G'}) \circ \phi_E = \phi_V \circ (s_G, t_G)$ *and* $\ell^E_{G'} \circ \phi_E = \ell^E_G$.

Definition 12 ([12]). *The category* **Graph** *has graphs as objects, parameterized over some global (and usually implicit) label set* $\mathcal{L}$, *and homomorphisms as arrows. The subcategory* **FinGraph** *restricts to graphs with finite* V *and* E.

2.2 Termination

The topic of termination dates back at least to Turing, and is studied in many different settings. For a systematic overview for term rewriting systems (not yet existent for graph transformation systems), see [34, Chapter 6]. Plump has shown that termination of graph rewriting is undecidable [29].

Definition 13. *Let* $R \subseteq A \times A$ *be given. An* infinite R-sequence *is a function* $f : \mathbb{N} \to A$ *such that for all* $i \in \mathbb{N}$, $(f(i), f(i+1)) \in R$.

Definition 14. *A binary relation* R *is* terminating *if there does not exist an infinite* R-sequence.

Definition 15 ([2,17]). *Let $R, S \subseteq A \times A$ be binary relations. Then R is terminating relative to S if every infinite $R \cup S$-sequence contains a finite number of R steps.*

For our purposes, it suffices to measure objects as natural numbers (instead of a general well-founded order).

Definition 16 (Measure). *A* measure *is a function $\mathbf{w} : A \to \mathbb{N}$. The measure $\mathbf{w}$ is* decreasing *for a binary relation $R \subseteq A \times A$ if for all $(x,y) \in R$, $\mathbf{w}(x) > \mathbf{w}(y)$, and it is* non-increasing *for R if for all $(x,y) \in R$, $\mathbf{w}(x) \geq \mathbf{w}(y)$.*

Proposition 17 ([2,17]). *Let $R, S \subseteq A \times A$ be binary relations. Assume that there exists a measure $\mathbf{w}$ that is decreasing for R and non-increasing for S. Then R is terminating relative to S. Consequently $R \cup S$ is terminating iff S is.* □

In a framework agnostic setting, a rule ρ is a mathematical object that induces a binary relation $\Rightarrow_\rho \subseteq A \times A$. We say that a rule or a system or rules is terminating, decreasing or non-increasing if the induced rewrite relations have the respective property (and analogously for relative termination). Note that also Proposition 17 can then be applied to systems of rules in place of relations.

3   PBPO$^+$

PBPO$^+$ is short for *Pullback-Pushout with strong matching*. It is obtained by strengthening the matching mechanism of PBPO [8] by Corradini et al.

We provide the necessary definitions and results on PBPO$^+$. See Sect. 5 for many examples of rules. For a gentler introduction to PBPO$^+$, with examples of rewrite steps, see especially the tutorial [23].

Definition 18 (PBPO$^+$ Rewriting [8,25]**).** *A PBPO$^+$ rule ρ is a diagram as shown on the left of:*

$$
\begin{array}{ccc}
L \xleftarrow{l} K \xrightarrow{r} R & \qquad & K \xrightarrow{r} R \\
\downarrow t_L \quad \text{PB} \quad \downarrow t_K & & {\scriptstyle !u} \searrow \ \text{PO} \ \downarrow w \\
L' \xleftarrow{l'} K' & & L \xrightarrow{m} G_L \xleftarrow{g_L} G_K \ \Big| \ g_R \to G_R \\
& & \Big\| \quad \text{PB} \ \downarrow \alpha \ \text{PB} \ \ \downarrow u' \swarrow {}^{t_K} \\
& & L \xrightarrow{t_L} L' \xleftarrow{l'} K'
\end{array}
$$

L *is the* lhs pattern *of the rule, L' its* (context) type *and t_L the* (context) typing *of L. Likewise for the* interface K*. R is the* rhs pattern *or* replacement *for L.*

Rule ρ, match morphism $m : L \to G_L$ and adherence morphism $\alpha : G_L \to L'$ induce a rewrite step $G_L \Rightarrow_\rho^\alpha G_R$ on arbitrary objects G_L and G_R if the properties indicated by the commuting diagram on the right hold, where $u : K \to G_K$ is the unique morphism satisfying $t_K = u' \circ u$ [25, Lemma 15].

Assumption 19. *We assume that for any rule ρ analyzed for termination, the pushout of t_K along r exists.*

Remark 20 ([25, Section 3]*).* If Assumption 19 holds for a rule ρ, then any ρ step defines (up to isomorphism) a diagram shown on the right, where the bold diagram is ρ ($t_L \circ l = l' \circ t_K$ a pullback and $t_R \circ r = r' \circ t_K$ a pushout). Our method uses the extra pushout to analyze how rewritten objects (the middle span) relate to the context types (the bottom span).

$$
\begin{array}{ccccc}
L & \xleftarrow{\;l\;} & K & \xrightarrow{\;r\;} & R \\
{\scriptstyle m}\downarrow & \text{PB} & {\scriptstyle u}\downarrow & \text{PO} & {\scriptstyle w}\downarrow \\
G_L & - {\scriptstyle g_L} - & G_K & - {\scriptstyle g_R} \to & G_R \\
{\scriptstyle t_L}\downarrow & \text{PB} & {\scriptstyle t_K}\downarrow & \text{PO} & {\scriptstyle t_R}\downarrow \\
L' & \xleftarrow{\;l'\;} & K' & \xrightarrow{\;r'\;} & R'
\end{array}
$$

Theorem 21 ([25, Theorem 73]). *Let* $\mathbf{C}$ *be a quasitopos, and let matches* m *be regular monic. For rewriting formalisms* $\mathcal{F}$ *and* $\mathcal{G}$*, let* $\mathcal{F} \prec \mathcal{G}$ *express that in* $\mathbf{C}$*, for any* $\mathcal{F}$ *rule* ρ*, there exists a* $\mathcal{G}$ *rule* τ *such that* $\Rightarrow^{\rho}_{\mathcal{F}} = \Rightarrow^{\tau}_{\mathcal{G}}$*. We have:*

$$
SqPO \prec AGREE \overset{\prec}{} \begin{array}{c} PBPO^{+} \\ \curlyvee \\ PBPO \end{array} \overset{\prec}{} DPO
$$

$\square$

Observe that $\prec$ is transitive. As the constructive proofs in [25] show, the procedures to encode the mentioned formalisms into PBPO^{+} are straightforward. We moreover conjecture SPO $\prec$ PBPO^{+} [25, Remark 26].

4   Decreasingness by Counting Weighted Elements

We start with an explanation of the general idea behind our termination approach. Given a set of rules $\mathcal{T}$, we seek to construct a measure $\mathbf{w}$ such that for all steps $G_L \Rightarrow_{\rho} G_R$ generated by a rule $\rho \in \mathcal{T}$, $\mathbf{w}(G_L) > \mathbf{w}(G_R)$. Then $\mathbf{w}$ is a decreasing measure for the rewrite relation generated by $\mathcal{T}$, so that $\mathcal{T}$ is terminating. We construct such a measure $\mathbf{w}$ by weighing objects as follows.

Definition 22 (Weight Functions). *Given a set of objects* $\mathbb{T}$*, weight function* $\mathbf{w} : \mathbb{T} \to \mathbb{N}$*, and class of morphisms* $\mathcal{A}(\mathbf{C})$*, we define the* tiling weight function

$$
\mathbf{w}^{\mathcal{A}}_{\mathbb{T}}(X) \quad = \quad \sum_{t \in \mathcal{A}(\mathbb{T}, X)} \mathbf{w}(dom(t))
$$

for objects $X \in \mathrm{Obj}(\mathbf{C})$*. In this context, we refer to the objects of* $\mathbb{T}$ *as* tiles. *(Note that* $\mathbf{w}^{\mathcal{A}}_{\mathbb{T}}$ *is well-defined if* $\mathbb{T}$ *is finite and* $\mathbf{C}$ *is* $\mathcal{A}$*-locally finite.)*

Example 23. Let $\mathbf{C} = \mathbf{FinGraph}$ with singleton label set $\mathcal{L}$, and G and arbitrary graph. Some basic examples of tile sets and parameters are as follows.

- Let $\bullet$ represent the graph consisting of a single node. If $\mathbb{T} = \{\bullet\}$, $\mathbf{w}(\bullet) = 1$, and $\mathcal{A}(\mathbf{C}) \in \{\mathrm{Hom}(\mathbf{C}), \mathrm{mono}(\mathbf{C}), \mathrm{rm}(\mathbf{C})\}$, then $\mathbf{w}^{\mathcal{A}}_{\mathbb{T}}(G) = |V_G|$.
- Let $\bullet \longrightarrow \bullet$ represent the graph consisting of a single edge with distinct endpoints. If $\mathbb{T} = \{\bullet \longrightarrow \bullet\}$, $\mathbf{w}(\bullet \longrightarrow \bullet) = 1$ and $\mathcal{A}(\mathbf{C}) = \mathrm{Hom}(\mathbf{C})$, then $\mathbf{w}^{\mathcal{A}}_{\mathbb{T}}(G) = |E_G|$. If instead $\mathcal{A}(\mathbf{C}) = \mathrm{Hom}(\mathbf{C})$, then $\mathbf{w}^{\mathcal{A}}_{\mathbb{T}}(G)$ counts the number of subgraph occurrences isomorphic to $\bullet \longrightarrow \bullet$ in G (loops are not counted). (See also Example 50.)

- If $\mathbb{T} = \{\bullet, \bullet \longrightarrow \bullet\}$, $\mathbf{w}(\bullet) = 2$, $\mathbf{w}(\bullet \longrightarrow \bullet) = 1$ and $\mathcal{A}(\mathbf{C}) = \mathrm{Hom}(\mathbf{C})$, then $\mathbf{w}_{\mathbb{T}}^{\mathcal{A}}(G) = 2 \cdot |V_G| + |E_G|$.

Our goal is to use $\mathbf{w}_{\mathbb{T}}^{\mathcal{A}}(\cdot)$ as a decreasing measure. This gives rise to two main challenges: finding a suitable $\mathbb{T}$ (if it exists), and determining whether $\mathbf{w}_{\mathbb{T}}^{\mathcal{A}}(\cdot)$ is decreasing. In this paper, we focus exclusively on the second problem, and show that the matter can be decided through a finite rule analysis.

Certain assumptions on $\mathcal{A}(\mathbf{C})$ will be needed. To prevent clutter and to help intuition, we state them now, valid for the remainder of this paper. In the individual proofs, we clarify which assumptions on $\mathcal{A}(\mathbf{C})$ are used.

Assumption 24. *We assume that $\mathcal{A}(\mathbf{C})$ satisfies $\mathrm{rm}(\mathbf{C}) \subseteq \mathcal{A}(\mathbf{C})$; is stable under pullback, composition ($g, f \in \mathcal{A}(\mathbf{C}) \implies g \circ f \in \mathcal{A}(\mathbf{C})$) and decomposition ($g \circ f \in \mathcal{A}(\mathbf{C}) \implies f \in \mathcal{A}(\mathbf{C})$); and that $\mathcal{A}$-factorizations exist.*

Note that $\mathrm{iso}(\mathbf{C}) \subseteq \mathcal{A}(\mathbf{C})$, because $\mathrm{iso}(\mathbf{C}) \subseteq \mathrm{rm}(\mathbf{C})$.

Proposition 25. *In any category, the class $\mathrm{Hom}(\mathbf{C})$ satisfies Assumption 24. Likewise for $\mathrm{mono}(\mathbf{C})$ and $\mathrm{rm}(\mathbf{C})$ in (quasi)toposes.* ⊛[1]

Now suppose that a rule ρ generates a rewrite step diagram. This defines a factorization $t_R = R \xrightarrow{w} G_R \xrightarrow{w'} R'$ (Remark 20). Any tiling of G_R can be partitioned into two using the following definition.

Definition 26. *For arrows $f : A \to B$ and sets S of arrows with codomain B we define the partitioning $S = S_{\cong}^f \uplus S_{\not\cong}^f$ where $S_{\cong}^f = \{g \in S \mid \langle f \mid g \rangle \in \mathrm{iso}(\mathbf{C})\}$ and $S_{\not\cong}^f = \{g \in S \mid \langle f \mid g \rangle \notin \mathrm{iso}(\mathbf{C})\}$.*

Intuitively, $\mathcal{A}(\mathbb{T}, G_R)_{\cong}^w$ contains all tilings that lie isomorphically in the pattern $w(R)$, and $\mathcal{A}(\mathbb{T}, R')_{\cong}^{t_R}$ the remaining tilings, which overlap partially or fully with the context. The remainder of this section is structured as follows.

We will start by centrally identifying some key assumptions and properties that we need in order to reason on the level of the rule (Sect. 4.1).

We then prove that there exists a domain-preserving bijection between $\mathcal{A}(\mathbb{T}, G_R)_{\cong}^w$ and $\mathcal{A}(\mathbb{T}, R')_{\cong}^{t_R}$, allowing us to determine $\mathbf{w}(\mathcal{A}(\mathbb{T}, G_R)_{\cong}^w)$ on the level of the rule (Sect. 4.2).

Determining $\mathbf{w}(\mathcal{A}(\mathbb{T}, G_R)_{\not\cong}^w)$ on the level of the rule is in general impossible, because usually G_R can have an arbitrary size. Instead, we give precise conditions, formulated on the level of the rule, that ensure that there exists a domain-preserving injection $\xi : \mathcal{A}(\mathbb{T}, G_R)_{\not\cong}^w \rightarrowtail \mathcal{A}(\mathbb{T}, G_L)$ across the rewrite step diagram, so that $\mathbf{w}(\xi \circ \mathcal{A}(\mathbb{T}, G_R)_{\not\cong}^w) = \mathbf{w}(\mathcal{A}(\mathbb{T}, G_R)_{\not\cong}^w)$ (Sect. 4.3). Such injections often exist in the usual categories of interest, because the context of G_R is roughly inherited from the left.

---

[1] The proofs for results marked with ⊛ can be found on arXiv [24].

The two results are then combined as follows. If we additionally find a tiling $\Delta \subseteq \mathcal{A}(\mathbb{T}, L)$ such that for the given match $m : L \to G_L$, $m \circ \Delta \subseteq \mathcal{A}(\mathbb{T}, G_L)$, $\mathbf{w}(m \circ \Delta) > \mathbf{w}(\mathcal{A}(\mathbb{T}, G_R)^w_{\cong})$ and $(m \circ \Delta) \cap (\xi \circ \mathcal{A}(\mathbb{T}, G_R)^w_{\ncong}) = \varnothing$, then

$$\begin{aligned}
\mathbf{w}^{\mathcal{A}}_{\mathbb{T}}(G_L) &\geq \mathbf{w}(m \circ \Delta) + \mathbf{w}(\xi \circ \mathcal{A}(\mathbb{T}, G_R)^w_{\ncong}) \\
&> \mathbf{w}(\mathcal{A}(\mathbb{T}, G_R)^w_{\cong}) + \mathbf{w}(\mathcal{A}(\mathbb{T}, G_R)^w_{\ncong}) \\
&= \mathbf{w}^{\mathcal{A}}_{\mathbb{T}}(G_R)
\end{aligned}$$

and we will have successfully proven that $\mathbf{w}^{\mathcal{A}}_{\mathbb{T}}(\cdot)$ is a decreasing measure. This is the main result of this section (Sect. 4.4).

4.1   Relating Rule and Step

In order to reason about steps on the level of rules, the following variant of adhesivity is needed. It does not yet occur in the literature.

Definition 27 (PBPO⁺-Adhesive).   *A pushout square $r' \circ t_K = t_R \circ r$ is (VK) if, whenever it lies at the bottom of a commutative cube shown on the right, where the top face is a pushout and the back faces are pullbacks, we have that the front faces are pullbacks.*

Corollary 28. *If $\mathbf{C}$ is rm-adhesive, pushouts $r' \circ t_K = t_R \circ r$ with $t_K \in \mathrm{rm}(\mathbf{C})$ are PBPO⁺-adhesive.*   □

Remark 29. Not all quasitoposes are PBPO⁺-adhesive: the counterexample by Johnstone et al. [16, Fig. 1], which shows that the category of simple graphs is not rm-adhesive, is also a counterexample for PBPO⁺-adhesivity. We ask: are there interesting PBPO⁺-adhesive categories that are not rm-adhesive?

The following equalities will prove crucial. Recall Notation 1.

Lemma 30. *Assume $\mathbf{C}$ has pullbacks. Let a rewrite step for a PBPO⁺ rule ρ be given. If square $r' \circ t_K = t_R \circ r$ is PBPO⁺-adhesive, then for any ρ-rewrite step and any $t : T \to G_R$*

1. $\langle g_R \mid t \rangle = \langle r' \mid w' \circ t \rangle$;
2. $u' \circ \langle t \mid g_R \rangle = \langle w' \circ t \mid r' \rangle$;
3. $\langle w \mid t \rangle = \langle t_R \mid w' \circ t \rangle$; and
4. $\langle t \mid w \rangle = \langle w' \circ t \mid t_R \rangle$.

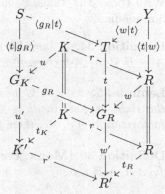

Proof. In the diagram on the right, the bottom face of the bottom cube is PBPO⁺-adhesive by assumption, its top face is a pushout, and its back faces are pullbacks in any category [25, Lemma 15]. Hence its front faces are pullbacks by PBPO⁺-adhesivity. Then all claims follow by composing pullback squares, using the pullback lemma.   □

Remark 31. Because every $t \in \mathcal{A}(T, G_R)$ defines an arrow $w' \circ t \in \mathrm{Hom}(T, R')$, we can overapproximate $\mathcal{A}(T, G_R)$ using $\mathrm{Hom}(T, R')$. The equalities of Lemma 30 will then be used as follows.

1. We will slide morphisms $t \in \mathcal{A}(\mathbb{T}, G_R)_{\not\cong}^w$ to the left. If $\langle g_R \mid t \rangle$ is invertible, then $g_L \circ \langle t \mid g_R \rangle \circ \langle g_R \mid t \rangle^{\leftarrow} : T \to G_L$ is an arrow towards the left. Lemma 30.1 implies that invertibility of $\langle g_R \mid t \rangle$ can be verified on the level of the rule.
2. Although we cannot deduce $\langle t \mid g_R \rangle$, Lemma 30.2 implies that we can at least deduce how it is mapped into K'.
3. Lemma 30.3 implies that it suffices to restrict the overapproximation of $\mathcal{A}(\mathbb{T}, G_R)_{\not\cong}^w$ to $\mathrm{Hom}(\mathbb{T}, R')_{\not\cong}^{t_R}$.
4. If $t \in \mathcal{A}(\mathbf{C})$, then $\langle t \mid w \rangle \in \mathcal{A}(\mathbf{C})$ by the pullback stability assumption. Thus, Lemma 30.4 implies that it suffices to restrict the overapproximation even further to $\{ f \in \mathrm{Hom}(\mathbb{T}, R')_{\not\cong}^{t_R} \mid \langle f \mid t_R \rangle \in \mathcal{A}(\mathbf{C}) \}$.

4.2   Determining $\mathbf{w}(\mathcal{A}(\mathbb{T}, G_R)_{\cong}^w)$

The weight of $\mathcal{A}(\mathbb{T}, G_R)_{\cong}^w$ can be determined under minimal assumptions.

Lemma 32. *Let the pullback on the right be given with $t_R \in \mathcal{A}(\mathbf{C})$. Let $\mathbf{C}$ be $\mathcal{A}$-locally finite and $\mathbb{T}$ a set of objects. Then $\chi(t) = w' \circ t$ is a domain-preserving bijection $\chi : \mathcal{A}(\mathbb{T}, G_R)_{\cong}^w \to \mathcal{A}(\mathbb{T}, R')_{\cong}^{t_R}$.*

$$
\begin{array}{ccc}
R & =\!=\!=\!= & R \\
\scriptstyle t_R \downarrow & \text{PB} & \downarrow \scriptstyle w \\
R' & \leftarrow\! w' \!- & G_R
\end{array}
$$

$\circledast$

Corollary 33. *If the conditions of Lemma 32 are met, then* $\mathbf{w}(\mathcal{A}(\mathbb{T}, G_R)_{\cong}^w) = \mathbf{w}(\mathcal{A}(\mathbb{T}, R')_{\cong}^{t_R})$. $\qquad\square$

4.3   Sliding Tiles Injectively

In this section we establish conditions for the existence of a domain-preserving injection $\xi : \mathcal{A}(\mathbb{T}, G_R)_{\not\cong}^w \rightarrowtail \mathcal{A}(\mathbb{T}, G_L)$. Intuitively, one can think of ξ as sliding tiles from right to left across the rewrite step diagram.

If $l' \in \mathrm{rm}(\mathbf{C})$, then ξ will be seen to exist rather straightforwardly. However, in general it suffices to require more weakly that l' preserves any tiles to be slid (and distinctly so). Definitions 34 and 36 help capture such a weaker requirement. With these definitions, ξ can be shown to exist even for non-trivial rules with non-monic l'.

Definition 34. *A morphism $g : B \to C$ preserves the $\mathcal{A}$-factorization of $f : A \to B$ if the $\mathcal{A}$-factorization $f = f'' \circ f'$ exists and $g \circ f'' \in \mathcal{A}(C)$.*

Lemma 35. *Assume $\mathbf{C}$ has pullbacks. Let the diagram on the right be given, with $x \in \mathcal{A}(\mathbf{C})$. If f preserves the $\mathcal{A}$-factorization of $g' \circ x$, then $f' \circ x \in \mathcal{A}(\mathbf{C})$.*

$$
\begin{array}{ccc}
A & \leftarrow\! f' \!- B \leftarrow\! x \sim & X \\
\scriptstyle g \downarrow & \text{PB} & \downarrow \scriptstyle g' \\
C & \leftarrow\! f \!- & D
\end{array}
$$

$\circledast$

Definition 36 (Monic For). *Morphism $h : B \to C$ is monic for morphisms $f, g : A \to B$ if $h \circ f = h \circ g$ implies $f = g$.*

Lemma 37. *Let the diagram on the right be given. If f is monic for $g' \circ x$ and $g' \circ y$, then f' is monic for x and y.* ⊛

$$\begin{array}{ccccc} A & \leftarrow f' - & B & \overset{\leftarrow x -}{\underset{\leftarrow y -}{}} & X \\ \downarrow g & & \downarrow & \text{PB} & \downarrow g' \\ C & \leftarrow f - & D & & \end{array}$$

The morphism $g_R : G_K \to G_R$ of the rewrite step may identify elements. So for the injection ξ from right to left to exist, we must be able to go into the inverse direction, without identifying tiles. To this end, the following lemma will prove useful.

Lemma 38. *If epimorphisms e and e' in diagram*

$$\begin{array}{ccccc} A' & -f' \to & B' & \leftarrow g' - & C' \\ \downarrow e & = & \downarrow h & = & \downarrow e' \\ A & -f \to & B & \leftarrow g - & C \end{array}$$

are split, then for right inverses $e^{\leftarrow}$ and $e'^{\leftarrow}$, $f'e^{\leftarrow} = g'e'^{\leftarrow} \implies f = g$. ⊛

We are now ready for the main theorem of this subsection. We recommend keeping the diagram for Lemma 30 alongside it.

Theorem 39. *Assume $\mathbf{C}$ has pullbacks. Let a rewrite rule ρ be given with square $r' \circ t_K = t_R \circ r$ PBPO$^+$-adhesive. Fix a class $\mathcal{A}(\mathbf{C})$ and a set of objects $\mathbb{T}$. Let $\Phi = \{f \in \mathrm{Hom}(\mathbb{T}, R')^{t_R}_{\ncong} \mid \langle f \mid t_R \rangle \in \mathcal{A}(\mathbf{C})\}$. If*

1. *for all $f \in \Phi$, $\langle r' \mid f \rangle$ is a split epimorphism; and*
2. *for some right inverse choice function $(\cdot)^{\leftarrow}$ and for all $f, g \in \Phi$,*
   (a) *l' preserves the $\mathcal{A}$-factorization of $\langle f \mid r' \rangle \circ \langle r' \mid f \rangle^{\leftarrow}$; and*
   (b) *l' is monic for $\langle f \mid r' \rangle \circ \langle r' \mid f \rangle^{\leftarrow}$ and $\langle g \mid r' \rangle \circ \langle r' \mid g \rangle^{\leftarrow}$.*

then for any rewrite step diagram induced by ρ, the function

$$\xi(t) = g_L \circ \langle t \mid g_R \rangle \circ \langle g_R \mid t \rangle^{\leftarrow}$$

defines an injection $\mathcal{A}(\mathbb{T}, G_R)^w_{\ncong} \rightarrowtail \mathcal{A}(\mathbb{T}, G_L)$.

Proof. We first argue that the use of $(\cdot)^{\leftarrow}$ in $\xi(t)$ is well-defined. Because $t \in \mathcal{A}(\mathbf{C})$, we have $\langle t \mid w \rangle \in \mathcal{A}(\mathbf{C})$ by pullback stability. And $\langle t \mid w \rangle = \langle w' \circ t \mid t_R \rangle$ by Lemma 30.4. Moreover, $\langle w \mid t \rangle \notin \mathrm{iso}(\mathbf{C})$, and $\langle w \mid t \rangle = \langle t_R \mid w' \circ t \rangle$ by Lemma 30.3. So $\langle w' \circ t \mid t_R \rangle \in \Phi$. Then by local assumption 1., $\langle r' \mid w' \circ t \rangle$ is a split epimorphism. And $\langle r' \mid w' \circ t \rangle = \langle g_R \mid t \rangle$ by Lemma 30.1. So $\langle g_R \mid t \rangle^{\leftarrow}$ is well-defined.

We next argue that $\xi(t) \in \mathcal{A}(\mathbb{T}, G_L)$. As established, $\langle r' \mid w' \circ t \rangle = \langle g_R \mid t \rangle$, and by Lemma 30.2, $u' \circ \langle t \mid g_R \rangle = \langle w' \circ t \mid r' \rangle$. So the diagram

$$\begin{array}{ccccc} G_L & \overset{g_L}{\longleftarrow} & G_K & \overset{\langle t \mid g_R \rangle \circ \langle g_R \mid t \rangle^{\leftarrow}}{\longleftarrow\!\!\!\rightsquigarrow} & T \\ \downarrow \alpha & \text{PB} & \downarrow u' & & \\ L' & \overset{l'}{\longleftarrow} & K' & \overset{\langle w' \circ t \mid r' \rangle \circ \langle r' \mid w' \circ t \rangle^{\leftarrow}}{\longleftarrow} & \end{array}$$

commutes, where the pullback square is given by the rewrite step. Moreover, $\langle t \mid g_R \rangle \circ \langle g_R \mid t \rangle^{\leftarrow} \in \mathcal{A}(\mathbf{C})$ (as indicated in the diagram) by stability under composition, using $\langle t \mid g_R \rangle \in \mathcal{A}(\mathbf{C})$ (by pullback stability and $t \in \mathcal{A}(\mathbf{C})$) and $\langle g_R \mid t \rangle^{\leftarrow} \in \mathrm{rm}(\mathbf{C}) \subseteq \mathcal{A}(\mathbf{C})$ (using Proposition 8 and Assumption 24). By local assumption (a) and the commuting triangle of the diagram, l' preserves the $\mathcal{A}$-factorization of $u' \circ \langle t \mid g_R \rangle \circ \langle g_R \mid t \rangle^{\leftarrow}$. So by Lemma 35, $\xi(t) \in \mathcal{A}(\mathbf{C})$ and consequently $\xi(t) \in \mathcal{A}(\mathbb{T}, G_L)$.

For injectivity of ξ, assume $\xi(t) = \xi(s)$ for $t, s \in \mathcal{A}(\mathbb{T}, G_R)_{\not\cong}^w$. By local assumption (b), Lemma 30.1 and Lemma 30.2, l' is monic for

$$\langle w' \circ t \mid r' \rangle \circ \langle r' \mid w' \circ t \rangle^{\leftarrow} = u' \circ \langle t \mid g_R \rangle \circ \langle g_R \mid t \rangle^{\leftarrow}$$

and

$$\langle w' \circ s \mid r' \rangle \circ \langle r' \mid w' \circ s \rangle^{\leftarrow} = u' \circ \langle s \mid g_R \rangle \circ \langle g_R \mid s \rangle^{\leftarrow}.$$

So by Lemma 37, g_L is monic for $\langle t \mid g_R \rangle \circ \langle g_R \mid t \rangle^{\leftarrow}$ and $\langle s \mid g_R \rangle \circ \langle g_R \mid s \rangle^{\leftarrow}$. Then because $\xi(t) = \xi(s)$, $\langle t \mid g_R \rangle \circ \langle g_R \mid t \rangle^{\leftarrow} = \langle s \mid g_R \rangle \circ \langle g_R \mid s \rangle^{\leftarrow}$. Then finally, $t = s$ by Lemma 38. □

4.4   The Main Result

We are now ready to prove the main result of this paper (Theorem 40) and its corollary (Corollary 42). We also show that in rather common settings, many technical conditions of the theorem are met automatically (Lemma 44 and Propositions 45 and 46). We close with a complementary lemma that establishes decreasingness for deleting rules (Lemma 47). Examples of applications will be given in Sect. 5.

Theorem 40 (Decreasingness by Element Counting). *Let $\mathcal{T}$ and $\mathcal{T}'$ be disjoint sets of $PBPO^+$ rules. Assume $\mathbf{C}$ has pullbacks and let $\mathcal{A}(\mathbf{C})$ be a class such that $\mathbf{C}$ is $\mathcal{A}$-locally finite. Let $\mathbb{T}$ be a set of objects and $\mathbf{w} : \mathbb{T} \to \mathbb{N}$ a weight function such that, for every $\rho \in \mathcal{T} \uplus \mathcal{T}'$, the following conditions hold:*

- *ρ's pushout square $r' \circ t_K = t_R \circ r$ is $PBPO^+$-adhesive; and*
- *$t_R \in \mathcal{A}(\mathbf{C})$; and*
- *set $\Phi_\rho = \{ f \in \mathrm{Hom}(\mathbb{T}, R')_{\not\cong}^{t_R} \mid \langle f \mid t_R \rangle \in \mathcal{A}(\mathbf{C}) \}$ meets the conditions of Theorem 39 for some right inverse choice function $(\cdot)^{\leftarrow}$; and*
- *there exists a set $\Delta_\rho \subseteq \mathcal{A}(\mathbb{T}, L)$ such that*
  - *for all $f \in \Phi_\rho$ and $t \in \Delta_\rho$, $l' \circ \langle f \mid r' \rangle \circ \langle r' \mid f \rangle^{\leftarrow} \neq t_L \circ t$;*
  - *t_L is monic for all $t, t' \in \Delta_\rho$; and*
  - *$\mathbf{w}(\Delta_\rho) > \mathbf{w}(\mathcal{A}(\mathbb{T}, R')_{\cong}^{t_R})$ if $\rho \in \mathcal{T}$ and $\mathbf{w}(\Delta_\rho) \geq \mathbf{w}(\mathcal{A}(\mathbb{T}, R')_{\cong}^{t_R})$ if $\rho \in \mathcal{T}'$.*

Then for any rewrite step with match $m \in \mathcal{A}(\mathbf{C})$, induced by a rule $\rho \in \mathcal{T} \uplus \mathcal{T}'$, we have $\mathbf{w}_{\mathbb{T}}^{\mathcal{A}}(G_L) > \mathbf{w}_{\mathbb{T}}^{\mathcal{A}}(G_R)$ if $\rho \in \mathcal{T}$ and $\mathbf{w}_{\mathbb{T}}^{\mathcal{A}}(G_L) \geq \mathbf{w}_{\mathbb{T}}^{\mathcal{A}}(G_R)$ if $\rho \in \mathcal{T}'$.

Proof. Let a step induced by a $\rho \in \mathcal{T} \uplus \mathcal{T}'$ be given.
By Corollary 33, $\mathbf{w}(\mathcal{A}(\mathbb{T}, G_R)_{\cong}^w) = \mathbf{w}(\mathcal{A}(\mathbb{T}, R')_{\cong}^{t_R})$.

By Theorem 39, we obtain an injection $\xi : \mathcal{A}(\mathbb{T}, G_R)_{\ncong}^{w} \rightarrowtail \mathcal{A}(\mathbb{T}, G_L)$ with $dom(\xi(t)) = dom(t)$, using the assumption on Φ_ρ. So $\mathbf{w}(\xi \circ \mathcal{A}(\mathbb{T}, G_R)_{\ncong}^{w}) = \mathbf{w}(\mathcal{A}(\mathbb{T}, G_R)_{\ncong}^{w})$.

Moreover, by $m \in \mathcal{A}(\mathbf{C})$ and stability under composition, we have $(m \circ \Delta_\rho) \subseteq \mathcal{A}(\mathbb{T}, G_L)$. And by t_L monic for all $t, t' \in \Delta_\rho$, we have m monic for all $t, t' \in \Delta_\rho$, and so $\mathbf{w}(m \circ \Delta_\rho) = \mathbf{w}(\Delta_\rho)$. It remains to show that $(m \circ \Delta_\rho)$ and $(\xi \circ \mathcal{A}(\mathbb{T}, G_R)_{\ncong}^{w})$ are disjoint. If for a $t' \in \Delta_\rho$ and $t \in \mathcal{A}(\mathbb{T}, G_R)_{\ncong}^{w}$, $m \circ t' = \xi(t)$, then $t_L \circ t' = \alpha \circ m \circ t' = \alpha \circ \xi(t) = \alpha \circ g_L \circ \langle t \mid g_R \rangle \circ \langle g_R \mid t \rangle^{\leftarrow} = l' \circ u' \circ \langle t \mid g_R \rangle \circ \langle g_R \mid t \rangle^{\leftarrow} = l' \circ \langle w' \circ t \mid r' \rangle \circ \langle r' \mid w' \circ t \rangle^{\leftarrow}$, using Lemma 30.(1–2) and $\alpha \circ g_L = l' \circ u$, which contradictingly implies $t' \notin \Delta_\rho$ by the definition of Δ_ρ and $w' \circ t \in \Phi$. Thus $\xi(t) \neq m \circ t'$.

In summary,

$$\mathbf{w}_{\mathbb{T}}^{\mathcal{A}}(G_L) \geq \mathbf{w}(m \circ \Delta_\rho) + \mathbf{w}(\xi \circ \mathcal{A}(\mathbb{T}, G_R)_{\ncong}^{w})$$
$$= \mathbf{w}(\Delta_\rho) + \mathbf{w}(\mathcal{A}(\mathbb{T}, G_R)_{\ncong}^{w})$$
$$\succ \mathbf{w}(\mathcal{A}(\mathbb{T}, R')_{\cong}^{t_R}) + \mathbf{w}(\mathcal{A}(\mathbb{T}, G_R)_{\ncong}^{w})$$
$$= \mathbf{w}(\mathcal{A}(\mathbb{T}, G_R)_{\cong}^{w}) + \mathbf{w}(\mathcal{A}(\mathbb{T}, G_R)_{\ncong}^{w})$$
$$= \mathbf{w}_{\mathbb{T}}^{\mathcal{A}}(G_R)$$

for $\succ = >$ if $\rho \in \mathcal{T}$ and $\succ = \geq$ if $\rho \in \mathcal{T}'$, completing the proof. $\square$

Remark 41. The requirement $m \in \mathcal{A}(\mathbf{C})$ puts a lower bound on what one can choose for $\mathcal{A}(\mathbf{C})$ in a termination proof. Usually two factors are relevant: the class of t_L, and match restrictions imposed by the setting. More precisely, let $X(\mathbf{C})$ and $Y(\mathbf{C})$ be classes of morphisms. If $t_L \in X(\mathbf{C})$, where $X(\mathbf{C})$ satisfies the decomposition property (meaning $m \in X(\mathbf{C})$ by $t_L = \alpha \circ m$), and the setting imposes $m \in Y(\mathbf{C})$, then the choice of $\mathcal{A}(\mathbf{C})$ must satisfy $X(\mathbf{C}) \cap Y(\mathbf{C}) \subseteq \mathcal{A}(\mathbf{C})$.

From Theorem 40 and Remark 41 the following is immediate.

Corollary 42 (Termination by Element Counting). *Let $\mathcal{T}$ and $\mathcal{T}'$ be disjoint sets of $PBPO^+$ rules. Let $\mathcal{A}(\mathbf{C})$ be a class such that for all rules $\rho \in (\mathcal{T} \uplus \mathcal{T}')$, $t_L(\rho) \in \mathcal{A}(\mathbf{C})$ or matching of ρ is restricted to a class $X \subseteq \mathcal{A}(\mathbf{C})$. If the conditions of Theorem 40 are met, then $\mathcal{T}$ terminates relative to $\mathcal{T}'$. Hence $\mathcal{T} \uplus \mathcal{T}'$ is terminating iff $\mathcal{T}'$ is.* $\square$

Remark 43 (Generalizing $\mathcal{A}(\mathbf{C})$). Theorem 40 and the results it depends on still hold if in instead of having $\mathcal{A}(\mathbf{C})$ globally fixed, a class $\mathcal{A}(\mathbf{C})$ is fixed for each individual $T \in \mathbb{T}$, and match morphism m is in the intersection of every class. This for instance allows counting some tiles monically, and others non-monically.

The following lemma implies that in many categories of interest, tilings of L and slid tiles never collide. For instance, in quasitoposes, pushouts along t_K are pullbacks if $t_K \in \mathrm{rm}(\mathbf{C})$ [15, Lemma A2.6.2].

Lemma 44. *Let a rule ρ be given. If pushout $r' \circ t_K = t_R \circ r$ is a pullback and $t_K \in \mathrm{mono}(\mathbf{C})$, then for all $t \in \mathrm{Hom}(T, R')_{\ncong}^{t_R}$ with $\langle r' \mid t \rangle$ a split epi, we have that for all $t' \in \mathrm{Hom}(T', L)$, $l' \circ \langle t \mid r' \rangle \circ \langle r' \mid t \rangle^{\leftarrow} \neq t_L \circ t'$.*

Proof. By contradiction. Assume that for some $t \in \mathrm{Hom}(T, R')^{t_R}_{\ncong}$ with $\langle r' \mid t \rangle$ a split epi and some $t' \in \mathrm{Hom}(T', L)$, $l' \circ \langle t \mid r' \rangle \circ \langle r' \mid t \rangle^{\leftarrow} = t_L \circ t'$. Then we have the commuting diagram

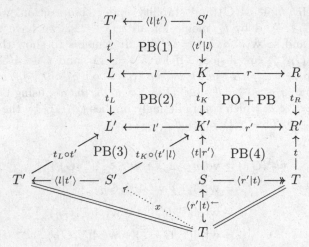

where

- squares PB(2) and PO + PB are given by ρ;
- squares PB(1) and PB(4) are constructed;
- square PB(3) follows from PB(1) and PB(2), using the pullback lemma;
- morphism x exists by the hypothesis and the pullback property; and
- $\langle r' \mid t \rangle \circ \langle r' \mid t \rangle^{\leftarrow} = 1_T$ by the right inverse property.

By a diagram chase we thus have $t = r' \circ t_K \circ \langle t' \mid l \rangle \circ x$. Then from

$$
\begin{array}{ccccccc}
R & \xleftarrow{\quad r \quad} & K & =\!=\!=\!=\!= & K & \xleftarrow{\langle t' \mid l \rangle \circ x} & T \\
\downarrow t_R & \mathrm{PB} & \curlyvee t_K & \mathrm{PB} & \| & \mathrm{PB} & \| \\
R' & \xleftarrow{\quad r' \quad} & K' & \xleftarrow{\ t_K\ } & K & \xleftarrow{\langle t' \mid l \rangle \circ x} & T \\
& & & \underbrace{\qquad\qquad\qquad\qquad}_{t} & & &
\end{array}
$$

and two applications of the pullback lemma, we have $\langle t_R \mid t \rangle \in \mathrm{iso}(\mathbf{C})$, contradicting $t \in \mathrm{Hom}(T, R')^{t_R}_{\ncong}$. $\qquad\square$

The two propositions below state further sufficient conditions for satisfying the termination method's preconditions. Many graph categories of interest meet these conditions.

Proposition 45. *If $\mathbf{C}$ is an rm-locally finite, rm-adhesive quasitopos, then* $\mathrm{rm}(\mathbf{C})$ *satisfies Assumption 24. $\mathbf{C}$ also has all pullbacks and all pushouts, and so in particular the required pushouts described in Assumption 19. If moreover $t_L(\rho) \in \mathrm{rm}(\mathbf{C})$, then $m, t_K, t_R \in \mathrm{rm}(\mathbf{C})$, ρ's pushout square is $PBPO^+$-adhesive, and t_L is monic for $\mathcal{A}(\mathbb{T}, L(\rho))$.*

Proof. A quasitopos has by definition all limits and colimits. That $\mathrm{rm}(\mathbf{C})$ satisfies Assumption 24 was stated in Proposition 25. If $t_L \in \mathrm{rm}(\mathbf{C})$, and $m, t_K \in \mathrm{rm}(\mathbf{C})$ by stability under decomposition and pullback, respectively. That $t_R \in \mathrm{rm}(\mathbf{C})$ subsequently follows from pushout stability in quasitoposes [15, Lemma A.2.6.2]. Because of the assumed rm-adhesivity and $t_K \in \mathrm{rm}(\mathbf{C})$, ρ's pushout square is PBPO$^+$-adhesive (Corollary 28). Finally, that t_L is monic for $\mathcal{A}(\mathbb{T}, L(\rho))$ follows trivially from the fact that t_L is monic. (See [4, Corollary 1] and [25, Proposition 36] for relevant summaries of quasitopos properties.)   □

As is well known, if $\mathcal{I}$ is small, then the functor category $[\mathcal{I}, \mathbf{Set}]$ is a topos, and many structures that are of interest to the graph transformation community can be defined in this manner (e.g., $\mathbf{Graph} \cong [\cdot \rightrightarrows \cdot, \mathbf{Set}]$). The following proposition assures us that such toposes are closed under finite restrictions. We are not aware of a similar principle for quasitoposes.

Proposition 46. *If $\mathcal{I}$ is finite and $\mathbf{C} \cong [\mathcal{I}, \mathbf{FinSet}]$, then $\mathbf{C}$ is a Hom-locally finite topos, and so for any $\mathcal{A}(\mathbf{C})$, an $\mathcal{A}$-locally finite rm-adhesive quasitopos.*

Proof. $\mathbf{C}$ is a topos [5, Example 5.2.7], and it is locally finite because $\mathbf{FinSet}$ is Hom-locally finite. Moreover, any topos is rm-adhesive [18], and any topos is a quasitopos.   □

Finally, we have the following general principle, which does not require any assumptions on t_K and t_R, nor any adhesivity assumptions.

Lemma 47 (Deleting Rules are Decreasing). *Assume $\mathbf{C}$ has pullbacks and is mono-locally finite. Suppose that for a PBPO$^+$ rule ρ, l' is monic, l is not epic, and r is iso; and that for any matches m for ρ, m is monic. Then ρ is decreasing for $\mathbb{T} = \{L\}$, $\mathbf{w}(L) > 0$, $\Delta_\rho = \{1_L\}$ and $\mathcal{A}(\mathbf{C}) = \mathrm{mono}(\mathbf{C})$.*   ⊛

5   Examples

We give a number of examples of applying Theorem 40 in category $\mathbf{C} = \mathbf{FinGraph}$ (Definition 12), each demonstrating new features. For each example, we will fix $\mathbb{T}$, $\mathbf{w}$ and $\mathcal{A}(\mathbf{C})$, and usually some properties of the relevant morphism sets (such as cardinalities) or related comments. The remaining details of the proofs are routine. Note that in $\mathbf{FinGraph}$ (and more generally in any topos), $\mathrm{rm}(\mathbf{C}) = \mathrm{mono}(\mathbf{C})$, and because the rules in examples satisfy $t_L \in \mathrm{mono}(\mathbf{C})$, we are in each case free to choose $\mathrm{mono}(\mathbf{C})$ or $\mathrm{Hom}(\mathbf{C})$ for $\mathcal{A}(\mathbf{C})$ (Remark 41).

Notation 48 (Visual Notation). *In our examples of rules, the morphisms $t_X : X \hookrightarrow X'$ ($X \in \{L, K, R\}$) of rules are regular monos (embeddings). We depict t_X by depicting the graph X', and then let solid, colored vertices and solid edges denote $t_X(X)$, with dotted blank vertices and dotted edges the remainder of X'. For example, in Example 49 below, subgraph L of $t_L : L \hookrightarrow L'$ is $\boxed{x} \longrightarrow \boxed{y}$.*
The vertices of graphs are non-empty sets $\{x_1, \ldots, x_n\}$ depicted by boxes $\boxed{x_1 \; \cdots \; x_n}$. When depicting the homomorphism $(r' = (r'_V, r'_E)) : K' \to R'$ we

*will choose the vertices of K' and R' in such a way that component r'_V is fully
determined by $S \subseteq r'_V(S)$ for all $S \in V_{K'}$. For example, for nodes $\{x\}, \{y\} \in V_{K'}$
of Example 49 below (in which morphism r' is implicit), $r'(\{x\}) = r'(\{y\}) =
\{x, y\} \in V_{R'}$. If component r'_E is not uniquely determined by r'_V, then let r'_E
preserve the relative positioning of the edges (although often, this choice will be
inconsequential). Morphism $l' : K' \to L'$ is depicted similarly.*

Example 49 (Folding an Edge). The rule

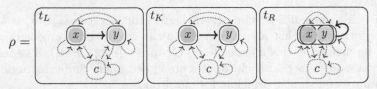

folds a non-loop edge $\bullet \longrightarrow \bullet$ into a loop $\bullet \circlearrowleft$ (in any context). Define tile set
$\mathbb{T} = \{\bullet \longrightarrow \bullet\}$ with $\mathbf{w}(\bullet \longrightarrow \bullet) = 1$ and fix $\mathcal{A}(\mathbf{C}) = \mathrm{mono}(\mathbf{C})$. Then $|\Phi_\rho| = 5$ (every
non-loop dotted edge in R', and the loop on c). For every $f \in \Phi_\rho$, $\langle r' \mid f \rangle$ is iso
and hence a regular epi with a unique right inverse. Because l' is monic (even
iso), the remaining details of Theorem 39 are immediate. Finally, for the only
choice of Δ_ρ, $\mathbf{w}(\Delta_\rho) = 1 > \mathbf{w}(\mathrm{mono}(\mathbb{T}, R')^{t_R}_{\cong}) = \mathbf{w}(\varnothing) = 0$. So ρ is terminating.

Example 50. Let the rewrite system $\mathcal{T}$ be given consisting of ρ and τ, respectively:

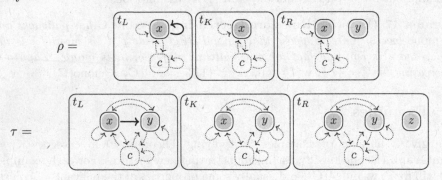

Rule ρ deletes a loop in any context, and adds a node; and rule τ, deletes a
non-loop edge in any context, and adds a node.

Because the r morphisms of ρ and τ are not iso, Lemma 47 cannot be applied.
But for $T = \bullet \longrightarrow \bullet$, $\mathbb{T} = \{T\}$, $\mathbf{w}(T) = 1$, $\mathcal{A}(\mathbf{C}) = \mathrm{Hom}(\mathbf{C})$, and Δ_{ρ_1} and Δ_{ρ_2}
the singleton sets containing the unique morphisms $T \to L(\rho)$ and $T \to L(\tau)$,
respectively, $\mathcal{T}$ is proven decreasing and thus terminating by Theorem 40. This
argument captures the natural argument: "the number of edges is decreasing".

An alternative argument lets $\mathbb{T} = \{T, T'\}$, with $T' = \bullet \circlearrowleft$, $\mathbf{w}(T) = \mathbf{w}(T') = 1$,
and $\mathcal{A}(\mathbf{C}) = \mathrm{mono}(\mathbf{C})$. Then Δ_ρ contains the unique mono $T' \rightarrowtail L(\rho)$ and Δ_τ
the unique mono $T \rightarrowtail L(\tau)$. This captures the argument: "the sum of loop and
non-loop edges is decreasing".

Remark 51 (Fuzzy Categories). In a fuzzy graph category $\mathbf{C}$, a graph is labeled using labels from a partial order $(\mathcal{L}, \leq)$ [20,25]. Unlike in **Graph**, $\mathrm{mono}(\mathbf{C}) \neq \mathrm{rm}(\mathbf{C})$: $\mathrm{mono}(\mathbf{C})$ contains all injective graph homomorphisms such that labels are non-decreasing ($\leq$), and $\mathrm{rm}(\mathbf{C})$ restricts to monomorphisms that preserve labels ($=$). In previous papers [22,23,25], we have shown that fuzzy graphs are useful structures for implementing relabeling mechanics for graph transformation.

In these categories, rules that change labels (but leave the structure of the graph unchanged) can be proven terminating by using $\mathcal{A}(\mathbf{C}) = \mathrm{rm}(\mathbf{C})$, but not always by using $\mathcal{A}(\mathbf{C}) = \mathrm{mono}(\mathbf{C})$. For instance, a rule that increases a loop edge label a into label $b > a$, is shown terminating by $\mathbb{T} = \{\cdot \circlearrowright a\}$ and $\mathcal{A}(\mathbf{C}) = \mathrm{rm}(\mathbf{C})$, but no proof exists for $\mathcal{A}(\mathbf{C}) = \mathrm{mono}(\mathbf{C})$, because $\cdot \circlearrowright b$ has strictly more monic elements than $\cdot \circlearrowright a$.

That the termination method is indeed applicable for these categories is immediate from a result in a recent paper by Rosset and ourselves [32]: fuzzy presheaves (which includes fuzzy graphs) are rm-adhesive quasitoposes.

Example 52. The rule

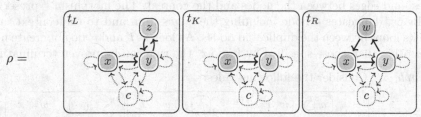

$$\rho =$$

is proven terminating by $\mathbb{T} = \{\cdot \rightleftarrows \cdot\}$, $\mathbf{w}(\cdot \rightleftarrows \cdot) = 1$ and $\mathcal{A}(\mathbf{C}) = \mathrm{mono}(\mathbf{C})$.

This example shows that applicability of our technique is not invariant under rule equivalence. Let ρ' be obtained from ρ by dropping the solid edge from x to y in K and K'. The rules ρ and ρ' are equivalent: they induce the same rewrite relation. However, our termination technique fails for ρ', because not all tiles can be intactly transferred. We pose the question: *can every rule τ be converted into an equivalent representative standard(τ) such that the method fails on standard(τ) iff it fails on all rules in the equivalence class?*

Example 53. Consider the rules ρ and τ, respectively:

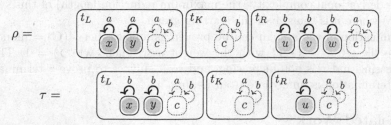

$$\rho =$$

$$\tau =$$

Intuitively, these rules model replacements in multisets over $\{a, b\}$. The elements of the multiset are modeled by nodes with loops that carry the label a or b, respectively. Rule ρ replaces two a's by three b's, and τ replaces two b's by one a.

These rules are terminating. To prove this, we use tiles $\mathbb{T} = \{\cdot\circlearrowright a, \cdot\circlearrowright b\}$ together with the weight assignment $\mathbf{w}(\cdot\circlearrowright a) = 5$ and $\mathbf{w}(\cdot\circlearrowright b) = 3$, and let $\mathcal{A}(\mathbf{C}) = \mathrm{mono}(\mathbf{C})$. The context is isomorphically preserved along l' and r', and partial overlaps with the pattern are not possible. So Theorem 39 is easily verified for Φ_ρ and Φ_τ. Then for the obvious largest choices of Δ_ρ and Δ_τ, we have $\mathbf{w}(\Delta_\rho) = 2 \cdot 5 = 10 > \mathbf{w}(\mathrm{mono}(\mathbb{T}, \rho(R'))_{\cong}^{\rho(t_R)}) = 3 \cdot 3 = 9$ for ρ and $\mathbf{w}(\Delta_\tau) = 2 \cdot 3 = 6 > \mathbf{w}(\mathrm{mono}(\mathbb{T}, \tau(R'))_{\cong}^{\tau(t_R)}) = 5$ for τ, completing the proof.

The above termination proof works also for vast generalizations of the rules. For instance, the first rule can be generalized to

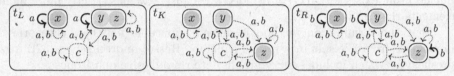

Observe that L' now allows an unbounded number of additional loops on the nodes, and edges between the nodes and the context. The morphism l' preserves the loops, duplicates a node including the edges from and to the context, and unfolds loops between the duplicated nodes. As long as l' and r' do not create new loops other than those specified by l and r, the rule can be proven terminating.

Example 54. Consider the following rules:

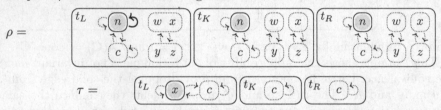

Rule ρ deletes an arbitrary loop, and in doing so, allows arbitrarily many bipartite graph components in the context to duplicate (such components can either be mapped onto node c or onto the right subgraph component). Note that this makes the rule non-deterministic. Rule τ deletes an arbitrary node including incident edges.

The derivational complexity (the maximum reduction length) of this system is $O(2^n)$ where n is the size of the starting graph.

Termination of the system can be proven as follows. Let $\mathcal{A}(\mathbf{C}) = \mathrm{mono}(\mathbf{C})$. Use the tile set $\mathbb{T} = \{\cdot\circlearrowright\}$ with the weight assignment $\mathbf{w}(\cdot\circlearrowright) = 1$. Then ρ is decreasing and τ is non-increasing, and so it suffices to prove τ terminating, whose termination is immediate from Lemma 47.

6   Related Work

We consider two closely related approaches by Bruggink et al. [6, 7] to be the most relevant to our method. Both approaches use weighted type graphs T to measure graphs G by means of counting weighted morphisms $G \to T$ (instead of weighted morphisms $T' \to G$ for tiles T'). So the general idea is dual to ours. Moreover, to our knowledge, these approaches are the only systematic termination methods in the algebraic tradition based on decreasing interpretations.

Both methods are defined for DPO in the category of edge-labeled multigraphs. The first approach [7] requires that l and r of DPO rules $L \xleftarrow{l} K \xrightarrow{r} R$, and matches $m : L \rightarrowtail G_L$, are monic. The second approach [6] has no such restrictions.

Because our method is applicable in a much broader setting, our method will prove rules terminating that are outside the scope of the methods by Bruggink et al. Nonetheless, it is interesting to ask how the approaches relate in settings where they are all defined.

On the one hand, although Examples 5 and 6 of [6] are within the scope of our method, our method cannot prove them terminating. The intuitive reason is that the examples terminate because of global properties, rather than local ones. On the other hand, Example 55 below defines a DPO rule that falls inside the scope of all three methods, and only our method can prove it terminating. In conclusion, within the restricted setting, the methods are incomparable.

Example 55. Consider the following DPO rule ρ in category **FinGraph**:

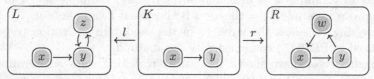

and assume matching is required to be monic. This requirement is often used in practice, because monic matching increases the expressiveness of DPO [14].

The approach in [7] cannot prove ρ terminating. For establishing termination (on all graphs), the weighted type graph T has to contain a node with a loop (called a *flower node*). The flower node ensures that every graph G can be mapped into T. Then, in particular, the technique requires a weight decrease (from L to R) for the case that the interface K is mapped onto the flower node. However, this makes L and R indistinguishable for the technique in [7].

Although matches are required to be monic, the method of [6] overapproximates for unrestricted matches by design. Observe that if matching is not monic, then graph L of ρ, but with x and y identified, rewrites to itself, meaning ρ is not terminating. As a consequence, the overapproximation of [6] causes it to fail in proving ρ terminating for the monic matching setting. (For the same reason, the method of [6] fails on the simpler top span of Example 49, which is a DPO rule, for the monic matching setting.)

Rule ρ can be proven terminating with our method as follows. Encode ρ into PBPO$^+$ using the standard encoding [25, Definition 71]. The resulting rule and a termination proof is given in Example 52.

Additional examples by Bruggink et al. that our method can prove are Example 4 of [7] (= Example 2 of [6]), and Example 4 of [6]. Additional examples that our method cannot prove are Example 1 and the example of Sect. 3.5 of [7]. However, unlike the earlier referenced Examples 5 and 6 of [6], these examples are in reach if our morphism counting technique can take into account antipatterns (Remark 56), because they terminate because of local properties.

Remark 56 (Antipatterns). A rule that matches an isolated node, and adds a loop, cannot be proven terminating with our method. For this, one must be able to count nodes *without* loops (an *antipattern*), which is currently unsupported. We believe that extending our method with support for such antipatterns is a natural first step for significantly strengthening it.

We discuss some additional related work. An early systematic termination criterion for hypergraph rewriting with DPO, is due to Plump, based on the concept of forward closures [27]. Both of the examples proven terminating with forward closures, Examples 3.8 and 4.1 of [27], can be handled with our method. The encoding and proof of Example 3.8 is available on arXiv [24, Example 38].

More recently, Plump formulated a modularity criterion for hypergraph rewriting using DPO [31]: the union of two terminating systems is terminating if there are no sequential critical pairs. Of this paper, our method can prove three out of four examples: Examples 3 (= Example 3.8 of [27]), 5 and 6. The modeling of Example 6 is available on arXiv [24, Example 39]. Our method cannot prove Example 4 (= the already discussed Example 5 of [7]). It would be interesting to assess the strength of the modularity criterion (especially if generalized to PBPO$^+$) combined with our method.

Bruggink et al. have shown that string rewriting rules are terminating on graphs iff they are terminating on cycles [7], making cycle rewriting techniques [33,36] applicable to graph transformation systems consisting of string rewrite rules. Similarly, in a previous paper [22], we have shown that particular PBPO$^+$ encodings of linear term rewrite rules are terminating on graphs iff they are terminating on terms.

There also exist a variety of methods that generalize TRS methods (such as simplification orderings) to term graphs [21,28,30] and drags [11].

7   Conclusion and Future Work

We have introduced a termination method for graph transformation systems that can be utilized across frameworks, and which is defined in a broad array of categories. Our examples and comparisons with related work show that the method adds considerable value to the study of termination for graph transformation.

Future work for strengthening the method includes solving the issues raised related to rule equivalence (Example 52) and antipatterns (Remark 56). Methods for finding $\mathbb{T}$, if it exists, and identifying useful sufficient conditions for the nonexistence of $\mathbb{T}$, would also be very useful. A possible metatheoretical direction for future research includes the question posed for PBPO$^+$-adhesivity (Remark 29). Finally, we plan to formally verify and implement our method.

Acknowledgments. We thank anonymous reviewers for many helpful suggestions. Both authors received funding from the Netherlands Organization for Scientific Research (NWO) under the Innovational Research Incentives Scheme Vidi (project. No. VI.Vidi.192.004).

References

1. Adámek, J., Herrlich, H., Strecker, G.E.: Abstract and Concrete Categories - The Joy of Cats. Dover Publications (2009)
2. Bachmair, L., Dershowitz, N.: Commutation, transformation, and termination. In: Siekmann, J.H. (ed.) CADE 1986. LNCS, vol. 230, pp. 5–20. Springer, Heidelberg (1986). https://doi.org/10.1007/3-540-16780-3_76
3. Barr, M., Wells, C.: Category theory for Computing Science. Prentice Hall, Hoboken (1990)
4. Behr, N., Harmer, R., Krivine, J.: Concurrency theorems for non-linear rewriting theories. In: Gadducci, F., Kehrer, T. (eds.) ICGT 2021. LNCS, vol. 12741, pp. 3–21. Springer, Cham (2021). https://doi.org/10.1007/978-3-030-78946-6_1
5. Borceux, F.: Handbook of Categorical Algebra: Volume 3, Sheaf Theory, Cambridge University Press, Cambridge (1994)
6. Bruggink, H.J.S., König, B., Nolte, D., Zantema, H.: Proving termination of graph transformation systems using weighted type graphs over semirings. In: Parisi-Presicce, F., Westfechtel, B. (eds.) ICGT 2015. LNCS, vol. 9151, pp. 52–68. Springer, Cham (2015). https://doi.org/10.1007/978-3-319-21145-9_4
7. Bruggink, H.J.S., König, B., Zantema, H.: Termination analysis for graph transformation systems. In: Diaz, J., Lanese, I., Sangiorgi, D. (eds.) TCS 2014. LNCS, vol. 8705, pp. 179–194. Springer, Heidelberg (2014). https://doi.org/10.1007/978-3-662-44602-7_15
8. Corradini, A., Duval, D., Echahed, R., Prost, F., Ribeiro, L.: The PBPO graph transformation approach. J. Logical Algebraic Methods Program. **103**, 213–231 (2019). https://doi.org/10.1016/j.jlamp.2018.12.003
9. Corradini, A., Duval, D., Echahed, R., Prost, F., Ribeiro, L.: Algebraic graph rewriting with controlled embedding. Theor. Comput. Sci. **802**, 19–37 (2020). https://doi.org/10.1016/j.tcs.2019.06.004
10. Corradini, A., Heindel, T., Hermann, F., König, B.: Sesqui-pushout rewriting. In: Corradini, A., Ehrig, H., Montanari, U., Ribeiro, L., Rozenberg, G. (eds.) ICGT 2006. LNCS, vol. 4178, pp. 30–45. Springer, Heidelberg (2006). https://doi.org/10.1007/11841883_4
11. Dershowitz, N., Jouannaud, J.-P.: Graph path orderings. In Proceedings of Conference on Logic for Programming, Artificial Intelligence and Reasoning (LPAR), vol, 57 of EPiC Series in Computing, pp 307–325. EasyChair (2018) https://doi.org/10.29007/6hkk
12. Ehrig, H., Ehrig, K., Prange, U., Taentzer, G.: Fundamentals of algebraic graph transformation. MTCSAES, Springer, Heidelberg (2006). https://doi.org/10.1007/3-540-31188-2
13. Ehrig, H., Pfender, M., Schneider, H.J.: Graph-Grammars: An Algebraic Approach. In: Proceedings of the 10th Annual Symposium on Switching and Automata Theory (SWAT), pp. 167–180. IEEE Computer Society (1973). https://doi.org/10.1109/SWAT.1973.11

14. Habel, A., Müller, J., Plump, D.: Double-pushout graph transformation revisited. Math. Struct. Comput. Sci. **11**(5), 637–688 (2001). https://doi.org/10.1017/S0960129501003425

15. Johnstone, P.T.: Sketches of an Elephant: A Topos Theory Compendium, Vol. 1. Oxford University Press (2002)

16. Johnstone, P.T., Lack, S., Sobociński, P.: Quasitoposes, Quasiadhesive Categories and Artin Glueing. In: Mossakowski, T., Montanari, U., Haveraaen, M. (eds.) CALCO 2007. LNCS, vol. 4624, pp. 312–326. Springer, Heidelberg (2007). https://doi.org/10.1007/978-3-540-73859-6_21

17. Klop. J.W.: Term Rewriting Systems: A tutorial. Bulletin of the European Association for Theoretical Computer Science (1987)

18. Lack, S., Sobociński., P.: Adhesive categories. In: Proceedings of Conference on Foundations of Software Science and Computation Structures (FOSSACS), vol. 2987, LNCS, pp. 273–288. Springer, Cham (2004). https://doi.org/10.1007/978-3-540-24727-2_20

19. Lack, S., Sobocinski, P.: Adhesive and quasiadhesive categories. RAIRO Theor. Informatics Appl. **39**(3), 511–545 (2005). https://doi.org/10.1051/ita:2005028

20. Mori, M., Kawahara, Y.: Fuzzy graph rewritings. 数理解析研究所講究録, **918**, 65–71 (1995)

21. Moser, G., Schett. M.A.: Kruskal's tree theorem for acyclic term graphs. In: Proceedings of Workshop on Computing with Terms and Graphs, TERMGRAPH, vol. 225, EPTCS, pp. 25–34 (2016). https://doi.org/10.4204/EPTCS.225.5

22. Overbeek, R., Endrullis. J.: From linear term rewriting to graph rewriting with preservation of termination. In: Proceedings of Workshop on Graph Computational Models (GCM), vol. 350, EPTCS, pp. 19–34 (2021). https://doi.org/10.4204/EPTCS.350.2

23. Overbeek, R., Endrullis. J.: A PBPO$^+$ graph rewriting tutorial. In Proceedings of Workshop on Computing with Terms and Graphs (TERMGRAPH), vol. 377, EPTCS, pp. 45–63. Open Publishing Association (2023). https://doi.org/10.4204/EPTCS.377.3

24. Overbeek, R., Endrullis, J.: Termination of graph transformation systems using weighted subgraph counting. CoRR, abs/2303.07812 (2023). https://doi.org/10.48550/arXiv.2303.07812

25. Overbeek, R., Endrullis, J., Rosset, A.: Graph rewriting and relabeling with PBPO$^+$: A unifying theory for quasitoposes. J. Log. Algebraic Methods Program. (2023). https://doi.org/10.1016/j.jlamp.2023.100873

26. Pierce, B.C.: Basic Category Theory for Computer Scientists. MIT Press, Cambridge (1991). https://doi.org/10.7551/mitpress/1524.001.0001

27. Plump, D.: On termination of graph rewriting. In: Nagl, M. (ed.) WG 1995. LNCS, vol. 1017, pp. 88–100. Springer, Heidelberg (1995). https://doi.org/10.1007/3-540-60618-1_68

28. Plump, D.: Simplification orders for term graph rewriting. In: Prívara, I., Ružička, P. (eds.) MFCS 1997. LNCS, vol. 1295, pp. 458–467. Springer, Heidelberg (1997). https://doi.org/10.1007/BFb0029989

29. Plump, D.: Termination of graph rewriting is undecidable. Fundam. Informaticae **33**(2), 201–209 (1998). https://doi.org/10.3233/FI-1998-33204

30. Plump. D.: Term Graph Rewriting. Handbook of Graph Grammars and Computing by Graph Transformation: Volume 2: Applications, Languages and Tools, pp. 3–61 (1999). https://www-users.york.ac.uk/djp10/Papers/tgr_survey.pdf

31. Plump, D.: Modular termination of graph transformation. In: Heckel, R., Taentzer, G. (eds.) Graph Transformation, Specifications, and Nets. LNCS, vol. 10800, pp. 231–244. Springer, Cham (2018). https://doi.org/10.1007/978-3-319-75396-6_13
32. Rosset, A., Overbeek, R., Endrullis. J.: Fuzzy presheaves are quasitoposes. In: Fernández, M., Poskitt, C.M. (eds.) ICGT 2023. LNCS, vol. 13961, pp. 102–122. Springer, Cham (2023). https://doi.org/10.1007/978-3-031-36709-0_6
33. Sabel, D., Zantema. H.: Termination of cycle rewriting by transformation and matrix interpretation. Log. Methods Comput. Sci. **13**(1) (2017). https://doi.org/10.23638/LMCS-13(1:11)2017
34. Terese (ed.) Term Rewriting Systems, volume 55 of Cambridge Tracts in Theoretical Computer Science. Cambridge University Press (2003)
35. Wyler. O.: Lecture Notes on Topoi and Quasitopoi. World Scientific Publishing Co. (1991)
36. Zantema, H., König, B., Bruggink, H.J.S.: Termination of cycle rewriting. In: Dowek, G. (ed.) RTA 2014. LNCS, vol. 8560, pp. 476–490. Springer, Cham (2014). https://doi.org/10.1007/978-3-319-08918-8_33

Fuzzy Presheaves are Quasitoposes

Aloïs Rosset[✉][ID], Roy Overbeek[ID], and Jörg Endrullis

Vrije Universiteit Amsterdam, Amsterdam, The Netherlands
{a.rosset,r.overbeek,j.endrullis}@vu.nl

Abstract. Quasitoposes encompass a wide range of structures, including various categories of graphs. They have proven to be a natural setting for reasoning about the metatheory of algebraic graph rewriting. In this paper we propose and motivate the notion of *fuzzy presheaves*, which generalises fuzzy sets and fuzzy graphs. We prove that fuzzy presheaves are rm-adhesive quasitoposes, proving our recent conjecture for fuzzy graphs. Furthermore, we show that simple fuzzy graph categories are quasitoposes.

Keywords: Quasitopos · Presheaf · Fuzzy set · Graph rewriting

The algebraic graph transformation tradition uses category theory to define graph rewriting formalisms. The categorical approach enables rewriting a large variety of structures, and allows meta-properties to be studied in uniform ways, such as concurrency, parallelism, termination, and confluence. Different formalisms have been developed and studied since 1973, such as DPO [14], SPO [22], DPU [4], SqPO [13], AGREE [11], PBPO [12], and PBPO$^+$ [31].

The study of meta-properties on a categorical level has given new motivation to study existing concepts, such as quasitoposes [45], and has given rise to new ones, such as different notions of adhesivity [20,21]. Quasitoposes are categories with rich structure. They encompass many examples of particular interest in computer science [19]. They have moreover been proposed as a natural setting for non-linear rewriting [5]. In addition, they provide a unifying setting: we have shown that PBPO$^+$ subsumes DPO, SqPO, AGREE and PBPO in quasitoposes [30, Theorem 73].[1]

Fuzzy sets generalise the usual notion of sets by allowing elements to have membership values in the unit interval [46] or more generally in a poset [16]. Similarly, fuzzy graphs [26] and fuzzy Petri nets [10,40] have been defined by allowing elements to have membership values. Having a range of possible membership is used in artificial intelligence for example to express a degree of certainty or strength of a connection. In graph rewriting, it can be used to restrict matching [26,33–35]. We have recently shown that for fuzzy graphs, where we see the membership value as a label, the order structure on the labelling set lends itself well to implementing relabelling mechanics and type hierarchies [30,31].

---

[1] We moreover conjecture that PBPO$^+$ subsumes SPO in quasitoposes [30, Rem. 26].

M. Fernández and C. M. Poskitt (Eds.): ICGT 2023, LNCS 13961, pp. 102–122, 2023.
https://doi.org/10.1007/978-3-031-36709-0_6

These features have proved useful, for example, for faithfully modelling linear term rewriting with graph rewriting [27].

In this paper, we do the following.

1. We propose the notion of a *fuzzy presheaf*, which is a presheaf endowed with a fuzzy structure, generalising fuzzy sets and fuzzy graphs.
2. We show that fuzzy presheaf categories are rm-adhesive quasitoposes, when the membership values are taken in a complete Heyting algebra (Theorems 15 and 41). We obtain as a corollary the known result that fuzzy sets form a quasitopos, prove our conjecture that fuzzy graphs form a quasitopos [30], and as a new result that fuzzy sets, graphs, hypergraphs and other fuzzy presheaves are rm-adhesive quasitoposes.
3. We examine the related question of whether simple fuzzy graphs form a quasitopos. Because simple graphs are not presheaves (nor rm-adhesive), our main theorem cannot be applied. However, we show that (directed and undirected) simple fuzzy graphs are quasitoposes by using topologies and separated elements.

An immediate practical application of our results is that the termination technique [29], recently developed for rm-adhesive quasitoposes is applicable to rewriting fuzzy presheaves. In addition, it follows that our already mentioned subsumption result for PBPO$^+$ holds for fuzzy presheaf categories [30, Theorem 73].

The paper is structured as follows. Section 1 introduces preliminary definitions of presheaves, Heyting algebra and quasitoposes. Section 2 proves that fuzzy presheaves are quasitoposes. Section 3 proves that fuzzy presheaves are rm-adhesive. Section 4 looks at examples and applications. Section 5 establishes that simple fuzzy graph categories are also quasitoposes. Section 6 discusses future work and concludes.

1   Preliminaries

We assume the reader is familiar with basic notions of category theory [2,23]. First, we recall the basic category of graphs that we consider.

Definition 1. *A* **graph** A *consists of a set of vertices* $A(V)$, *a set of edges* $A(E)$, *and source and target functions* $A(s), A(t) : A(E) \to A(V)$. *A* **graph homomorphism** $f : A \to B$ *is a pair of functions* $f_V : A(V) \to B(V)$ *and* $f_E : A(E) \to B(E)$ *respecting sources and targets, i.e.,* $(B(s), B(t)) \cdot f_E = (f_V \times f_V) \cdot (A(s), A(t))$. *They form the category* Graph.

Definition 2. *A* **presheaf** *on a category* I *is a functor* $F : I^{\mathrm{op}} \to$ Set. *A morphism of presheaves is a natural transformation between functors. The category of presheaves on* I *is denoted by* Set$^{I^{\mathrm{op}}}$ *or* $\hat{I}$.

The category Graph is in fact the presheaf category on $I^{\mathrm{op}} = E \overset{s}{\underset{t}{\rightrightarrows}} V$. We will showcase how many other categories of graphs can be seen as presheaf categories in Example 42.

In this article, we study presheaves endowed with a fuzzy structure, generalising fuzzy sets and fuzzy graphs. Let us therefore recall those two definitions.

Definition 3 ([46]). *Given a poset* $(\mathcal{L}, \leqslant)$, *an* $\mathcal{L}$**-fuzzy set** *is a pair* (A, α) *consisting of a set* A *and a* **membership function** $\alpha : A \to \mathcal{L}$. *A morphism* $f : (A, \alpha) \to (B, \beta)$ *of fuzzy sets is a function* $f : A \to B$ *such that* $\alpha \leqslant \beta f$, *i.e.,* $\alpha(a) \leqslant \beta f(a)$ *for all* $a \in A$. *They form the category* FuzzySet($\mathcal{L}$). *Membership functions are represented in grey. We express that* $\alpha \leqslant \beta f$ *by the diagram on the right, which we call* $\leqslant$**-commuting**.

$$A \xrightarrow{\;f\;} B$$
$$\alpha \searrow \;\leqslant\; \swarrow \beta$$
$$\mathcal{L}$$

Definition 4 ([31, **Def. 76**]). *A* **fuzzy graph** (A, α) *consists of a graph* A *and two membership functions* $\alpha_V : A(V) \to \mathcal{L}(V)$, *and* $\alpha_E : A(E) \to \mathcal{L}(E)$, *where* $\mathcal{L}(V)$ *and* $\mathcal{L}(E)$ *are posets. A* **fuzzy graph homomorphism** $f : (A, \alpha) \to (B, \beta)$ *is a graph homomorphism* $f : A \to B$ *such that* $\alpha_V \leqslant \beta_V f_V$ *and* $\alpha_E \leqslant \beta_E f_E$. *They form the category* FuzzyGraph.

Remark 5. Some authors require that fuzzy graphs satisfy $\mathcal{L}(V) = \mathcal{L}(E)$ and $\alpha_E(e) \leqslant \alpha_V(s(e)) \wedge \alpha_V(t(e))$ for all edges $e \in A(E)$ [37]. For survey works on fuzzy graphs, see e.g. [25,32].

The main goal of this article is to demonstrate that the categories of presheaves endowed with a fuzzy structure are quasitoposes. We thus recall what a quasitopos is, along with the required notions of regular monomorphisms, (regular) subobjects, (regular) subobject classifiers, and (locally) cartesian closedness.

Definition 6 ([1, **Def. 7.56**]). *A monomorphism* $m : A \rightarrowtail B$ *is called* **regular**, *denoted* $m : A \hookrightarrow B$, *if it is the equaliser of two parallel morphisms* $B \rightrightarrows C$.

Definition 7 ([2, **Def. 8.15**]). *Monomorphisms into an object* C *form a class denoted* Mono(C). *There is a preorder on this class:* $m \leqslant m'$ *if there exists* u *such that* $m = m'u$. *The equivalence relation* $m \simeq m' := (m \leqslant m' \wedge m \geqslant m')$ *is equivalent to having* $m = m'u$ *for some isomorphism* u. *The equivalence classes are the* **subobjects** *of* C. *If each representative of a subobject is a special kind of monomorphism, for instance a regular monomorphism, then we talk of* **regular subobjects**.

$$A \cdots\xrightarrow{\;u\;} B$$
$$m \searrow \;\;\swarrow m'$$
$$C$$

Definition 8 ([45, **Def. 14.1**]). *Let* C *be a category with finite limits, and* $\mathcal{M}$ *be a class of monomorphisms. An* $\mathcal{M}$**-subobject classifier** *in* C *is a monomorphism* True $: 1 \rightarrowtail \Omega$ *such that for every monomorphism* $m : A \rightarrowtail B$ *in* $\mathcal{M}$, *there exists a unique* $\chi_A : B \to \Omega$ *such that the diagram on the right is a pullback square. When unspecified,* $\mathcal{M}$ *is all monomorphisms.*

$$\begin{array}{ccc} A & \xrightarrow{\;!\;} & 1 \\ m \downarrow & & \downarrow \text{True} \\ B & \xrightarrow{\;\chi_A\;} & \Omega \end{array}$$

Definition 9 ([45, **Def. 9.1, Def. 18.1**]). *A category* C *is* **cartesian closed** *if it has finite products and any two objects* $A, B \in$ C *have an exponential object* $B^A \in$ C. *In more details, there must exist a functor* $-^A :$ C $\to$ C *that is right adjoint to the product functor:* $(- \times A) \dashv (-^A)$. *A category* C *is* **locally cartesian closed** *if all its slice categories* C$/C$ *are cartesian closed.*

Definition 10 ([19, **Def. 9**]). *A category is a* **quasitopos** *if*

- *every finite limit and finite colimit exists,*
- *it has a regular-subobject classifier, and*
- *it is locally cartesian closed.*

When studying fuzzy sets and fuzzy structure in general, we work with a poset $(\mathcal{L}, \leqslant)$. The properties of that poset can be important to derive certain results. We therefore recall a few of those properties in the next definition.

Definition 11. *We recall some order theory notions.*

- *A* **lattice** *is a poset $(\mathcal{L}, \leqslant)$ that has a binary join $\vee$ and binary meet $\wedge$.*
- *A lattice is* **bounded** *if it has a greatest element $\top$ and a least element $\bot$.*
- *A lattice is* **complete** *if all its subsets have both a join and a meet.*
- *A* **Heyting algebra** *is a bounded lattice with, for all elements a and b, an element $a \Rightarrow b$ such that for all c*

$$a \wedge b \leqslant c \iff a \leqslant b \Rightarrow c \tag{1}$$

Each definition has an appropriate notion of morphism. A poset morphism is a monotone function. A lattice morphism respects the binary join and the binary meet. A bounded lattice morphism additionally respects the greatest and the least element. A Heyting algebra morphism additionally respects the implication. A complete lattice respects all joins and all meets. They thus form categories, Poset, Lattice, HeytAlg, CompHeytAlg *etc.*

A poset/lattice/Heyting algebra can be seen as a thin category, i.e., given objects a, b there is at most one morphism from a to b denoted $a \leqslant b$.[2] Meet and join correspond to binary product and coproduct. Least and greatest elements correspond to initial and a terminal objects. A bounded lattice is thus finitely bicomplete. A lattice is complete in the usual sense if and only if it is complete as a category.[3] Heyting algebras are cartesian closed categories because (1) is the Hom-set adjunction requirement asking $a \Rightarrow b$ to be an exponential object. The counit of the adjunction tells us that we have *modus ponens* [24, I.8, Prop. 3]:

$$a \wedge (a \Rightarrow b) = a \wedge b. \tag{2}$$

The Yoneda embedding is needed in some proofs, so let us recall its definition.

Definition 12. *Let I be a locally small category.*

- *The* **contravariant** Hom-*functor $I(-, i) : I^{\mathrm{op}} \to$ Set is defined on objects $j \in I^{\mathrm{op}}$ as the set of I-morphisms $I(j, i)$, and on I^{op}-morphisms $\kappa^{\mathrm{op}} : j \to k$ as the precomposition $I(j, i) \xrightarrow{-\circ\kappa} I(k, i) : (j \xrightarrow{\iota} i) \mapsto (k \xrightarrow{\kappa} j \xrightarrow{\iota} i)$.*
- *The (contravariant)* **Yoneda embedding** $y : I \to \mathrm{Set}^{I^{\mathrm{op}}}$ *is defined on objects $y(i \in I) = I(-, i)$ as the contravariant Hom-functor, and on morphisms $y(j \xrightarrow{\iota} i)$ as the natural transformation that postcomposes by ι, i.e., for every $k \in I$:*

$$I(k, j) \xrightarrow{\iota \circ -} I(k, i) : (k \xrightarrow{\kappa} j) \mapsto (k \xrightarrow{\kappa} j \xrightarrow{\iota} i).$$

---

[2] See e.g. [2, Def. 6.7], [45, Def. 24.1], or [24, 8.1].

[3] Note that completeness and cocompleteness are equivalent for lattices [2, Def. 6.8].

2   Fuzzy Presheaves are Quasitoposes

We introduce the notion of a *fuzzy presheaf*, which generalises fuzzy sets and fuzzy graphs. This allows to obtain in only one definition the additional concepts of fuzzy undirected graphs, fuzzy hypergraphs, fuzzy k-uniform hypergraphs, and plenty more. For examples see Example 42.

Let us denote by incl : Set → Poset the inclusion functor, which gives the coarsest possible partial order, i.e., the reflexive relation.

Definition 13. *Given a category I and a functor $\mathcal{L} : I^{\mathrm{op}} \to$ Poset, a (**poset**) $\mathcal{L}$-**fuzzy presheaf** is a pair (A, α) consisting of a presheaf $A : I^{\mathrm{op}} \to$ Set and a lax natural transformation α : incl $\cdot A \Rightarrow \mathcal{L}$, i.e., a natural transformation whose naturality squares $\leqslant$-commute, as shown on the right for each $\iota : j \to i$. A morphism of $\mathcal{L}$-fuzzy presheaves $f : (A, \alpha) \to (B, \beta)$ is a natural transformation $f : A \to B$ such that $\alpha \leqslant \beta f$, i.e., $\alpha_i \leqslant \beta_i f_i$ for all $i \in I$. They form a category* FuzzyPresheaf$(I, \mathcal{L})$.

$$
\begin{array}{ccc}
A(i) & \xrightarrow{A(\iota)} & A(j) \\
\alpha_i \downarrow & \leqslant & \downarrow \alpha_j \\
\mathcal{L}(i) & \xrightarrow[\mathcal{L}(\iota)]{} & \mathcal{L}(j)
\end{array}
$$

The terminology *lax* natural transformation comes from Poset being a 2-category, where a 2-morphism between two 1-morphisms f and g is simply $f \leqslant g$. In our simple case, a natural transformation is *lax*, if its naturality squares commute up to a 2-morphism, i.e., $\leqslant$-commute.

Taking $\mathcal{L}$ to be the terminal functor, i.e., each $\mathcal{L}(i)$ is a singleton set, makes us retrieve the usual definition of presheaf. For Graph, i.e., the presheaf category on $I^{\mathrm{op}} = E \overset{s}{\underset{t}{\rightrightarrows}} V$, asking for $\mathcal{L}(E) = \mathcal{L}(V)$ and for $\mathcal{L}(s) = \mathcal{L}(t) = $ id gives the fuzzy graphs as in Remark 5.

Definition 13 is about *poset* fuzzy presheaves, because each $\mathcal{L}(i)$ is a poset and each $\mathcal{L}(\iota)$ is a poset morphism. We similarly have a definition of lattice fuzzy presheaves, bounded lattice fuzzy presheaves, complete lattice fuzzy presheaves, Heyting algebra fuzzy presheaves, etc.

Notice how the morphisms $\mathcal{L}(\iota)$ create dependencies between the membership values. In the case of posets with bottom, the dependency can be removed by choosing every $\mathcal{L}(\iota)$ to be the constant map to $\bot$. That way, we obtain fuzzy graphs as in Definition 4, where vertex and edge labels are independent. However this fails in some other cases, for instance with bounded lattices. Then, morphisms are required to send $\top$ to $\top$ and must respect $\vee$, which makes the mapping of $\mathcal{L}(i) \setminus \{\top\}$ to bottom a non-valid bounded lattice morphism.

Remark 14. Another way to define fuzzy presheaves is as follows. First, define a *poset fuzzy set* as a function $\alpha : A \to \mathcal{L}(A)$ from a set A to a poset $\mathcal{L}(A)$, and a *poset fuzzy set morphism* as a pair $(f, \mathcal{L}(f))$ of a function $f : A \to B$ and of a poset morphism $\mathcal{L}(f) : \mathcal{L}(A) \to \mathcal{L}(f)$ such that $\mathcal{L}(f) \cdot \alpha \leqslant \beta \cdot f$. Call the category thus formed PosetFuzzySet. It is the lax comma 2-category $(\mathrm{id}_{\mathsf{Set}} \Downarrow U)$ where U : Poset → Set is the forgetful functor. Define then a *poset fuzzy presheaf* as a functor $I^{\mathrm{op}} \to$ PosetFuzzySet, and morphisms as natural transformations. Call this category PosetFuzzyPresheaf(I). This approach is similar to the one of Spivak [39], who defines *fuzzy simplicial sets* as functors $\Delta^{\mathrm{op}} \to$ FuzzySet$(\mathcal{L})$, where Δ is the simplex category.

The category $\mathsf{PosetFuzzyPresheaf}(I)$ differs from $\mathsf{FuzzyPresheaf}(I,\mathcal{L})$, for instance with fuzzy graphs, in that every graph G has its own two membership/labeling sets $\mathcal{L}(G(E))$ and $\mathcal{L}(G(V))$. Poset fuzzy graph morphisms give rise to a cube like on the right. This is not desirable in the applications we are interested in, see e.g. Examples 44 and 46.

Take a category I and a functor $\mathcal{L} : I^{\mathrm{op}} \to \mathsf{Poset}$. $\mathsf{PosetFuzzyPresheaf}(I)$ contains $\mathsf{FuzzyPresheaf}(I,\mathcal{L})$ as a subcategory. Theorem 15 gives a sufficient condition for $\mathsf{FuzzyPresheaf}(I,\mathcal{L})$ to be a quasitopos. It remains open whether $\mathsf{PosetFuzzyPresheaf}(I)$ is also a quasitopos. However, a positive answer to this question does not immediately imply our result. Indeed, knowing that a category is a quasitopos does not immediately imply that a subcategory is one too. See, for example, the proof in Sect. 5 that simple (fuzzy) graphs form a quasitopos.

The rest of this section is dedicated to proving the next theorem, which says that when considering complete Heyting algebra fuzzy presheaves, they turn out to form a quasitopos.

Theorem 15. *Given a small category I and a functor $\mathcal{L} : I^{\mathrm{op}} \to \mathsf{CompHeytAlg}$, then $\mathsf{FuzzyPresheaf}(I,\mathcal{L})$ is a quasitopos.*

In the rest of this section I is a small category and $\mathcal{L} : I^{\mathrm{op}} \to \mathsf{CompHeytAlg}$ is a functor. We give a brief high-level overview of the proof. We tackle the properties of being a quasitopos one by one (Definition 10):

- Section 2.1: Finite limits and colimits exist for presheaves, and it suffices to combine them with a terminal (respectively initial) structure to have finite limits (respectively finite colimits) for fuzzy presheaves.
- Section 2.2: We modify the subobject classifier for presheaves by giving all its elements full membership. This gives us the regular-subobject classifier for fuzzy presheaves.
- Section 2.3: We combine how exponential objects look like for fuzzy sets and presheaves to obtain exponential objects for fuzzy presheaves, resulting in cartesian closedness.
- Section 2.4: Lastly, to show that we have a locally cartesian closed category, we show that every slice of a fuzzy presheaf category gives another fuzzy presheaf category, extending a result from the literature [2, Lemma 9.23].

Remark 16. The proof of Theorem 15 stays true even when we don't require the complete Heyting algebras morphisms $\mathcal{L}(\iota)$ to exist. Indeed, the constructions of (co)limits, exponential objects, and of the subobject classifier Ω are the same; it only adds each time an extra step of verification that a diagram $\leqslant$-commutes.

2.1 Finite Limits and Colimits

The goal of this subsection is to prove the following.

Proposition 17. FuzzyPresheaf$(I, \mathcal{L})$ *has all finite limits and colimits.*

Proof. To show existence of finite limits, it suffices to show the existence of a terminal object and of pullbacks [7, Prop. 2.8.2]. Those are the same as in $\mathsf{Set}^{I^{\mathrm{op}}}$, paired with a final structure for the membership function [41, §3.1.1].

The terminal object is $1(i) = \{\,\cdot^\top\}$ for each $i \in I$, with the single element having full membership $\top \in \mathcal{L}(i)$. For 1 to be a fuzzy presheaf, we must have $\mathcal{L}(\iota) \cdot \top \leqslant \top \cdot 1(\iota) = \top$, for every $\iota : j \to i$, which trivially holds.

Given two morphisms $f : (A, \alpha) \to (C, \gamma)$ and $g : (B, \beta) \to (C, \gamma)$, their pullback $(A \times_C B, \delta)$ is also defined pointwise. Its membership function δ must be maximal satisfying $\delta \leqslant \alpha g'$ and $\delta \leqslant \beta f'$ and is therefore the meet of both conditions: $\delta_i(d) := \alpha_i g_i'(d) \wedge \beta_i f_i'(d)$. For $A \times_C B$ to be a fuzzy presheaf, we must have $\mathcal{L}(\iota) \cdot \delta_i \leqslant \delta_j \cdot (A \times_C B)(\iota)$ for each $\iota : j \to i$. Indeed, for $d \in A(i) \times_{C(i)} B(i)$:

$$
\begin{aligned}
\mathcal{L}(\iota)(\delta_i(d)) &= \mathcal{L}(\iota)\big(\alpha_i g_i'(d) \wedge \beta_i f_i'(d)\big) && (\text{def. } \delta)\\
&= \big(\mathcal{L}(\iota) \cdot \alpha_i \cdot g_i'\big)(d) \wedge \big(\mathcal{L}(\iota) \cdot \beta_i \cdot f_i'\big)(d) && (\mathcal{L}(\iota) \text{ lattice homom.})\\
&\leqslant \big(\alpha_j \cdot A(\iota) \cdot g_i'\big)(d) \wedge \big(\beta_j \cdot B(\iota) \cdot f_i'\big)(d) && ((A,\alpha),(B,\beta) \text{ fuz. presh.})\\
&= \big(\alpha_j \cdot g_j' \cdot (A \times_C B)(\iota)\big)(d)\\
&\quad \wedge \big(\beta_j \cdot f_j' \cdot (A \times_C B)(\iota)\big)(d) && (f', g' \text{ nat. transf.})\\
&= \big(\delta_j \cdot (A \times_C B)(\iota)\big)(d). && (\text{def. } \delta)
\end{aligned}
$$

The proof that FuzzyPresheaf$(I, \mathcal{L})$ has all finite colimits is dual.   □

2.2   Regular-Subobject Classifier

The goal of this subsection is to prove the following.

Proposition 18. FuzzyPresheaf$(I, \mathcal{L})$ *has a regular-subobject classifier.*

Example 19. We first look at some examples of regular monomorphisms

- In Set, a monomorphism is a subset inclusion (up to isomorphism), and all monomorphisms are regular [1, 7.58(1)].
- In $\mathsf{Set}^{I^{\mathrm{op}}}$, a morphism $m : A \to B$ is (regular) monic if and only if every $m_i : A(i) \to B(i)$, where $i \in I$, is (regular) monic. All monomorphisms are therefore regular for the same reason as in Set.
- In FuzzySet$(\mathcal{L})$, $m : (A, \alpha) \hookrightarrow (B, \beta)$ is a regular monomorphism if it is the equaliser of some functions $f_1, f_2 : B \rightrightarrows C$. Because $m : (A, \beta m) \to (B, \beta)$ also equalises f_1 and f_2, the universal property of the equaliser implies $\beta m \leqslant \alpha \leqslant \beta m$ and thus $a = \beta m$. When $A \subseteq B$, α is the restriction $\beta|_A$ [41, 3.1.3].

Fuzzy presheaves being pointwise fuzzy sets, we have the following.

Lemma 20. *A regular monomorphism $m : (A, \alpha) \hookrightarrow (B, \beta)$ between fuzzy presheaves is a natural transformation $m : A \to B$ such that each $m_i : (A(i), \alpha_i) \hookrightarrow (B(i), \beta_i)$ is regular monic, in particular $\alpha_i = \beta_i m_i$.*

$$A(i) \xrightarrow{m_i} B(i)$$
$$\alpha_i \searrow \;\; = \;\; \swarrow \beta_i$$
$$\mathcal{L}(i)$$

Notation 21. *We denote the pullback of a morphism $m : A \to C$ along another morphism $f : B \to C$ by $f^{\leftharpoonup} m$.*

Example 22. We continue Example 19 and look at the (regular-)subobject classifiers of the aforementioned categories.

– In Set, we have $\mathsf{True} : \{\cdot\} \to \Omega := \{0, 1\} : \cdot \mapsto 1$. For $m : A \rightarrowtail B$, the characteristic function $\chi_A : B \to \{0, 1\}$ is 1 on $m(A) \subseteq B$ and 0 on the rest.
– In FuzzySet($\mathcal{L}$) [41, 3.1.3], we have $\Omega := \{ 0^\top, 1^\top \}$ with both elements having full membership. The square being a pullback forces $\alpha = \beta m$. Hence Ω classifies only regular subobjects.

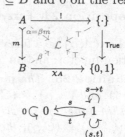

– In Graph [44], the terminal object is $\{ \varsigma \cdot \}$, the classifying object Ω is shown on the right, and $\mathsf{True} : \varsigma \cdot \mapsto 1 \circlearrowleft s \to t$. Given a graph H and a subgraph $G \subseteq H$, the characteristic function χ_G sends vertices in G to 1 and edges in G to $s \to t$.

Vertices not in G are sent to 0; and edges not in G are sent to (s, t) if both endpoints are in G, to 0 if neither endpoint is in G, and to s or t if only the source or target is in G, respectively.

When I is small, $\mathsf{Set}^{I^{op}}$ has a subobject classifier [24]. By smallness of I, the class of the subobjects of a presheaf $B \in \hat{I}$ (also called subpresheaves) is necessarily a set. This allows us to define a functor $\mathsf{Sub}_{\hat{I}} : \hat{I}^{op} \to \mathsf{Set}$:

– $\mathsf{Sub}_{\hat{I}}(B) := \{\text{subobjects } m : A \rightarrowtail B\}$,
– $\mathsf{Sub}_{\hat{I}}(f : B \to C)$ sends a subobject $(A \xrightarrow{n} C)$ of C to $f^{\leftharpoonup} n$, the pullback of n along f, which is a subobject of B.

The classifying object is the presheaf $\Omega := \mathsf{Sub}_{\hat{I}}(y(-)) : I^{op} \to \mathsf{Set}$. For $i \in I$, $\Omega(i)$ is thus the set of all subpresheaves of $y(i)$, which are sometimes called *sieves*. Let $\mathsf{True}_i : \{\cdot\} \to \Omega(i) : \cdot \mapsto y(i)$, because $y(i)$ is indeed a subpresheaf of itself. We prove that True classifies subobjects. Pick a subpresheaf $m : A \rightarrowtail B$, and suppose w.l.o.g. $A(i) \subseteq B(i)$ for each $i \in I$. We want to define the natural transformation $\chi_A : B \to \Omega$. For $i \in I$ and $b \in B(i)$, then $(\chi_A)_i(b) \in \Omega(i)$ must be a subpresheaf of $y(i) = I(-, i)$. On $j \in I$, let

$$(\chi_A)_i(b)(j) := \{\iota : j \to i \mid B(\iota)(b) \in A_j\}. \tag{3}$$

Lemma 23. *We have the following properties:*

1. χ_A *is natural,*
2. χ_A *characterises the subpresheaf $A \subseteq B$, i.e., gives a pullback square, and*

3. χ_A is unique in satisfying Item 2.   ∎[4]

In FuzzyPresheaf$(I, \mathcal{L})$, we consider the same construction and give full membership value to Ω, analogously to the fuzzy set regular-subobject classifier. Explicitly, for $i \in I$, let $\omega_i : \Omega(i) \to \mathcal{L}(i) : (A \rightarrowtail y(i)) \mapsto \top$. Notice that (Ω, ω) is a fuzzy presheaf since for each $\iota \in I(j,i)$, we trivially have $\mathcal{L}(\iota) \cdot \omega_i \leqslant \omega_j \cdot \Omega(\iota) = \top$. By Lemma 23, (Ω, ω) is the regular-subobject classifier that we seek and Proposition 18 is hence proved.

2.3 Cartesian Closed

The goal of this section is to prove the following.

Proposition 24. FuzzyPresheaf$(I, \mathcal{L})$ *is cartesian closed.*

Let (A, α) be a fuzzy presheaf. We construct a product functor, an exponentiation functor, and an adjunction between them: $- \times (A, \alpha) \dashv -^{(A,\alpha)}$.

Definition 25. *The product of two fuzzy presheaves* (C, γ) *and* (A, α) *is* $(C \times A, \gamma\pi_1 \wedge \alpha\pi_2)$. *The membership function is written* $\gamma \wedge \alpha$ *for short. Given* $f : (C', \gamma') \to (C, \gamma)$, *then* $f \times (A, \alpha) := f \times \mathrm{id}_A$.

$$C' \times A \xrightarrow{f \times \mathrm{id}_A} C \times A$$
$$\gamma' \wedge \alpha \searrow \quad \overset{\leqslant}{\swarrow} \gamma \wedge \alpha$$
$$\mathcal{L}$$

Lemma 26. *The* $- \times (A, \alpha)$ *functor is well-defined on morphisms.*

Proof. Because $\gamma' \leqslant \gamma f$, it follows that $\gamma' \wedge \alpha \leqslant (\gamma \wedge \alpha) \circ (f \times \mathrm{id}_A)$. Moreover, a product is a limit, and thus by Sect. 2.1, $(C \times A, \gamma\pi_1, \alpha\pi_2)$ is indeed a fuzzy presheaf and satisfies $\mathcal{L}(\iota) \cdot (\gamma_i\pi_1 \wedge \alpha_i\pi_2) \leqslant (\gamma_j\pi_1 \wedge \alpha_j\pi_2) \cdot (A(\iota) \times B(\iota))$.   □

Example 27. To construct exponential objects of fuzzy presheaves, we first look at simpler examples.

– In Set, given two sets A, B, then $B^A := \mathsf{Set}(A, B)$.
– In FuzzySet$(\mathcal{L})$ [41, p. 81], given fuzzy sets $(A, \alpha : A \to \mathcal{L}), (B, \beta : B \to \mathcal{L})$, then $(B, \beta)^{(A,\alpha)} = (B^A, \theta)$ has the same carrier as in Set, and θ is given below. Logically speaking, the truth value of $f \in B^A$ is the minimum truth-value of "truth preservation by f", i.e., "if a then $f(a)$".

$$\theta(f : A \to B) := \bigwedge_{a \in A} (\alpha(a) \Rightarrow \beta f(a)).$$

– In Graph [44], given two graphs A, B then B^A is the graph
  – with vertex set $B^A(V) := \mathsf{Graph}(\{\cdot\} \times A, B)$,
  – with edge set $B^A(E) := \mathsf{Graph}(\{s \to t\} \times A, B)$,
  – the source of $m : \{s \to t\} \times A \to B$ is $m(s, -) : \{\cdot\} \times A \to B$, and
  – the target function is analogous.

---

[4] The symbol ∎ denotes that the proof is in the extended version on arXiv [38]. A sketch of the proof is in [24, p.37-38].

– In $\hat{I}$ [2, Section 8.7], given $A, B : I^{\mathrm{op}} \to \mathsf{Set}$, then $B^A = \hat{I}(y(-) \times A, B)$ is:
  – on $i \in I$, the set of all natural transformations $y(i) \times A \Rightarrow B$.
  – on $\iota : j \to i$, the precomposition by $y(\iota) \times \mathrm{id}_A$.

$$y(i) \times A \xrightarrow{m} B \quad \longmapsto \quad y(j) \times A \xrightarrow{y(\iota) \times \mathrm{id}_A} y(i) \times A \xrightarrow{m} B \qquad (4)$$

Notice how this generalises the Graph case: for $I^{\mathrm{op}} = E \underset{t}{\overset{s}{\rightrightarrows}} V$, we indeed have $y(V) \cong \{\cdot\}$ and $y(E) \cong \{s \to t\}$.

Definition 28. *Given fuzzy presheaves (A, α) and (B, β), define $(B, \beta)^{(A, \alpha)}$ as $\left(\hat{I}(y(-) \times A, B), \theta\right)$. The carrier is the same as in $\hat{I}$ above: on $i \in I$, it is the set of natural transformations $y(i) \times A \Rightarrow B$ and on $\iota : j \to i$ it is the precomposition by $y(\iota) \times \mathrm{id}_A$. The membership function is, for each $i \in I$:*

$$\theta_i\left(y(i) \times A \xrightarrow{m} B\right) := \bigwedge_{a \in A(i)} \left(\alpha_i a \Rightarrow \beta_i m_i(\mathrm{id}_i, a)\right). \qquad (5)$$

Given $(B, \beta) \xrightarrow{g} (B', \beta')$, the morphism $g^{(A, \alpha)}$ is defined by postcomposition by g:

$$\hat{I}(y(i) \times A, B) \xrightarrow{g_i^{(A, \alpha)} := g \circ -} \hat{I}(y(i) \times A, B')$$
$$\underset{\theta_i}{\searrow} \ \mathcal{L}(i) \ \underset{\theta_i'}{\swarrow} \qquad (6)$$

Lemma 29. *The exponentiation functor is well-defined on morphisms.*

Proof. The mapping $g^{(A, \alpha)}$ is clearly natural in $i \in I$. Indeed, $g_i^{(A, \alpha)}$ is defined by postcomposition by g (6), $\hat{I}(y(\iota) \times A, B)$ is precomposition by $y(\iota) \times \mathrm{id}_A$ (4), and pre- and postcomposition always commute.

Fix an arbitrary $i \in I$. We need to verify that (6) $\leqslant$-commutes, i.e., $\theta_i \leqslant \theta_i' g_i^{(A, \alpha)}$. Take some $m : y(i) \times A \to B$. We prove $\theta_i(m) \leqslant \theta_i' g_i^{(A, \alpha)}(m)$. By hypothesis $g : (B, \beta) \to (B', \beta')$, so $\beta_i \leqslant \beta_i' g_i$. For any $a' \in A(i)$, we have $m_i(\mathrm{id}_i, a') \in B(i)$.

$\Rightarrow \forall a' \in A(i) : \beta_i m_i(\mathrm{id}_i, a') \leqslant \beta_i' g_i m_i(\mathrm{id}_i, a')$

$\Rightarrow \forall a' \in A(i) : \alpha_i(a') \wedge \beta_i m_i(\mathrm{id}_i, a') \leqslant \beta_i' g_i m_i(\mathrm{id}_i, a')$

$\overset{(2)}{\Leftrightarrow} \forall a' \in A(i) : \alpha_i(a') \wedge (\alpha_i(a') \Rightarrow \beta_i m_i(\mathrm{id}_i, a')) \leqslant \beta_i' g_i m_i(\mathrm{id}_i, a')$

$\overset{(1)}{\Leftrightarrow} \forall a' \in A(i) : (\alpha_i(a') \Rightarrow \beta_i m_i(\mathrm{id}_i, a')) \leqslant (\alpha_i(a') \Rightarrow \beta_i' g_i m_i(\mathrm{id}_i, a'))$

$\Rightarrow \forall a' \in A(i) : \bigwedge_{a \in A(i)} (\alpha_i(a) \Rightarrow \beta_i m_i(\mathrm{id}_i, a)) \leqslant (\alpha_i(a') \Rightarrow \beta_i' g_i m_i(\mathrm{id}_i, a'))$

$\Rightarrow \bigwedge_{a \in A(i)} (\alpha_i(a) \Rightarrow \beta_i m_i(\mathrm{id}_i, a)) \leqslant \bigwedge_{a' \in A(i)} (\alpha_i(a') \Rightarrow \beta_i' g_i m_i(\mathrm{id}_i, a'))$

$\overset{(5) \wedge (6)}{\Rightarrow} \theta_i(m) \leqslant \theta_i' g_i^{(A, \alpha)}(m)$.

Lastly, for $(B, \beta)^{(A,\alpha)}$ to be a fuzzy presheaf, we check for each $\iota : j \to i$ that $\mathcal{L}(\iota) \cdot \theta_i \leqslant \theta_j \cdot (B, \beta)^{(A,\alpha)}(\iota)$. Take $m : y(i) \times A \Rightarrow B$.

$$
\begin{aligned}
\mathcal{L}(\iota)(\theta_i(m)) &= \mathcal{L}(\iota)\big(\bigwedge_{a \in A(i)} \alpha_i a \Rightarrow \beta_i m_i(\mathrm{id}_i, a)\big) && \text{(def. } \theta \text{ (5))} \\
&= \bigwedge_{a \in A(i)} \mathcal{L}(\iota)(\alpha_i a) \Rightarrow \mathcal{L}(\iota)\big(\beta_i m_i(\mathrm{id}_i, a)\big) && (\mathcal{L}(\iota) \text{ CompHeytAlg hom.)} \\
&\leqslant \bigwedge_{a \in A(i)} \big(\alpha_j \cdot A(\iota)\big)(a) \Rightarrow \big(\beta_j \cdot B(\iota) \cdot m_i\big)(\mathrm{id}_i, a) && ((A,\alpha), (B,\beta) \text{ fuz. presh.)} \\
&= \bigwedge_{a \in A(i)} \big(\alpha_j \cdot A(\iota)\big)(a) \Rightarrow \big(\beta_j \cdot m_j \cdot (y(i)(\iota) \times A(\iota))\big)(\mathrm{id}_i, a) && (m \text{ nat.)} \\
&= \bigwedge_{a' \in \mathrm{im}\, A(\iota)} \alpha_j a' \Rightarrow \big(\beta_j \cdot m_j \cdot (y(i)(\iota) \times \mathrm{id}_{A(j)})\big)(\mathrm{id}_j, a') && \text{(index change)} \\
&= \bigwedge_{a' \in \mathrm{im}\, A(\iota)} \alpha_j a' \Rightarrow \big(\beta_j \cdot m_j \cdot (y(\iota) \times \mathrm{id}_{A(j)})\big)(\mathrm{id}_j, a') && \text{(Def. 12 of } y) \\
&\leqslant \bigwedge_{a' \in A(j)} \alpha_j a' \Rightarrow \big(\beta_j \cdot m_j \cdot (y(\iota) \times \mathrm{id}_{A(j)})\big)(\mathrm{id}_j, a') && \text{(bounding index)} \\
&= \theta_j\big(m_j \cdot (y(\iota) \times \mathrm{id}_{A(j)})\big) && \text{(def. } \theta \text{ (5))} \\
&= \theta_j\big((B,\beta)^{(A,\alpha)}(\iota)(m)\big) && \text{(def. } (B,\beta)^{(A,\alpha)}(\iota))
\end{aligned}
$$

This concludes the proof.   □

Lemma 30. *Given a fuzzy presheaf* (A, α), *then* $- \times (A, \alpha) \dashv -^{(A,\alpha)}$.   ■

As a consequence, Proposition 24 is now proved.

2.4   Locally Cartesian Closed

The goal of this section is to prove the following.

Proposition 31. FuzzyPresheaf$(I, \mathcal{L})$ *is locally cartesian closed.*

Because we consider slice categories, let us fix an arbitrary fuzzy presheaf (D, δ) for the rest of this section. We aim to prove that FuzzyPresheaf$(I,\mathcal{L})/(D,\delta)$ is cartesian closed. Here are the details of this category.

- Its objects are of the form $((A, \alpha), p)$ where (A, α) is a fuzzy presheaf and $p : (A, \alpha) \to (D, \delta)$ is a fuzzy presheaf morphism to the fixed object (D, δ), hence satisfying $\alpha \leqslant \delta p$. We denote this object by (A, α, p) for simplicity.
- A morphism $f : (A, \alpha, p) \to (B, \beta, q)$ is a natural transformation $f : A \to B$ satisfying $p = qf$. For p, q and f to be also well-defined in the base category FuzzyPresheaf$(I, \mathcal{L})$, we must also have $\alpha \leqslant \delta p$, $\beta \leqslant \delta q$ and $\alpha \leqslant \beta f$.

$$
\begin{array}{cc}
\begin{array}{ccc}
A & \xrightarrow{\ p\ } & D \\
& \searrow^{\alpha} \ \leqslant \ \swarrow_{\delta} & \\
& \mathcal{L} &
\end{array}
&
\begin{array}{ccc}
& \nearrow^{p} \quad D \quad \nwarrow^{q} & \\
A & \xrightarrow[\ f\ ]{\ \ } & B \\
& \searrow_{\alpha} \ \ \swarrow^{\delta} \ \searrow_{\beta} & \\
& \mathcal{L} &
\end{array}
\end{array}
$$

To prove locally cartesian closedness, we extend on the technique used in [2, Lemma 9.23]. We demonstrate that the slice category FuzzyPresheaf$(I,\mathcal{L})/(D,\delta)$

is equivalent to *another* category of fuzzy presheaves $\mathsf{FuzzyPresheaf}(J, \tilde{\mathcal{L}})$ for some other J and $\tilde{\mathcal{L}}$. Because cartesian closedness is preserved by categorical equivalence [24, Ex. I.4], and because $\mathsf{FuzzyPresheaf}(J, \tilde{\mathcal{L}})$ is cartesian closed by Proposition 24, then Proposition 31 follows immediately. The category J is the category of elements of D, so let us recall its definition.

Definition 32 ([2, p. 203]). *Given a presheaf $D \in \mathsf{Set}^{I^{\mathrm{op}}}$, its* **category of elements** *is denoted* $\mathrm{el}(D)$ *or* $\int_I D$ *and contains the following:*

- *objects are pairs (i, d) where $i \in I$ and $d \in D(i)$, and*
- *morphisms $\iota : (j, e) \to (i, d)$ are I-morphisms $\iota : j \to i$ such that $D(\iota)(d) = e$.*

Example 33. Recall that Graph is the presheaf category on $I^{\mathrm{op}} = E \overset{s}{\underset{t}{\rightrightarrows}} V$. Given a graph D, the objects of its category of elements are $\{(V, v) \mid v \in D(V)\} \cup \{(E, e) \mid e \in D(E)\}$. Each edge $e \in D(E)$ induces two $\mathrm{el}(D)$-morphisms, one going onto the source of e, $(E, e) \to (V, s(e))$, and one onto its target $(E, e) \to (V, t(e))$. All morphisms of $\mathrm{el}(D)$ are obtained this way.

Lemma 34. *Let I be a small category and $(\mathcal{L}(i))_{i \in I}$ a family of complete Heyting algebras. For any object $(D, \delta) \in \mathsf{FuzzyPresheaf}(I, \mathcal{L})$, we have an equivalence of categories*

$$\mathsf{FuzzyPresheaf}(I, \mathcal{L})\big/_{(D, \delta)} \simeq \mathsf{FuzzyPresheaf}(\mathrm{el}(D), \tilde{\mathcal{L}})$$

where $\tilde{\mathcal{L}}$ is defined on objects $(i, d) \in \mathrm{el}(D)^{\mathrm{op}}$ as $\tilde{\mathcal{L}}(i, d) := \mathcal{L}(i)_{\leqslant \delta_i(d)}$, and on morphisms $\iota : (j, e) \to (i, d)$ as $\mathcal{L}(\iota)\big|_{\mathcal{L}(i)_{\leqslant \delta_i(d)}} : \mathcal{L}(i)_{\leqslant \delta_i(d)} \to \mathcal{L}(j)_{\leqslant \delta_j(e)}$. ∎

By Lemma 34 and Proposition 24, then Proposition 31 is proved. By Propositions 17, 18, 31, the proof of Theorem 15 is complete.

Remark 35. If I is a finite category, then every finite presheaf category $\mathsf{FinSet}^{I^{\mathrm{op}}}$ is a topos [8, Ex. 5.2.7]. Analogously, if I is finite we also have that every finite fuzzy presheaf category is a quasitopos.

3 Fuzzy Presheaves are Rm-Adhesive

In this section we show that fuzzy presheaves are rm-adhesive. To state and motivate the result we need some definitions.

Definition 36 ([20]). *A pushout square is* **Van Kampen** **(VK)** *if, whenever it lies at the bottom of a commutative cube, like $ABCD$ on the diagram, where the back faces $FBAE$, $FBCG$ are pullbacks, then*

the front faces are pullbacks $\iff$ the top face is pushout.

A pushout square is **stable** *(under pullback) if, whenever it lies at the bottom of such a cube, then* *the front faces are pullbacks* $\Longrightarrow$ *the top face is pushout.*

Definition 37 ([15]). *A category* C *is*

- **adhesive** *if pushouts along monomorphisms exist and are VK;*
- **rm-adhesive**[5] *if pushouts along regular monomorphisms exist and are VK;*
- **rm-quasiadhesive** *if pushouts along regular monomorphisms exist, are stable, and are pullbacks.*

Toposes are adhesive [20, Proposition 9] (and thus rm-adhesive), and quasitoposes are rm-quasiadhesive [15]. Not all quasitoposes are rm-adhesive. The category of simple graphs is a counterexample [19, Corollary 20].

Coproducts are a generalisation of the notion of disjoint unions in Set. However, a notion of *union* can still be defined. In a quasitopos, there is an elegant description of binary union of regular subobjects [19, Proposition 10(iii)].

Definition 38. *The* binary union *of two regular subobjects* $f : A \hookrightarrow C$ *and* $g : B \hookrightarrow C$ *in a quasitopos is obtained as the pushout of the pullback of* f *and* g.

$$D := A \times_C B \qquad\qquad A \sqcup_D B \cdots\cdots h \cdots\cdots C \qquad\qquad (7)$$

Proposition 39 ([15, **Proposition 2.4**], [6, **Proposition 3**]). *In a quasitopos, if* f *or* g *of* (7) *is regular monic, then* h *is monic.* □

To prove FuzzyPresheaf$(I, \mathcal{L})$ rm-adhesive, we use the following theorem.

Theorem 40 ([19, **Theorem 21**]). *If* C *is a quasitopos, then* C *is rm-adhesive iff the class of regular subobjects is closed under binary union.* □

Theorem 41. *If* I *is a small category and* $(\mathcal{L}(i))_{i \in I}$ *is a family of complete Heyting algebras, then* FuzzyPresheaf$(I, \mathcal{L})$ *is rm-adhesive.*

Proof. By Theorem 15, FuzzyPresheaf$(I, \mathcal{L})$ is a quasitopos, and so by Theorem 40, it suffices to show that the class of regular subobjects is closed under binary union. By the reasoning given in Example 19, it suffices to show this property pointwise, i.e., on the level of FuzzySet$(\mathcal{L})$.

So let two regular subobjects $f : A \hookrightarrow C$ and $g : B \hookrightarrow C$ of FuzzySet$(\mathcal{L})$ be given. In (7), f' and g' are regular monic by (general) pullback stability, and then f'' and g'' are regular monic by pushout stability in quasitoposes [18, Lemma A.2.6.2]. Moreover, by Proposition 39, h is monic.

Now consider an arbitrary element $y \in A \sqcup_D B$. By general pushout reasoning for Set, y is in the image of f'' or g'', say w.l.o.g. $y = f''(a)$ for some $a \in A$. Because $f = hf''$ and f is regular (i.e., membership-preserving), a and $h(y)$ have the same membership value. The membership value of y is sandwiched between the ones of a and of $h(y)$ and all three are thus equal. This means that h preserves the membership, i.e., is regular. □

---

[5] Also known as *quasiadhesive*.

4   Examples and Applications

In this section, we showcase the wide variety of structures that presheaf categories cover. We can therefore add labels from a complete Heyting algebra to all of those categories, and have an rm-adhesive quasitopos.

Example 42. We describe multiple categories of graph as presheaf categories $\mathsf{Set}^{I^{\mathrm{op}}}$ and specify I^{op} for each.

- *Multigraphs*: We allow parallel edges, i.e., with same source and target.
  - Directed: denoted Graph in this paper; $I^{\mathrm{op}} = E \underset{t}{\overset{s}{\rightrightarrows}} V$.
  - Undirected [44, §8]: each edge e requires another one $\mathsf{sym}(e)$ in the other direction, which is equivalent to having an undirected edge. Take $I^{\mathrm{op}} = {}^{\mathsf{sym}}\!\subsetneq E \underset{t}{\overset{s}{\rightrightarrows}} V$ where $\mathsf{sym} \cdot \mathsf{sym} = \mathrm{id}_E$, $s \cdot \mathsf{sym} = t$, and $t \cdot \mathsf{sym} = s$.
  - Directed reflexive [44, §8]: Each vertex v has a specific loop $\mathsf{refl}_v \subsetneq v$ preserved by graph homomorphisms. We can equivalently omit those reflexive loops and consider that an edge mapped onto a reflexive loop is now sent to the vertex instead, giving *degenerate* graphs. This is the presheaf category on $I^{\mathrm{op}} = E \underset{\underset{t}{\longrightarrow}}{\overset{\overset{s}{\longrightarrow}}{\leftarrow \mathsf{refl} -}} V$ where $s \cdot \mathsf{refl} = t \cdot \mathsf{refl} = \mathrm{id}_V$.
  - Undirected reflexive [44, §8]: with both sym and refl in I^{op}, we must add the equation $\mathsf{sym} \cdot \mathsf{refl} = \mathsf{refl}$.
- *(Multi) Hypergraphs*: An *hyperedge* can contain any number of vertices.
  - For hyperedges to be ordered lists $\langle v_1, \ldots, v_m \rangle$ of vertices, let I^{op} have objects V and E_m for each $m \in \mathbb{N}_{\geqslant 1}$, and let $I^{\mathrm{op}}(E_m, V) = \{s_1, \ldots, s_m\}$. Note that homomorphisms preserve arities.
  - For hyperedges to be unordered sets $\{v_1, \ldots, v_m\}$, take the convention that all possible ordered lists containing the same vertices must exist. For that, add m symmetry arrows, i.e., $I^{\mathrm{op}}(E_m, E_m) = \{\mathsf{sym}_1, \ldots, \mathsf{sym}_m\}$ for each $m \in \mathbb{N}_{\geqslant 1}$, with the equation $\mathsf{sym}_j \cdot \mathsf{sym}_j = \mathrm{id}_{E_m}$, the three equations $s_1 \cdot \mathsf{sym}_1 = s_2$, $s_2 \cdot \mathsf{sym}_1 = s_1$, $s_i \cdot \mathsf{sym}_1 = s_i$ for $i \neq 1,2$ that show how sym_1 commutes the first two component, and similar equations for $\mathsf{sym}_2, \ldots, \mathsf{sym}_m$.
  - To consider *m-uniform* hyperedges, i.e., having exactly m vertices in all of them, for m fixed, let I^{op} have as objects only E_m and V in the previous examples.
  - We can extend the previous hyperedge examples to admit targets [22, Example 3.4] by having for each $m, n \in \mathbb{N}_{\geqslant 1}$ an object $E_{m,n}$ with $I^{\mathrm{op}}(E_{m,n}, V) = \{s_0, \ldots, s_{m-1}\} \cup \{t_0, \ldots, t_{n-1}\}$.
  - Here is an alternative definition of a hypergraph: Let G be a hypergraph if it has three sets $G(R), G(E), G(V)$ and two functions $f : G(R) \to G(E)$ and $g : G(R) \to G(V)$. We read $r \in G(R)$ as meaning that $g(r)$ is a vertex incident to the hyperedge $f(r)$. Note that a vertex can be incident to same hyperedge multiple times and that morphisms do not preserve arities here. This is the presheaf for $I^{\mathrm{op}} = E \overset{f}{\longleftarrow} R \overset{g}{\longrightarrow} V$.

Example 43. As pointed out in [3], the category of Petri nets is not a topos because it is not cartesian closed. However, the authors show that the category of *pre-nets*, which are Petri nets where the input and the output of a transition are ordered, is equivalent to a presheaf category.

Example 44 ([30, Example 81]). Let us illustrate the main reason for our interest in fuzzy structures: practical graph relabeling when seeing the membership values as labels. Let $(\mathcal{L}, \leqslant)$ be a complete lattice. The diagram

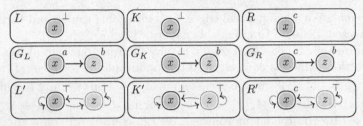

depicts a rewrite step from a graph G_L to a graph G_R using the graph rewriting formalism PBPO$^+$. The top and bottom row are part of a PBPO$^+$ rewrite rule ρ, and the middle row is an application of ρ to graph G_L, producing graph G_R. Graph G_K is an intermediate graph in the rewrite process. As can be seen, the effect of the rule is that the label a on vertex x is replaced by label c. Fuzzy graphs allow expressing label bounds in rules that make such relabeling possible. For detailed information on how this works and how this compares to existing relabeling approaches, see the cited example and [30, Section 7.2].

Remark 45. Recall from Definition 13 that an $\mathcal{L}$-fuzzy presheaf (A, α) must satisfy $\mathcal{L}(\iota) \cdot \alpha_i \leqslant \alpha_j \cdot A(\iota)$ for all $\iota : j \rightarrow i$. In practice, that condition gives a constraint on what can be called a fuzzy presheaf, but it has no impact on fuzzy presheaves morphisms. In the case of relabeling graphs, as discussed in Example 44, it restricts the amount of fuzzy graphs on which rewriting can be performed. However, that a given fuzzy graph satisfies the constraint or not, the output of the PBPO$^+$ rewrite step stays the same, i.e., the rewriting dynamics are unchanged. Indeed, pullbacks and pushouts are the same with and without the constraints, as pointed out in Remark 16.

The original example of constraint for fuzzy graphs was by Rosenfeld [37]. When the membership values are taken in the unit interval, it makes sense that if an edge has membership $x \in [0, 1]$, its endpoints must be at least as much part of the graph, i.e., have membership $y \geqslant x$. Potential applications of restricting the amount of valid fuzzy graphs in graph rewriting is left for future work.

Example 46. Here is an example of how the fuzzy structure can be used in hypergraph rewriting. Assume we are in a distributed system setting, with $\mathbb{R} \supseteq \{\boxtimes, \multimap\}$ a globally fixed set of resources. A system state $S = \langle \mathbb{P}, holds, \mathbb{C}, connect \rangle$ consists of: a set of processes $\mathbb{P}$, a resource assignment $holds : \mathbb{P} \rightarrow \mathcal{P}(\mathbb{R})$, a set of ternary connections $\mathbb{C}$, and a connection assignment $connect : \mathbb{C} \rightarrow \mathbb{P} \times \mathbb{P} \times \mathbb{P}$.

Suppose that in this system, a process p can transmit $\boxtimes$ to a process q if, for some $C \in \mathbb{C}$ and $r \in \mathbb{P}$: (i) p, q and r are distinct, (ii) $\boxtimes \in holds(p)$, (iii) $\boxtimes \notin holds(q)$, (iv) $\multimap \in holds(r)$, and (v) $connect(C) = (p, q, r)$. Executing the transmission changes the state by removing $\boxtimes$ from $holds(p)$, adding $\boxtimes$ to $holds(q)$, removing $\multimap$ from $holds(r)$ and removing C from $\mathbb{C}$. We can thus think of r as a mediator for the transmission, which provides $\multimap$ as a required resource that is consumed.

We can model states as objects of the fuzzy presheaf $\mathsf{FuzzyPresheaf}(I, \mathcal{L})$ where: $I^{\mathrm{op}} = \mathbb{C} \underset{-t}{\overset{-s}{\underset{-m}{\rightrightarrows}}} \mathbb{P}$ and $\mathbb{P}$ is labeled from the subset lattice $\mathcal{L} = (\mathcal{P}(\mathbb{R}), \subseteq)$, which is a complete Heyting algebra. Then the transmission described above can be informally defined as a hypergraph transformation rule:

$$p^{P \uplus \{\boxtimes\}} \xrightarrow{\quad\quad} q^{Q \setminus \{\boxtimes\}} \qquad\qquad\qquad p^{P} \qquad\qquad q^{Q \cup \{\boxtimes\}}$$
$$r^{R \uplus \{\multimap\}} \qquad\qquad\Longrightarrow\qquad\qquad r^{R}$$

which is formally definable with a single PBPO^{+} rule similar to the one in Example 44, so that it is applicable in any context.

Example 47. Other concrete examples of modellings using fuzzy multigraphs and fuzzy multigraph rewriting using PBPO^{+} include:

- Binary decision diagrams reduction [28, Section 5].
- Processes that consume data from FIFO channels [30, Example 82].
- Terms and linear term rewriting [27, Section 4].

5   Fuzzy Simple Graphs Form a Quasitopos

In Example 42, we described different multigraph categories as presheaf categories. Hence, adding a fuzzy structure to these categories results in quasitoposes by Theorem 15. This raises the question whether simple fuzzy graph categories are quasitoposes, and if so, whether it can be shown using Theorem 15.

First, simple graphs do not arise as presheaf categories, meaning Theorem 15 cannot be applied. Here is one way to observe this: presheaves are toposes and in toposes, all monomorphisms are regular. However, monomorphisms in directed simple graphs [19, p. 315] and in undirected simple graphs [36, Lemma 3.7.1] are regular only if they reflect edges,[6] which is not always the case.

Nevertheless, we show in this section that directed and undirected simple fuzzy graph categories are quasitoposes, by using a different technique. In more detail, given a *(Lawvere-Tierney) topology* τ on a topos, the τ-separated elements form a subcategory which is a quasitopos [17, Theorem 10.1]. Vigna [44] has used this fact to prove that the (directed/undirected) simple graphs are exactly the $\neg\neg$-separated elements of the respective (directed/undirected) multigraph category, where $\neg\neg$ is a topology. We extend the approach by Vigna, showing that directed and undirected simple fuzzy graphs are quasitoposes.

---

[6] A graph homomorphism $f : G \to H$ *reflects edges* if, for every $v, w \in G(V)$, every edge between $f_V(v)$ and $f_V(w)$ has an f-preimage.

Because we start from fuzzy multigraphs, which form a quasitopos, we need to work with the more general definition of a *topology on a quasitopos* (Definition 49). The result that the separated elements form a quasitopos still holds true with this more general definition of a topology [45, Theorem 43.6]. Recall Notation 21 and Definition 7 if needed for the notations used.

Remark 48. Alternatively, Johnstone et al. show that Artin Glueing [9] can be used to prove that *directed* simple graphs form a quasitopos. Given a functor $T : \mathsf{C} \to \mathsf{D}$, denote by $\mathsf{C} \mathbin{/\!\!/} T$ the full subcategory of the slice category C/T consisting of the monomorphisms $X \rightarrowtail TY$. Then directed simple graphs can be seen as $\mathsf{Set} \mathbin{/\!\!/} T$ for $TX = X \times X$. Because Set is a quasitopos and T preserves pullbacks, we have by [9, Theorem 16] that $\mathsf{Set} \mathbin{/\!\!/} T$, i.e., directed simple graphs, is a quasitopos.

Definition 49 ([45, **Def. 41.1**]). *A* **topology** τ *on a quasitopos* E *is a family of mappings* $\mathrm{Mono}(A) \to \mathrm{Mono}(A)$ *for each* $A \in \mathsf{E}$. *It sends every monomorphism* $m : A_0 \rightarrowtail A$ *to another monomorphism* $\tau m : A_1 \rightarrowtail A$ *with same codomain* $A \in \mathsf{E}$, *such that*

(i) If $m \leqslant m'$, *then* $\tau m \leqslant \tau m'$.
(ii) $m \leqslant \tau m$.
(iii) $\tau\tau m \simeq \tau m$.
(iv) For all $f : B \to A$, *we have* $\tau(f^{\leftarrow} m) \simeq f^{\leftarrow}(\tau m)$.
(v) If m *is regular, then so is* τm.

Notice that axioms $(i) - (iii)$ says that τ is a closure operation. For instance in Set, given a subset $A_0 \subseteq A$, (ii) requires that $A_0 \subseteq \tau(A_0) \subseteq A$. By (iv) and (v) the closure operation must commute with pullbacks and preserve regularity.

Example 50 ([45, *Ex. 41.2*]). We give two basic examples of topologies. The third example is more advanced; examples of it in a few categories are given in the next lemma.

- The *trivial topology* $\tau(A_0 \overset{m}{\rightarrowtail} A) := (A \overset{\mathrm{id}_A}{=\!=} A)$.
- The *discrete topology* $\tau(A_0 \overset{m}{\rightarrowtail} A) := (A_0 \overset{m}{\rightarrowtail} A)$.
- Let $m : A_0 \rightarrowtail A$. Consider the (necessarily monic) morphism from the initial object $0 \rightarrowtail A$ and factorise it as (epi mono, regular mono) $\overline{o_A} \circ e : 0 \twoheadrightarrow A_1 \hookrightarrow A$. Let $\neg m$ be the exponential object $(\overline{o_A} : A_1 \hookrightarrow A)^{(m:A_0 \rightarrowtail A)}$ in the slice category E/A. Then $\neg\neg$ is a topology called the *double negation*.

Lemma 51. *We detail the* $\neg\neg$ *topology in a few categories. In each situation, we consider subobjects of an object* A *or* (A, α).

- *In* Set, $\neg A_0 = A \setminus A_0$. *Therefore,* $\neg\neg = \mathrm{id}$.
- *In* $\mathsf{FuzzySet}(\mathcal{L})$, $\neg(A_0, \alpha_0) = (A \setminus A_0, \alpha)$. *Therefore,* $\neg\neg$ *is the identity on the subset but replaces its membership function* α_0 *with* α, *the one of* A.
- *In* Graph, $\neg A_0$ *is the largest subgraph of* A *totally disconnected from* A_0. *Hence,* $\neg\neg A_0$ *is the largest subgraph of* A *induced by* $A_0(V)$, *i.e., every edge from* A *with source and target in* A_0 *is added in* $\neg\neg A_0$.

 – *In* FuzzyGraph, $\neg\neg$ *acts like in* Graph *on the underlying graph, i.e., it adds all edges that had source and target in the subgraph, and it replaces the membership function of the subobject with the one of the main object.*

The notions of density and separatedness from standard topology, can also be defined for (quasi)topos topologies. To illustrate what density is, recall that $\mathbb{Q}$ is called dense in $\mathbb{R}$ because closing it under the standard metric topology, i.e., adding an arbitrary small open interval around each rational number, gives the whole space $\mathbb{R}$. In standard topology, recall also that every separated space B (a.k.a. Hausdorff space), has the property that functions $f : A \to B$ are fully determined by the images on any dense subsets of A.

Definition 52 ([45, **Def. 41.4, 42.1**]). *Given a topology τ in a quasitopos* E:

 – *a monomorphism $A_0 \overset{m}{\rightarrowtail} A$ is τ-**dense** if $\tau(A_0 \overset{m}{\rightarrowtail} A) = (A \overset{\mathrm{id}_A}{=\!=\!=} A)$,*

 – *an object B is called τ-**separated** if for every τ-dense subobject $m : A_0 \rightarrowtail A$ and every morphism $f : A_0 \to B$ there exists at most one factorisation $g : A \to B$ of f through m.*

$$\begin{array}{ccc} & A_0 & \\ {\scriptstyle m}\Big\downarrow & & \searrow^{f} \\ A & \cdots\!\cdots\!\!\rightarrow & B \\ & {\scriptstyle g} & \end{array}$$

The $\neg\neg$-separated graphs are precisely the simple graphs.

Lemma 53 ([44, **Thms. 3 & 4**]). *In* Graph, *a subgraph $A_0 \subseteq A$ is $\neg\neg$-dense if it contains all the vertices $A_0(V) = A(V)$. As a consequence, a graph B is $\neg\neg$-separated if it has no parallel edges.*

Similarly, the $\neg\neg$-separated fuzzy graphs are the simple fuzzy graphs.

Lemma 54. *In the category of (directed and undirected) fuzzy graphs, a subgraph $(A_0, \alpha_0) \subseteq (A, \alpha)$ is $\neg\neg$-dense if it contains all the vertices $A_0(V) = A(V)$. Hence, a fuzzy graph (B, β) is $\neg\neg$-separated if it has no parallel edges.* ∎

Corollary 55. *Directed and undirected simple fuzzy graphs form quasitoposes.*

Proof. Immediate by Lemma 54 and [45, Theorem 43.6]. □

6 Conclusion

In this paper, we introduced the concept of fuzzy presheaves and proved that they form rm-adhesive quasitoposes. Furthermore, we showed that simple fuzzy graphs, both directed and undirected, also form quasitoposes.

There are several directions for future work. Having the poset/lattice/Heyting algebras morphisms $\mathcal{L}(\iota)$ was not of much use for the practical applications that we had in mind. The applicability of putting such constraints on the membership values/labels can be explored further.

Theorem 15 says that having the image of the functor $\mathcal{L}$ in CompHeytAlg is a sufficient condition for fuzzy presheaves to form a quasitopos. Whether that is also a necessary condition is left for future work. The existence of top elements $\top$

comes from the terminal fuzzy presheaf, but the existence of bottom elements $\perp$ is already more difficult to argue, because of the initial fuzzy presheaf being the empty presheaf. This might be a lead to find a counterexample. That question seems, already on the level of fuzzy sets, to be yet unanswered, with only partial results as of now, such as [45, Proposition 71.4].

To obtain fuzzy presheaves, we have added a pointwise fuzzy structure to presheaves. More generally, different notions of fuzzy categories have been defined [42,43]. It is natural to wonder if a fuzzy (quasi)topos is also a quasitopos.

Finally, a more abstract question is whether having a fuzzy structure is a particular instance of a more abstract categorical construction. As discussed after Definition 13, a diagram $\leqslant$-commuting can sometimes be expressed as a commuting diagram in a (lax) 2-categorical setting. The more comma category-like approach of Remark 14 is another lead for future developing the field of fuzzy sets within category theory. For comma categories, Artin Glueing gives a nice criterion for obtaining new quasitoposes [19]. We therefore wonder if more abstract uniform theorems for obtaining quasitoposes can be expressed and proved.

Acknowledgments. We thank Helle Hvid Hansen for discussions and valuable suggestions. The authors received funding from the Netherlands Organization for Scientific Research (NWO) under the Innovational Research Incentives Scheme Vidi (project. No. VI.Vidi.192.004).

References

1. Adámek, J., Herrlich, H., Strecker, G.E.: Abstract and Concrete Categories: The Joy of Cats. Wiley, New York (1990)
2. Awodey, S.: Category Theory. Oxford Logic Guides. Ebsco Publishing (2006)
3. Baez, J.C., Genovese, F., Master, J., Shulman, M.: Categories of nets. In: 36th Annual Symposium on Logic in Computer Science, pp. 1–13. IEEE (2021). https://doi.org/10.1109/LICS52264.2021.9470566
4. Bauderon, M.: A uniform approach to graph rewriting: the pullback approach. In: Nagl, M. (ed.) WG 1995. LNCS, vol. 1017, pp. 101–115. Springer, Heidelberg (1995). https://doi.org/10.1007/3-540-60618-1_69
5. Behr, N., Harmer, R., Krivine, J.: Concurrency theorems for non-linear rewriting theories. In: Gadducci, F., Kehrer, T. (eds.) ICGT 2021. LNCS, vol. 12741, pp. 3–21. Springer, Cham (2021). https://doi.org/10.1007/978-3-030-78946-6_1
6. Behr, N., Harmer, R., Krivine, J.: Concurrency theorems for non-linear rewriting theories. CoRR abs/2105.02842 (2021). https://arxiv.org/abs/2105.02842
7. Borceux, F.: Handbook of Categorical Algebra. 1, Encyclopedia of Mathematics and its Applications, vol. 50. Cambridge University Press, Cambridge (1994). Basic category theory
8. Borceux, F.: Handbook of Categorical Algebra. 3, Encyclopedia of Mathematics and its Applications, vol. 52. Cambridge University Press, Cambridge (1994). Categories of sheaves
9. Carboni, A., Johnstone, P.T.: Connected limits, familial representability and Artin glueing. Math. Struct. Comput. Sci. **5**(4), 441–459 (1995). https://doi.org/10.1017/S0960129500001183

10. Cardoso, J., Valette, R., Dubois, D.: Fuzzy petri nets: an overview. IFAC Proc. Volumes **29**(1), 4866–4871 (1996). https://doi.org/10.1016/S1474-6670(17)58451-7. 13th World Congress of IFAC, 1996
11. Corradini, A., Duval, D., Echahed, R., Prost, F., Ribeiro, L.: AGREE – algebraic graph rewriting with controlled embedding. In: Parisi-Presicce, F., Westfechtel, B. (eds.) ICGT 2015. LNCS, vol. 9151, pp. 35–51. Springer, Cham (2015). https://doi.org/10.1007/978-3-319-21145-9_3
12. Corradini, A., Duval, D., Echahed, R., Prost, F., Ribeiro, L.: The PBPO graph transformation approach. J. Log. Algebraic Methods Program. **103**, 213–231 (2019). https://doi.org/10.1016/j.jlamp.2018.12.003
13. Corradini, A., Heindel, T., Hermann, F., König, B.: Sesqui-pushout rewriting. In: Corradini, A., Ehrig, H., Montanari, U., Ribeiro, L., Rozenberg, G. (eds.) ICGT 2006. LNCS, vol. 4178, pp. 30–45. Springer, Heidelberg (2006). https://doi.org/10.1007/11841883_4
14. Ehrig, H., Pfender, M., Schneider, H.J.: Graph-grammars: an algebraic approach. In: 14th Annual Symposium on Switching and Automata Theory, pp. 167–180. IEEE Computer Society (1973). https://doi.org/10.1109/SWAT.1973.11
15. Garner, R., Lack, S.: On the axioms for adhesive and quasiadhesive categories. Theory Appl. Categories **27**, 27–46 (2012)
16. Goguen, J.: *L*-fuzzy sets. J. Math. Anal. Appl. **18**(1), 145–174 (1967). https://doi.org/10.1016/0022-247X(67)90189-8
17. Johnstone, P.T.: On a topological topos. Proc. London Math. Soc. **s3–38**(2), 237–271 (1979). https://doi.org/10.1112/plms/s3-38.2.237
18. Johnstone, P.T.: Sketches of An Elephant: A Topos Theory Compendium. Oxford Logic Guides. Oxford University Press, New York (2002)
19. Johnstone, P.T., Lack, S., Sobociński, P.: Quasitoposes, quasiadhesive categories and artin glueing. In: Mossakowski, T., Montanari, U., Haveraaen, M. (eds.) CALCO 2007. LNCS, vol. 4624, pp. 312–326. Springer, Heidelberg (2007). https://doi.org/10.1007/978-3-540-73859-6_21
20. Lack, S., Sobociński, P.: Adhesive categories. In: Walukiewicz, I. (ed.) FoSSaCS 2004. LNCS, vol. 2987, pp. 273–288. Springer, Heidelberg (2004). https://doi.org/10.1007/978-3-540-24727-2_20
21. Lack, S., Sobociński, P.: Adhesive and quasiadhesive categories. RAIRO Theor. Inform. Appl. **39** (2005). https://doi.org/10.1051/ita:2005028
22. Löwe, M.: Algebraic approach to single-pushout graph transformation. Theor. Comput. Sci. **109**(1&2), 181–224 (1993). https://doi.org/10.1016/0304-3975(93)90068-5
23. Mac Lane, S.: Categories for the Working Mathematician, vol. 5. Springer, New York (1971). https://doi.org/10.1007/978-1-4757-4721-8
24. Mac Lane, S., Moerdijk, I.: Sheaves in Geometry and Logic. Springer-Verlag, New York (1994). A first introduction to topos theory, Corrected reprint of the 1992 edition
25. Mathew, S., Mordeson, J.N., Malik, D.S.: Fuzzy Graph Theory, vol. 363. Springer, Cham (2018). https://doi.org/10.1007/978-3-319-71407-3
26. Mori, M., Kawahara, Y.: Fuzzy graph rewritings. 数理解析研究所講究録 **918**, 65–71 (1995)
27. Overbeek, R., Endrullis, J.: From linear term rewriting to graph rewriting with preservation of termination. In: Proceedings of Workshop on Graph Computational Models (GCM). EPTCS, vol. 350, pp. 19–34 (2021). https://doi.org/10.4204/EPTCS.350.2

28. Overbeek, R., Endrullis, J.: A PBPO$^+$ graph rewriting tutorial. In: Proceedings of Workshop on Computing with Terms and Graphs (TERMGRAPH). EPTCS, vol. 377, pp. 45–63. Open Publishing Association (2023). https://doi.org/10.4204/EPTCS.377.3

29. Overbeek, R., Endrullis, J.: Termination of graph transformation systems using weighted subgraph counting. In: Graph Transformation - 16th International Conference, ICGT 2023. LNCS, vol. 13961. Springer (2023). https://doi.org/10.48550/arXiv.2303.07812

30. Overbeek, R., Endrullis, J., Rosset, A.: Graph rewriting and relabeling with PBPO$^+$: a unifying theory for quasitoposes. J. Log. Algebr. Methods Program. (2023). https://doi.org/10.1016/j.jlamp.2023.100873

31. Overbeek, R., Endrullis, J., Rosset, A.: Graph rewriting and relabeling with PBPO$^+$. In: ICGT (2020). https://doi.org/10.1007/978-3-030-78946-6_4

32. Pal, M., Samanta, S., Ghorai, G.: Modern Trends in Fuzzy Graph Theory. Springer, Singapore (2020). https://doi.org/10.1007/978-981-15-8803-7

33. Parasyuk, I.N., Ershov, S.V.: Transformations of fuzzy graphs specified by FD-grammars. Cybern. Syst. Anal. **43**, 266–280 (2007). https://doi.org/10.1007/s10559-007-0046-6

34. Parasyuk, I.N., Yershov, S.V.: Categorical approach to the construction of fuzzy graph grammars. Cybern. Syst. Anal. **42**, 570–581 (2006). https://doi.org/10.1007/s10559-006-0094-3

35. Parasyuk, I.N., Yershov, S.V.: Transformational approach to the development of software architectures on the basis of fuzzy graph models. Cybern. Syst. Anal. **44**, 749–759 (2008). https://doi.org/10.1007/s10559-008-9048-2

36. Plessas, D.J.: The categories of graphs. Ph.D. thesis, The University of Montana (2011). https://www.scholarworks.umt.edu/etd/967/. Dissertations & Professional Papers. 967

37. Rosenfeld, A.: Fuzzy graphs. In: Fuzzy Sets and Their Applications to Cognitive and Decision Processes, pp. 77–95. Elsevier (1975)

38. Rosset, A., Overbeek, R., Endrullis, J.: Fuzzy presheaves are quasitoposes. CoRR abs/2301.13067 (2023). https://doi.org/10.48550/arXiv.2301.13067

39. Spivak, D.I.: Metric realization of fuzzy simplicial sets (2009)

40. Srivastava, A.K., Tiwari, S.P.: On categories of fuzzy petri nets. Adv. Fuzzy Sys. **2011** (2011). https://doi.org/10.1155/2011/812040

41. Stout, L.N.: The logic of unbalanced subobjects in a category with two closed structures. In: Rodabaugh, S.E., Klement, E.P., Höhle, U. (eds.) Applications of Category Theory to Fuzzy Subsets. Theory and Decision Library, pp. 73–105. Springer, Dordrecht (1992). https://doi.org/10.1007/978-94-011-2616-8_4

42. Sugeno, M., Sasaki, M.: L-fuzzy category. Fuzzy Sets Syst. **11**(1), 43–64 (1983). https://doi.org/10.1016/S0165-0114(83)80068-2

43. Syropoulos, A., Grammenos, T.: A Modern Introduction to Fuzzy Mathematics. Wiley, Hoboken (2020)

44. Vigna, S.: A guided tour in the topos of graphs (2003). https://doi.org/10.48550/ARXIV.MATH/0306394

45. Wyler, O.: Lecture Notes on Topoi and Quasitopoi. World Scientific (1991)

46. Zadeh, L.A.: Fuzzy sets. Inf. Control **8**(3), 338–353 (1965). https://doi.org/10.1016/S0019-9958(65)90241-X

Mechanised DPO Theory: Uniqueness of Derivations and Church-Rosser Theorem

Robert Söldner[✉] and Detlef Plump

Department of Computer Science, University of York, York, UK
{rs2040,detlef.plump}@york.ac.uk

Abstract. We demonstrate how to use the proof assistant Isabelle/HOL to obtain machine-checked proofs of two fundamental theorems in the double-pushout approach to graph transformation: the uniqueness of derivations up to isomorphism and the so-called Church-Rosser theorem. The first result involves proving the uniqueness of pushout complements, first established by Rosen in 1975. The second result formalises Ehrig's and Kreowski's proof of 1976 that parallelly independent direct derivations can be interchanged to obtain derivations ending in a common graph. We also show how to overcome Isabelle's limitation that graphs in locales must have nodes and edges of pre-defined types.

Keywords: Double-pushout graph transformation · Isabelle/HOL · Uniqueness of derivations · Church-Rosser theorem

1 Introduction

Formal methods help to mitigate the risk of software defects by rigorously verifying the correctness of software against their specification. It involves the use of mathematical techniques to prove that a software implementation meets a set of specifications and requirements. These methods include techniques such as model checking, theorem proving, and symbolic interpretation [13].

Interactive theorem provers have been used to rigorously prove mathematical theorems such as the Four Colour Theorem [9], the Prime Number Theorem [1], and the Kepler Conjecture [11]. Moreover, specific algorithms and software components have been successfully verified against their specifications, such as the seL4 Microkernel Verification [14] and the development and formal verification of the CompCert compiler [16]. These achievements demonstrate the potential of interactive theorem provers in verifying complex theories and systems.

Our research aims to rigorously prove fundamental results in the double-pushout approach to graph transformations [6], which aligns with our long-term goal of verifying specific programs in the graph programming language GP 2 [4,21] using the Isabelle proof assistant [18]. This paper extends earlier work [24] and builds the foundations for our long-term goal.

While we are not the first to apply interactive theorem provers to graph transformation, previous approaches such as Strecker's [25] do not address the

© The Author(s), under exclusive license to Springer Nature Switzerland AG 2023
M. Fernández and C. M. Poskitt (Eds.): ICGT 2023, LNCS 13961, pp. 123–142, 2023.
https://doi.org/10.1007/978-3-031-36709-0_7

vast amount of theoretical results available for the double-pushout approach, see Sect. 6 for details. To the best of our knowledge, we are the first to formalise basic results in double-pushout graph transformation.

As usual in the double-pushout approach, we want to abstract from node and edge identifiers. Therefore, we introduce type variables, for each graph independently, that denote the type of the nodes and edges. Using different type variables for different graphs allows to use Isabelle's typechecker during development to prevent accidental mixing of node or edge identifiers from different graphs. However, a problem then is that Isabelle does not allow to quantify over new type variables within locale definitions. This poses a challenge to formalise the universal property of pushouts and pullbacks. We overcome this problem by requiring, in locales only, that nodes and edges are natural numbers.

To prove the uniqueness of direct derivations, we break down the proof into two parts. Firstly, we show the uniqueness of the pushout complement, and subsequently show that the pushout object is also unique given an isomorphic pushout complement. In the first part we follow Lack and Sobocinski's proof for adhesive categories [15], but replace adhesiveness with the pushout characterisation by the reduced chain condition [8] together with composition and decomposition lemmata for pushouts and pullbacks.

To prove the Church-Rosser theorem, we follow the original proof of Ehrig and Kreowski [7] and rely again on the pushout characterisation by the reduced chain condition. Additionally, we exploit composition and decomposition lemmata for pushouts and pullbacks, and also a lemma that allows to switch from pullbacks to pushouts and vice versa.

We believe that our proofs could be generalised from graphs to adhesive categories, but this is not our goal. Firstly, we want to avoid the overhead necessary to deal with an abstract class of categories, such as van Kampen squares. Secondly, our development addresses both abstract concepts such as pushouts and pullbacks, and set-theoretic constructions for such concepts. Giving corresponding constructions for all adhesive categories would not be feasible. Our goal is to lay the foundations for verifying graph programs in the language GP 2, a language whose syntax and semantics are built upon the double-pushout approach to graph transformation. Our ultimate goal is to provide both interactive and automatic tool support for formal reasoning on programs in graph transformation languages such as GP 2. An implementation of a proof assistant for GP 2 will have to use the concrete definitions of graphs, attributes, rules, derivations etc. Hence our formalisation focuses on concepts for transforming graphs. We are not saying that formalising the theory of adhesive categories is not interesting, but a large part of such an development would be a distraction from our main goal.

The rest of the paper is structured as follows. The next section is a brief introduction to Isabelle/HOL and the constructs used in our work. Section 3 describes the basics of DPO graph transformation and our formalisation in Isabelle. We show the uniqueness of direct derivations in Sect. 4, which entails a proof of the uniqueness of pushout complements. Section 5 introduces the so-called

Church-Rosser theorem which states that parallelly independent direct derivations can be interchanged to obtain derivations ending in a common graph. Section 6 provides a brief overview of related work. Finally, in Sect. 7, we summarise our findings and discuss future work.

2   Isabelle/HOL

Isabelle is a generic, interactive theorem prover based on the so-called *LCF* approach. It is based on a small (meta-logical) proof kernel, which is responsible for checking all proofs. This concept provides high confidence in the prover's soundness. Isabelle/HOL refers to the higher-order logic instantiation, which is considered to be the most established calculus within the Isabelle distribution [20]. HOL is strongly typed with support for polymorphic, higher-order functions [3].

In Isabelle/HOL, type variables are denoted by a leading apostrophe. For example, a term f of type 'a is denoted by f :: 'a. Our formalisation is based on *locales*, a mechanism for writing and structuring parametric specifications. The locale's context comprises a set of parameters $(x_1 \ldots x_n)$, assumptions $(A_1 \ldots A_m)$ and concluding theorem $\bigwedge x_1 \ldots x_n.\; [\![A_1; \ldots; A_m]\!] \implies C$. This logical view offers a compelling way of combining and enhancing contexts, resulting in a clear and maintainable representation. A detailed introduction can be found in [2].

The special syntax $[\![A_1; \ldots; A_n]\!] \implies P$ is syntactic sugar for $A_1 \implies \ldots \implies A_n \implies C$ and can be read as: if A_1 and $\ldots$ and A_n hold, C is implied. Furthermore, we use *intelligible semi-automated reasoning* (Isar) which is Isabelle's language of writing structured proofs [26]. In contrast to *apply-scripts*, which execute deduction rules in a linear manner, Isar follows a structured approach resulting in increased readability and maintainability [18]. A general introduction to Isabelle/HOL can be found in [18].

3   DPO Graph Transformation in Isabelle

We work with *directed labelled graphs* where the *label alphabet* consists of a set of node labels and a set of edge labels from which a graph can be labelled. In our Isabelle/HOL formalisation, we do not specify a concrete label alphabet and leave it unspecified. As a result, stated properties are satisfied for all possible label alphabets.

Our definition of graphs allows for parallel edges and loops, and we do not consider variables as labels.

Definition 1 (Graph ⟲). A *graph* $G = (V, E, s, t, l, m)$ over the alphabet $\mathcal{L}$ is a system where V is the finite set of nodes, E is the finite set of edges, $s, t \colon E \to V$ functions assigning the source and target to each edge, $l \colon V \to \mathcal{L}_V$ and $m \colon E \to \mathcal{L}_E$ are functions assigning a label to each node and edge.   □

We represent this definition in Isabelle/HOL as a record type together with a locale. A discussion of this approach can be found in earlier work [24].

```
record ('v,'e,'l,'m) pre_graph =
  nodes   :: "'v set"
  edges   :: "'e set"
  source  :: "'e ⇒ 'v"
  target  :: "'e ⇒ 'v"
  node_label :: "'v ⇒ 'l"
  edge_label :: "'e ⇒ 'm"
```

In contrast to previous work [24], the locale required the built-in `countable` typeclass for the node and edge identifiers to overcome the limitation of Isabelle/HOL of introducing new type variables within the definition of a locale. This approach is similar to the one discussed in [23], in which the authors utilise Isabelle's `statespace` command to establish a local environment comprising axioms for the conversion of concrete types to natural numbers. Instead, we use the injective `to_nat` function through the use of the `countable` typeclass to facilitate the conversion of types.

```
locale graph =
  fixes G :: "('v::countable,'e::countable,'l,'m) pre_graph"
  assumes
    finite_nodes: "finite V_G" and
    finite_edges: "finite E_G" and
    source_integrity: "e ∈ E_G ⟹ s_G e ∈ V_G" and
    target_integrity: "e ∈ E_G ⟹ t_G e ∈ V_G"
```

The injectivity also implies the existence of the inverse, `from_nat`. We use this technique to translate arbitrary identifiers for nodes and edges into a generic representation based on natural numbers.

Definition 2 (Graphs over naturals 🔗). A graph whose nodes and edges are natural numbers is called a *natural graph*.

In Isabelle/HOL, we create a type synonym `ngraph`, which specialises the existing `pre_graph` structure to `nat`, Isabelle's built-in type of natural numbers, for both identifiers.

```
type_synonym ('l,'m) ngraph = "(nat,nat,'l,'m) pre_graph"
```

The use of Isabelle/HOL's built-in functions `to_nat` and `from_nat` allows us to convert between both representations. We define the conversion function `to_ngraph` from `ngraph` to the parameterised `pre_graph` structure as follows:

```
definition to_ngraph
            :: "('v::countable,'e :: countable,'l,'m) pre_graph
               ⇒ ('l,'m) ngraph" where
  ⟨to_ngraph G ≡ ⦇nodes = to_nat ` V_G
              ,edges = to_nat ` E_G
              ,source = λe. to_nat (s_G (from_nat e))
              ,target = λe. to_nat (t_G (from_nat e))
              ,node_label = λv. l_G (from_nat v)
              ,edge_label = λe. m_G (from_nat e)⦈⟩
```

The node and edge identifiers are mapped using `to_nat` while for the source (target) function, we first translate the natural number back into the parameterised, then apply the original source (target) function, and finally convert it back to a natural number.

A *graph morphism* $f: G \to H$ is a structure preserving mapping from a source graph G to a target graph H. It consists of a pair of mappings $f = (f_V: V_G \to V_H, f_E: V_E \to V_H)$ for nodes and edges, respectively. We say a morphism f is *injective* (*surjective*), if f_V and f_E are injective (surjective). A morphism which is injective and surjective is called *bijective*. We call G and H *isomorphic*, denoted by $G \cong H$, if a bijective morphism exists. The *morphism composition* of $f: F \to G$ and $g: G \to H$, which we denote by the usual infix notation $g \circ f$, is the pairwise composition of the node and edge mappings $g \circ f = (g_V \circ f_V, g_E \circ f_E)$. If a morphism is uniquely identified by its source and target, we sometimes omit the name and write $F \to G \to H$ to stand for the composition $g \circ f$. In our formalisation, we denote morphism composition using the $\circ_\rightarrow$ symbol to prevent a naming clash with Isabelle's built-in function composition. Different from our earlier investigation [24], we generalise *rules* to use injective morphisms instead of inclusions.

Definition 3 (Rule ⌐). A *rule* $(L \leftarrow K \to R)$ consists of graphs L, K and R over $\mathcal{L}$ together with injective morphisms $K \to L$ and $K \to R$.   □

Another difference is the introduction of a new record type `pre_rule` to be consistent with our formalisation of graphs and morphisms. The left-hand, interface, and right-hand side of a rule can be accessed using the corresponding record accessor functions `lhs`, `interf`, and `rhs`. This approach reduces the amount of parameters we have to pass, resulting in better readability.

```
record ('v₁,'e₁,'v₂,'e₂,'v₃,'e₃,'l,'m) pre_rule =
  lhs    :: "('v₁,'e₁,'l,'m) pre_graph"
  interf :: "('v₂,'e₂,'l,'m) pre_graph"
  rhs    :: "('v₃,'e₃,'l,'m) pre_graph"
```

The `rule` locale now relies on the two injective morphisms $K \to L$ and $K \to R$. We populate the `countable` typeclass restriction to allow conversion of graphs to graphs over naturals. In order to leverage Isabelle's typechecker, each graph in a rule is allowed to have a different type for the node and edge identifiers.

```
locale rule =
  k: injective_morphism "interf r" "lhs r" b +
  r: injective_morphism "interf r" "rhs r" b'
  for r :: "('v₁::countable,'e₁::countable
          ,'v₂::countable,'e₂::countable
          ,'v₃::countable,'e₃::countable
          ,'l,'m) pre_rule" and b b'
```

In general, this design choice supports the development process, as the typechecker is rejecting certain terms. For example, we are not able to express membership of an item $x \in K_V \implies x \in L_V$ as the type of node identifiers in K is v_2 while L is of type v_1. From now on, we omit explicit type and typeclass restrictions within the shown code examples to enhance readability.

Definition 4 (Dangling condition ⌐). Let $b': K \to L$ be an injective graph morphism. An injective graph morphism $g: L \to G$ satisfies the *dangling condition* if no edge in $E_G - g_E(E_L)$ is incident to a node in $g_V(V_L - b'_V(V_K))$.   □

Definition 5 (Pushout ⬙). Given graph morphisms $b\colon A \to B$ and $c\colon A \to C$, a graph D together with graph morphisms $f\colon B \to D$ and $g\colon C \to D$ is a *pushout* of $A \to B$ and $A \to C$ if the following holds (see Fig. 1a):

1. Commutativity: $f \circ b = g \circ c$, and
2. Universal property: For all graph morphisms $p\colon B \to H$ and $t\colon C \to H$ such that $p \circ b = t \circ c$, there is a unique morphism $u\colon D \to H$ such that $u \circ f = p$ and $u \circ g = t$.

We call D the *pushout object* and C the *pushout complement*. □

Different from earlier work [24], we use graphs over naturals to formalise the universal property in order to overcome the mentioned limitation of Isabelle's locale mechanism. An important property is that pushouts are unique up to isomorphism, as formalised and proved in [24].

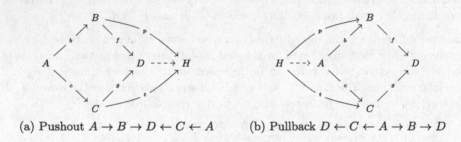

(a) Pushout $A \to B \to D \leftarrow C \leftarrow A$ (b) Pullback $D \leftarrow C \leftarrow A \to B \to D$

Fig. 1. Pushout and Pullback diagrams

Theorem 1 (Uniqueness of pushouts [6] ⬙). *Let $b\colon A \to B$ and $c\colon A \to C$ together with D induce a pushout as depicted in Fig. 1a. A graph H together with morphisms $p\colon B \to H$ and $t\colon C \to H$ is a pushout of b and c if and only if there is an isomorphism $u\colon D \to H$ such that $u \circ f = p$ and $u \circ g = t$.* □

The transformation of graphs by rules gives rise to *direct derivations*.

Definition 6 (Direct derivation ⬙). Let G and H be graphs, $r = \langle L \leftarrow K \to R \rangle$ and $g\colon L \to G$ an injective morphism satisfying the dangling condition. Then G *directly derives* H by r and g, denoted by $G \Rightarrow_{r,g} H$, as depicted in the double pushout diagram in Fig. 2. □

Note that the injectivity of the matching morphism $g\colon L \to K$ leads to a DPO approach that is more expressive than in the case of arbitrary matches [10].

In our earlier work [24], we used the `direct_derivation` locale to represent the operational view using gluing and deletion. Instead, here we use the categorical definition relying on pushouts. The operational definition is available within the `direct_derivation_construction` locale. We do not repeat its definition here, as it is not relevant for the current discussion.

```
locale direct_derivation =
  r: rule r b b' +
  gi: injective_morphism "lhs r" G g +
  po1: pushout_diagram "interf r" "lhs r" D G b d g c +
  po2: pushout_diagram "interf r" "rhs r" D H b' d f c'
  for r b b' G g D d c H f c'
begin
```

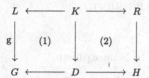
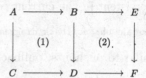

Fig. 2. Double-pushout diagram   **Fig. 3.** Composite commutative diagram

A pullback is dual to the concept of pushout.

Definition 7 (Pullback ⧉). Given graph morphism $f\colon B \to D$ and $g\colon C \to D$, a graph A together with graph morphisms $b\colon A \to B$ and $c\colon A \to C$ is a *pullback* of $C \to D \leftarrow B$ if the following holds (see Fig. 1b):

1. Commutativity: $f \circ b = g \circ c$, and
2. Universal property: For all graph morphisms $p\colon H \to B$ and $t\colon H \to C$ such that $f \circ p = g \circ t$, there is a unique morphism $u\colon H \to A$ such that $b \circ u = p$ and $c \circ u = t$. □

Definition 8 (Pullback construction [6] ⧉). Let $f\colon B \to D$ and $g\colon C \to D$ be graph morphisms. Then the following defines a graph A (see Fig. 1b), the pullback object of f and g:

1. $A = \{\langle x, y\rangle \in B \times C \mid f(x) = g(y)\}$ for nodes and edges, respectively
2. $s_A(\langle x, y\rangle) = \langle s_B(x), s_C(y)\rangle$ for $\langle x, y\rangle \in E_B \times E_C$
3. $l_A(\langle x, y\rangle) = l_B(x)$ for $\langle x, y\rangle \in V_B \times V_C$
4. $b\colon A \to B$ and $c\colon A \to C$ are defined by $b(\langle x, y\rangle) = x$ and $c(\langle x, y\rangle) = y$ □

We formalise the pullback construction in the `pullback_construction` locale, assuming morphisms $f\colon B \to D$ and $g\colon C \to D$. Our definition of the pullback object A is as follows:

```
abbreviation V where
  ‹V ≡ {(x,y). x ∈ V_B ∧ y ∈ V_C ∧ f_V x = g_V y}›
abbreviation E where
  ‹E ≡ {(x,y). x ∈ E_B ∧ y ∈ E_C ∧ f_E x = g_E y}›
fun s where ‹s (x,y) = (s_B x, s_C y)›
fun t where ‹t (x,y) = (t_B x, t_C y)›
fun l where ‹l (x,_) = l_B x›
fun m where ‹m (x,_) = m_B x›

definition A where
  ‹A ≡ (|nodes = V, edges = E, source = s, target = t
      , node_label = l, edge_label = m|)›
```

The next lemma shows that this construction leads to a graph that is indeed the pullback object of Fig. 1b.

Lemma 1 (Correctness of pullback construction ⌕). *Let* $f \colon B \to D$ *and* $g \colon C \to D$ *be graph morphisms and let graph* A *and graph morphisms* b *and* c *be defined as in Definition 8. Then the square in Fig. 1b is a pullback diagram.*

We use the `sublocale` command, instead of `interpretation`, to make these facts persistent in the current context via the `pb` identifier.

> `sublocale` *pb:* `pullback_diagram A B C D b c f g`

Similar to pushouts, pullbacks are unique up to isomorphism.

Theorem 2 (Uniqueness of pullbacks ⌕). *Let* $f \colon B \to D$ *and* $g \colon C \to D$ *together with* A *induce a pullback as depicted in Fig. 1b. A graph* H *together with morphisms* $p \colon H \to B$ *and* $t \colon H \to C$ *is a pullback of* f *and* g *if and only if there is an isomorphism* $u \colon H \to U$ *such that* $b \circ u = p$ *and* $c \circ u = t$.

The proof is dual to the uniqueness of the pushout (cf. Theorem 1) and we omit it due to space limitations. Essential properties for the forthcoming proofs in Sect. 4 and Sect. 5 are the composition and decomposition of pushouts and pullbacks.

Lemma 2 (Pushout/Pullback composition and decomposition). *Given the commutative diagram in Fig. 3, then the following statements are true:*

1. If (1) *and* (2) *are pushouts, so is* (1) + (2) ⌕
2. If (1) *and* (1) + (2) *are pushouts, so is* (2) ⌕
3. If (1) *and* (2) *are pullbacks, so is* (1) + (2) ⌕
4. If (2) *and* (1) + (2) *are pullbacks, so is* (1) ⌕   □

For example, item 3. of Lemma 2 is expressed in Isabelle/HOL as follows:

```
lemma pullback_composition:
  assumes
    1: <pullback_diagram A B C D f g g' f'> and
    2: <pullback_diagram B E D F e g' e'' e'>
  shows <pullback_diagram A E C F (e o→ f) g e'' (e' o→ f')>
```

The proof is analogous to [6, Fact 2.27]. In the next section, we show that direct derivations have a unique (up to isomorphism) result.

4   Uniqueness of Direct Derivations

The uniqueness of direct derivations is an important property when reasoning about rule applications. This section does not rely on the adhesiveness of the category of graphs, instead we base our proof on the characterisation of graph pushouts in [8]. Before stating the theorem, we introduce additional facts, mainly about pushouts and pullbacks, which are used within the proof of Theorem 4.

In general, pushouts along injective morphisms are also pullbacks.

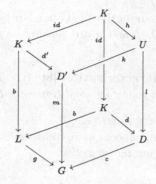

Fig. 4. Commutative cube based on direct derivation [15]

Lemma 3 (Injective pushouts are pullbacks [6] ⊡**).** *A pushout diagram as depicted in Fig. 1a is also a pullback if $A \to B$ is injective.*

The proof relies on the pullback construction (cf. Definition 8) and the fact that pullbacks are unique (cf. Theorem 2). Furthermore, pushouts and pullbacks preserve injectivity (surjectivity) in the sense that the opposite morphisms of the corresponding diagram (see Fig. 1) is also injective (surjective).

Lemma 4 (Preservation of injective and surjective morphisms [6] ⊡**).** *Given a pushout diagram in Fig. 1a, if $A \to B$ is injective (surjective), so is $C \to D$. Given a pullback diagram in Fig. 1b, if $C \to D$ is injective (surjective), so is $A \to B$.*

Certain forms of commutative diagrams give raise to pullbacks. This property is used in the proof of the uniqueness of the pushout complement (cf. Theorem 4).

Lemma 5 (Special pullbacks [6] ⊡**).** *The commutative diagram in Fig. 6 is a pullback if m is injective.*

Definition 9 (Reduced chain-condition [8]**).** The commutative diagram in Fig. 5 satisfies the *reduced chain-condition*, if for all $b' \in B$ and $c' \in C$ with $f(b') = g(c')$ there is $a \in A$ such that $b(a) = b'$ and $c(a) = c'$.

We show that pullbacks satisfy the reduced chain-condition.

Lemma 6 (Pullbacks satisfy the reduced chain-condition ⊡**).** *Each pullback diagram as depicted in Fig. 1b satisfies the reduced chain-condition.*

We state this lemma in Isabelle as follows:

```
lemma reduced_chain_condition_nodes:
  fixes x y
  assumes <x ∈ V_B> <y ∈ V_C> <f_V x = g_V y>
  shows <∃a ∈ V_A. (b_V a = x ∧ c_V a = y)>
```

Our proof relies on the pullback construction (cf. Definition 8) and the fact, that pullbacks are unique (cf. Theorem 2).

Definition 10 (Jointly surjective). Given injective graph morphisms $f: B \rightarrow D$ and $g: C \rightarrow D$. f and g are *jointly surjective*, if each item in D has a preimage in B or C.

Theorem 3 (Pushout characterization [8] ⌕**).** *The commutative diagram in Fig. 5 is a pushout, if the following conditions are true:*

1. *The morphisms b, c, f, g are injective.*
2. *The diagram satisfies the reduced chain-condition.*
3. *The morphisms g, f are jointly surjective.*

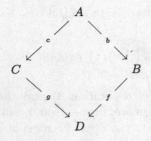

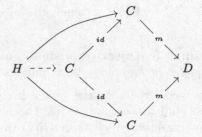

Fig. 5. Commutative diagram   **Fig. 6.** Special pullback diagram

The following theorem implies the uniqueness of the pushout complement, which is known to hold if the morphism $K \rightarrow L$ in the applied rule is injective, even if the matching morphism $L \rightarrow G$ is non-injective [22]. In our case, both morphisms are injective.

Theorem 4 (Uniqueness of direct derivations ⌕**).** *Let* $(1)+(2)$ *and* $(3)+(4)$ *be direct derivations as depicted in Fig. 9. Then* $D \cong D'$ *and* $H \cong H'$.

This theorem is stated in Isabelle/HOL within the locale `direct_derivation`. Within the `assumes` part, the second direct derivation, which we call `dd2`, is introduced.

```
theorem uniqueness_direct_derivation:
  assumes
    dd2: <direct_derivation r b b' G g D' d' m H' f' m'>
  shows <(∃u. bijective_morphism D D' u) ∧
         (∃u. bijective_morphism H H' u)>
```

The uniqueness proof of direct derivations (see Fig. 9) is performed in two phases. Firstly, we show the uniqueness of the pushout complement, which was first shown by Rosen [22]. Subsequently, we show that given a bijection between D and D', the pushout object is also unique up to isomorphism.

Proof (Theorem 4). The first phase of our proof closely follows Lack and Sobocin-ski [15], except for the final step, where the authors rely on adhesiveness. We finish the proof by relying on the pushout characterization (cf. Theorem 3). Given the two pushout diagrams (1) and (3) in Fig. 9 with injective $K \to D$ and $K \to D'$. To show the existence of a bijection between D and D', we construct the commutative cube in Fig. 4 with (1) as the bottom face and (3) as the front-left face, and show that l and k are bijections. For the latter, we show that the back-right and top faces are pushouts. (In [15], this is shown by adhesiveness, while we argue with the pushout characterization of Theorem 3.) The front-right face is a pullback construction (cf. Definition 8) which we tell Isabelle by interpretation of the `pullback_construction` locale.

```
interpret fr: pullback_construction D G D' c m ..
```

We use Isabelle's shorthand notation `..` for the standard tactic, to discharge the proof obligations which follow from the assumptions. Note that the pullback object together with the two morphisms is specified within the locale. Subsequent code will reference the pullback object by `fr.A`, the morphism l by `fr.b` and k by `fr.c` (see Fig. 4). (The identifiers within locales are given by the definition. As a result, the pullback object of the front-right face is referred to as `A` rather than the interpretation parameter K.) From Lemma 5, in our formalization referenced by `fun_algrtr_4_7_2`, we know that the back-left face is a pullback.

```
interpret bl: pullback_diagram "interf r" "interf r"
                "interf r" "lhs r" idM idM b b
  using fun_algrtr_4_7_2[OF r.k.injective_morphism_axioms]
  by assumption
```

To show that the back-right face is a pullback, we start with the front-left face. As the front left face is a pushout and m is injective, n' is too (cf. Lemma 4). Since pushouts along injective morphisms are also pullbacks (cf. Lemma 3), the front-left face is also a pullback. Using the pullback composition (cf. Lemma 2), the back face is pullback.

```
interpret backside: pullback_diagram "interf r" D' "interf r" G
                <d' o→ idM> idM m <g o→ b>
  using pullback_composition[OF bl.pullback_diagram_axioms
                          dd2.pb1.flip_diagram]
  by assumption
```

We define $h \colon K \to U$ using both, the d and d' morphisms as $h\,x = (d\,x, d'\,x)$ and subsequently prove the morphism properties.

```
define h where
  <h ≡ (|node_map = λv. (dV v, d'V v)
       ,edge_map = λe. (dE e, d'E e)|)>
```

We follow by showing that the top and bottom face commutes, i.e., $d' \circ id = k \circ h$ and $g \circ b = c \circ d$, respectively. This establishes the fact, that the right-side of the cube is a pullback. Using the pullback decomposition (cf. Lemma 2), the back-right face is a pullback. To approach the top-face, we start by showing it is a pullback, and subsequently it is also a pushout. Since m is injective, from Lemma 3 we know the bottom-face is also a pullback. Using the pullback composition (cf. Lemma 2), the bottom and back-left face is a pullback. By commutativity of the bottom face $g \circ b = c \circ d$ and back-right face $l \circ h = d \circ id$, the front-right and top face is a pullback. By the pullback decomposition (cf. Lemma 2) we can show that the top face is a pullback. We show this pullback is also a pushout by using the pushout characterization (cf. Theorem 3). Therefore, we need to show that h is injective, which follows from the construction above of h.

```
interpret h: injective_morphism "interf r" fr.A h
proof
  show <inj_on hV Vinterf r>
    using d_inj.inj_nodes
    by (simp add: h_def inj_on_def)
next
  show <inj_on hE Einterf r>
    using d_inj.inj_edges
    by (simp add: h_def inj_on_def)
qed
```

Joint surjectivity of k and d' follows from the pullback construction and the reduced-chain condition (cf. Lemma 9) of the front-left and top face. Note, that the reduced-chain condition holds for all pullbacks. Finally, we need to show that k and l are bijections. Since the top face is a pushout and the $C \to C$ morphism is a bijection, by Lemma 4, so is k.

```
interpret k_bij: bijective_morphism fr.A D' fr.c
  using top.b_bij_imp_g_bij[OF r.k.G.idm.bijective_morphism_axioms]
  by assumption
```

To show l is a bijection, we show that the back-right face is a pushout by using the pushout characterization. The bijectivity of l follows from the fact pushouts preserve bijections. We follow by defining the morphism $u: D \to D'$ as the composition of the l^{-1} and k. The inverse of l is obtained by using a lemma within our formalization of bijective morphisms, stating the existence of the inverse (ex_inv) using the obtain keyword:

```
obtain linv where linv:<bijective_morphism D fr.A linv>
  and <∧v. v ∈ VD⟹ fr.b o→linvV v = v>
    <∧e. e ∈ ED⟹ fr.b o→linvE e = e>
  and <∧v. v ∈ Vfr.A⟹ linv o→ fr.b V v = v>
    <∧e. e ∈ Efr.A⟹ linv o→ fr.b E e= e>
  by (metis l_bij.ex_inv)
```

We finish the first phase by defining the morphism $u: D \to D'$ as $k \circ l^{-1}$ and subsequently prove that morphism composition preserves bijections (using the already proven bij_comp_is_bij lemma).

```
define u where <u ≡ fr.c o→ linv>

interpret u: bijective_morphism D D' u
  using bij_comp_bij_is_bij[OF linv k_bij.bijective_morphism_axioms]
  by (simp add: u_def)
```

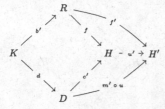

Fig. 7. Construction of u'

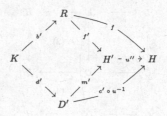

Fig. 8. Construction of u''

The second phase is to show the existence of an isomorphism $H \to H'$. We start by obtaining $u' : H \to H'$ and $u'' : H' \to H$ and show they are inverses.

We use the universal property of the pushout depicted in Fig. 7, which requires us to show commutativity: $f' \circ b' = m' \circ u \circ d$. So we substitute $u \circ d = d'$ into the commutativity equation of pushout (4) ($f' \circ b' = m' \circ d'$) in Fig. 9. We get $u \circ d = d'$ as follows: $u \circ d \overset{(1)}{=} k \circ l^{-1} \circ d \overset{(2)}{=} k \circ l^{-1} \circ l \circ h \overset{(3)}{=} k \circ h \overset{(4)}{=} d'$. Here, (1) is justified by the definition of u, (2) by the definition of l and h (which makes the back-right face in Fig. 4 commute), (3) by inverse cancellation, and finally (4) by the definitions of k and h (similarly to step (2)). We obtain $u'' : H' \to H$ by using the universal property of the pushout depicted in Fig. 8. We show the commutativity $f \circ b = c' \circ u^{-1} \circ d'$ by substituting $u^{-1} \circ d' = d$ into the commutativity equation of pushout (2) ($f \circ b' = c' \circ d$) in Fig. 9: $u^{-1} \circ d' \overset{(5)}{=} u^{-1} \circ u \circ d \overset{(6)}{=} d$. Here, (5) is justified by the above proven equation $u \circ d = d'$, and (6) follows from cancellation of inverses. The final steps are to show $u' \circ u'' = id$ and $u'' \circ u' = id$. To show the first equation, we start with $f' = u' \circ u'' \circ f'$ and $m' = u' \circ u'' \circ m'$, which we get from the definitions of u' and u''. Using the universal property of pushout (4) in Fig. 9 together with H', f', m', we conclude that the identity is the unique morphism $H' \to H'$ that makes the triangles commute. If $u' \circ u''$ makes the triangles commute as well, it is equal to the identity morphism. The first triangle commutes because $u' \circ u'' \circ f' \overset{(7)}{=} u' \circ f \overset{(8)}{=} f'$. Here, (7) and (8) are justified by the corresponding construction of u' and u'' (see the triangles in Fig. 7 and Fig. 8). For the second triangle, we start by using the commutativity of the bottom triangle in Fig. 7 and composing to the right with u^{-1}: $u' \circ c' \circ u^{-1} = m' \circ u \circ u^{-1}$. By cancellation of inverses we get $u' \circ c' \circ u^{-1} = m'$ and by substituting $c' \circ u^{-1}$ using the commutativity of the bottom triangle in Fig. 8, we prove that $u' \circ u'' \circ m' = m'$. Showing that $u'' \circ u' = id$ follows analogously and is omitted to save space. $\qquad \square$

With the uniqueness of direct derivations (cf. Theorem 4), we get the uniqueness of the pushout complement (Fig. 10).

Corollary 1 (Uniqueness of pushout complements ⤤). *Given a pushout as depicted in Fig. 1a where $A \to B$ is injective. Then the graph D is unique up to isomorphism.*

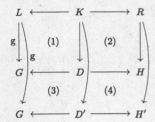

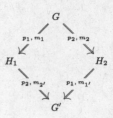

Fig. 9. Uniqueness of direct derivations   **Fig. 10.** Church-Rosser Theorem

We omit the proof to conserve space. The upcoming section introduces the so-called Church-Rosser Theorem, which states that parallell independent direct derivations have the diamond property.

5   Church-Rosser Theorem

The Church-Rosser Theorem refers to the idea that two graph transformation rules can be applied independently of each other, either sequentially or in parallel, without changing the final result. We follow the independence characterization of direct derivations given in [6].

Definition 11 (Parallel independence [6] ☑**).** The two direct derivations $G \Rightarrow_{p_1,m_1} H_1$ and $G \Rightarrow_{p_2,m_2} H_2$ in Fig. 11 are *parallel independent* if there exists morphisms $L_1 \to D_2$ and $L \to D_1$ such that $L_1 \to D_2 \to G = L_1 \to G$ and $L_2 \to D_1 \to G = L_2 \to G$.   □

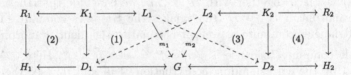

Fig. 11. Parallel independence

```
locale parallel_independence =
  p₁: direct_derivation r₁ b₁ b₁' G g₁ D₁ m₁ c₁ H₁ f₁ h₁ +
  p₂: direct_derivation r₂ b₂ b₂' G g₂ D₂ m₂ c₂ H₂ f₂ h₂
  for r₁ b₁ b₁' G g₁ D₁ m₁ c₁ H₁ f₁ h₁
      r₂ b₂ b₂' g₂ D₂ m₂ c₂ H₂ f₂ h₂ +
  assumes
    i: <∃i. morphism (lhs r₁) D₂ i
          ∧ (∀v ∈ V_lhs r₁· c₂ o_→ iV v = g₁V v)
          ∧ (∀e ∈ E_lhs r₁· c₂ o_→ iE e = g₁E e)> and
    j: <∃j. morphism (lhs r₂) D₁ j
          ∧ (∀v ∈ V_lhs r₂· c₁ o_→ jV v = g₂V v)
          ∧ (∀e ∈ E_lhs r₂· c₁ o_→ jE e = g₂E e)>
```

Theorem 5 (Church-Rosser Theorem [7] ⊡). *Given two parallel indepen-dent direct derivations $G \Rightarrow_{p_1,m_1} H_1$ and $G \Rightarrow_{p_2,m_2} H_2$, there is a graph G' together with direct derivations $H_1 \Rightarrow_{p_2,m'_2} G'$ and $H_2 \Rightarrow_{p_1,m'_1} G'$.*

Actually, we have shown more, namely that $G \Rightarrow_{p_1,m_1} H_1 \Rightarrow_{p_2,m'_2} G'$ and $G \Rightarrow_{p_2,m_2} H_2 \Rightarrow_{p_1,m'_1} G'$ are sequentially independent. We express this theo-rem in Isabelle/HOL within the `parallel_independence` locale as follows:

```
theorem (in parallel_independence) church_rosser:
   shows <∃g' D' m' c' H f' h' g'' D'' m'' c'' H f'' h''.
      direct_derivation r₂ b₂ b₂' H₂ g' D' m' c' H f' h'
    ∧ direct_derivation r₁ b₁ b₁' H₁ g'' D'' m'' c'' H f'' h''>
```

Proof (Theorem 5). We closely follow the original proof by Ehrig and Kre-owksi [7] where in a first stage, the pushouts (1)–(4) of Fig. 11 are vertically decomposed into pushouts (11) + (11), (21) + (22), (31) + (32), and (41) + (42), as depicted in Fig. 12. In a second stage, these pushouts are rearranged as in Fig. 13 and the new pushout (5) is constructed. Subsequently, we prove the two vertical pushouts (11) and (12). The pushouts (31) and (32) follow analogously and are not shown to conserve space.

We start by constructing the pullback (12) which we bind to the symbol `c12`, allowing later references, using our `pullback_construction` locale.

```
interpret "c12": pullback_construction D₁ G D₂ c₁ c₂ ..
```

The existence of $K_1 \to D$ follows from the universal property, and $D \to D_2$ from the construction of the pullback (12):

```
obtain j₁ where <morphism (interf r₁) c12.A j₁>
   and <∧v. v∈V interf r₁ ⟹ c12.b o→ j₁V v = m₁V v>
      <∧e. e∈E interf r₁ ⟹ c12.b o→ j₁E e = m₁E e>
   and <∧v. v∈V interf r₁ ⟹ c12.c o→ j₁V v = i₁ o→ b₁V v>
      <∧e. e∈E interf r₁ ⟹ c12.c o→ j₁E e = i₁ o→ b₁E e>
   using c12.pb.universal_property_exist_gen[OF p₁.r.k.G.graph_axioms
           wf_b₁i₁.morphism_axioms p₁.po1.c.morphism_axioms a b]
   by fast
```

From the fact that (1) = (11) + (12), we know (11) + (12) is a pushout and since $K_1 \to L_1$ is injective, it is also a pullback (cf. Lemma 3). By pullback decom-position (cf. Lemma 2), (11) is a pullback. We use the pushout characterization (cf. Theorem 3) to show it is also a pushout, which requires us to show injectivity of all morphisms, reduced-chain condition, and joint surjectivity of $D \to D_2$ and $L_1 \to D_2$. The injectivity of $K_1 \to L_1$ is given, $D \to D_2$ follows from pushout (1) and the injectivity of $K_1 \to L_1$ (cf. Lemma 4). To show the injectivity of $L_1 \to D_2$, we use the parallel independence (cf. Definition 11) $L_1 \to D_2 \to G = L_1 \to G$ and the injectivity of $L_1 \to G$. To show injectivity of $K_1 \to D$, we use the triangle $L_1 \to D_2 \to G = L_1 \to G$ obtained by the universal property of pullback (12) and the injectivity of both, $L_1 \to G$ and $D_2 \to G$. The reduced-chain condition follows by Lemma 6. To show the joint surjectivity of $D \to D_2$ and $L_1 \to D_2$ (that is each x in D_2 has a preimage in either D or L_1). Let y be the image of x in G. We apply the joint surjectivity of pushout (11) + (12) to y, that is y has a preimage in either D_1 or L_1. In the former case (y has a preimage z in D_1): from the pullback construction (cf. Definition 8), we get the common preimage of z and x in D which shows the former case. In the latter case, y has a preimage in L_1 via D_2. Since $D_2 \to G$ is injective, that preimage is mapped via x which means, x has a preimage in L_1. This shows the latter case.

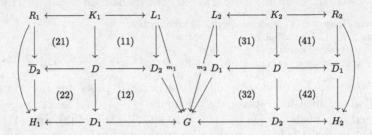

Fig. 12. Vertical pushout decomposition of Fig. 11

The pushouts (21), (41) are constructed using the **gluing** locale (see [24] for a detailed description).

```
interpret "c21": gluing "interf r1" c12.A "rhs r1" j1 b1' ..
interpret "c41": gluing "interf r2" c12.A "rhs r2" j2 b2' ..
```

The existence of $\overline{D}_2 \to H_1$ and $\overline{D}_1 \to H_2$ follows from the universal property of pushout (21) and (41), respectively. Pushouts (22) and (42) are obtained using the pushout decomposition (see Lemma 2). This finishes the first stage of the proof. The second stage rearranges the pushouts, as depicted in Fig. 13, such

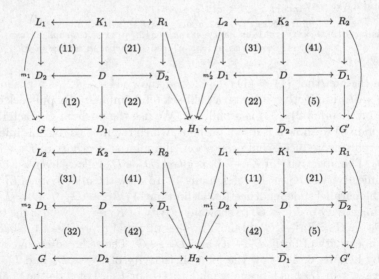

Fig. 13. Rearranged pushouts of Fig. 12

that we obtain two direct derivations $H_1 \Rightarrow_{p_2, m'_2} G'$ and $H_2 \Rightarrow_{p_1, m'_1} G'$. Here, we compose the pushouts from stage 1 (see Lemma 2). Exemplary, the pushouts (31) and (22) are composed in Isabelle/HOL.

```
interpret "31+22": pushout_diagram "interf r₂" c21.H "lhs r₂" H₁
                                   "c21.c o→ j₂" b₂ s₁ "h₁ o→ i₂"
    using pushout_composition[OF
        "31.flip_diagram" "22.flip_diagram" ] by assumption
```

The final pushout (5) is constructed and the pushouts are rearranged and vertically composed as depicted in Fig. 13. Isabelle is able to discharge the goal at this point automatically as we instantiated all required locales. This finishes the second stage of the proof.

6   Related Work

Isabelle/HOL was used by Strecker for interactive reasoning about graph transformations [25]. A major difference to our work is that he introduces a form of graph transformation that does not fit with any of the established approaches such as the double-pushout approach. As a consequence, his framework cannot draw on existing theory. Another difference is that [25] focuses on verifying first-order properties of some form of graph programs while the current paper is concerned with formalising and proving fundamental results of the DPO theory. Strecker's formalisation fixes node and edge identifiers as natural numbers, while we keep them abstract. Similar to our development, Isabelle's locale mechanism is employed.

Our formalisation of graphs follows the work of Noschinski [19], where records are used to group components and locales to enforce properties such as the well-formedness of graphs or morphisms. The main objective of [19] is to formalise and prove fundamental results of classical graph theory, such as Kuratowski's theorem.

da Costa Cavalheiro et al. [5] use the Event-B specification method and its associated theorem prover to reason about double-pushout and single-pushout graph transformations, where rules can have attributes and negative application conditions. Event-B is based on first-order logic and typed set theory. Different from our approach, [5] gives only a non-formalised proof for the equivalence between the abstract definition of pushouts and the set-theoretic construction. In contrast, we formalise both the abstract and the operational view and prove their correspondence using Isabelle/HOL. As Event-B is based on first-order logic, the properties that can be expressed and verified are quite limited. For it is known that non-local properties of finite graphs cannot be specified in first-order logic [17]. This restriction does not apply to our formalisation as we can make full use of higher-order logic.

7   Conclusion

In this paper, we formalise and prove in Isabelle/HOL two fundamental results in the theory of double-pushout graph transformation, namely the uniqueness of direct derivations and the so-called Church-Rosser theorem. Furthermore, we describe an approach to overcome the restriction of introducing new type variables within locale definitions.

Drawing on our experience so far, we plan to simplify the formalisation by reducing the number of type variables needed. The idea is to employ a single global type variable for both node and edge identifiers within graphs and morphisms. As a consequence, each graph in our commutative diagrams (such as pushouts or pullbacks) would share this ID-type whereas currently, each occurring graph may have a different ID-type. This would significantly decrease the number of parameters in our formalisation, and would also eliminate the need to work around Isabelle's restriction on introducing new type variables within locales. To implement this idea, we will need to revise our pushout and pullback constructions.

Our next objective is to extend the current approach to encompass attributed DPO graph transformation in the sense of [12]. Ultimately, we want to build a GP2 proof assistant within Isabelle/HOL that allows to interactively verify individual graph programs. Such a tool may use, for example, the proof calculus presented in [27].

Acknowledgements. We are grateful to Brian Courthoute and Annegret Habel for discussions on the topics of this paper.

References

1. Avigad, J., Donnelly, K., Gray, D., Raff, P.: A formally verified proof of the prime number theorem. ACM Trans. Comput. Log. **9**(1), 2 (2007). https://doi.org/10.1145/1297658.1297660
2. Ballarin, C.: Tutorial to locales and locale interpretation (2021). https://isabelle.in.tum.de/doc/locales.pdf
3. Brucker, A.D., Herzberg, M.: A formal semantics of the core DOM in Isabelle/HOL. In: Companion Proceedings of the The Web Conference 2018, WWW 2018, pp. 741–749. International World Wide Web Conferences Steering Committee (2018). https://doi.org/10.1145/3184558.3185980
4. Campbell, G., Courtehoute, B., Plump, D.: Fast rule-based graph programs. Sci. Comput. Program. **214** (2022). https://doi.org/10.1016/j.scico.2021.102727
5. da Costa Cavalheiro, S.A., Foss, L., Ribeiro, L.: Theorem proving graph grammars with attributes and negative application conditions. Theoret. Comput. Sci. **686**, 25–77 (2017). https://doi.org/10.1016/j.tcs.2017.04.010
6. Ehrig, H., Ehrig, K., Prange, U., Taentzer, G.: Fundamentals of Algebraic Graph Transformation. MTCSAES, Springer, Heidelberg (2006). https://doi.org/10.1007/3-540-31188-2
7. Ehrig, H., Kreowski, H.-J.: Parallelism of manipulations in multidimensional information structures. In: Mazurkiewicz, A. (ed.) MFCS 1976. LNCS, vol. 45, pp. 284–293. Springer, Heidelberg (1976). https://doi.org/10.1007/3-540-07854-1_188
8. Ehrig, H., Kreowski, H.J.: Pushout-properties: an analysis of gluing constructions for graphs. Math. Nachr. **91**, 135–149 (1979)

9. Gonthier, G.: The four colour theorem: engineering of a formal proof. In: Kapur, D. (ed.) ASCM 2007. LNCS (LNAI), vol. 5081, p. 333. Springer, Heidelberg (2008). https://doi.org/10.1007/978-3-540-87827-8_28

10. Habel, A., Müller, J., Plump, D.: Double-pushout graph transformation revisited. Math. Struct. Comput. Sci. **11**(5), 637–688 (2001). https://doi.org/10.17/S0960129501003425

11. Hales, T., et al.: A formal proof of the Kepler conjecture. In: Forum of Mathematics, Pi, vol 5 (2015). https://doi.org/10.1017/fmp.2017.1

12. Hristakiev, I., Plump, D.: Attributed graph transformation via rule schemata: Church-Rosser theorem. In: Milazzo, P., Varró, D., Wimmer, M. (eds.) STAF 2016. LNCS, vol. 9946, pp. 145–160. Springer, Cham (2016). https://doi.org/10.1007/978-3-319-50230-4_11

13. Huth, M., Ryan, M.D.: Logic in Computer Science - Modelling and Reasoning about Systems, 2nd edn. Cambridge University Press, Cambridge (2004)

14. Klein, G., et al.: seL4: formal verification of an OS kernel. In: Proceedings Symposium on Operating Systems Principles (SOSP 2009), pp. 207–220. Association for Computing Machinery (2009). https://doi.org/10.1145/1629575.1629596

15. Lack, S., Sobociński, P.: Adhesive categories. In: Walukiewicz, I. (ed.) FoSSaCS 2004. LNCS, vol. 2987, pp. 273–288. Springer, Heidelberg (2004). https://doi.org/10.1007/978-3-540-24727-2_20

16. Leroy, X.: Formal verification of a realistic compiler. Commun. ACM **52**(7), 107–115 (2009). https://doi.org/10.1145/1538788.1538814

17. Libkin, L.: Elements of Finite Model Theory. Texts in Theoretical Computer Science. Springer, Cham (2004). https://doi.org/10.1007/978-3-662-07003-1

18. Nipkow, T., Klein, G.: Concrete Semantics: with Isabelle/HOL. Springer, Cham (2014). https://doi.org/10.1007/978-3-319-10542-0. http://concrete-semantics.org/

19. Noschinski, L.: A graph library for Isabelle. Math. Comput. Sci. **9**(1), 23–39 (2014). https://doi.org/10.1007/s11786-014-0183-z

20. Paulson, L.C., Nipkow, T., Wenzel, M.: From LCF to Isabelle/HOL. Formal Aspects Comput. **31**(6), 675–698 (2019). https://doi.org/10.1007/s00165-019-00492-1

21. Plump, D.: Reasoning about graph programs. In: Proceedings with Terms and Graphs (TERMGRAPH 2016). Electronic Proceedings in Theoretical Computer Science, vol. 225, pp. 35–44 (2016). https://doi.org/10.4204/EPTCS.225.6

22. Rosen, B.K.: Deriving graphs from graphs by applying a production. Acta Informatica **4**, 337–357 (1975)

23. Schirmer, N., Wenzel, M.: State spaces - the locale way. In: Proceedings International Workshop on Systems Software Verification (SSV 2009). Electronic Notes in Theoretical Computer Science, vol. 254, pp. 161–179 (2009). https://doi.org/10.1016/j.entcs.2009.09.065

24. Söldner, R., Plump, D.: Towards Mechanised proofs in Double-Pushout graph transformation. In: Proceedings International Workshop on Graph Computation Models (GCM 2022). Electronic Proceedings in Theoretical Computer Science, vol. 374, pp. 59–75 (2022). https://doi.org/10.4204/EPTCS.374.6

25. Strecker, M.: Interactive and automated proofs for graph transformations. Math. Struct. Comput. Sci. **28**(8), 1333–1362 (2018). https://doi.org/10.1017/S096012951800021X

26. Wenzel, M.: Isar — a generic interpretative approach to readable formal proof documents. In: Bertot, Y., Dowek, G., Théry, L., Hirschowitz, A., Paulin, C. (eds.) TPHOLs 1999. LNCS, vol. 1690, pp. 167–183. Springer, Heidelberg (1999). https://doi.org/10.1007/3-540-48256-3_12

27. Wulandari, G.S., Plump, D.: Verifying graph programs with monadic second-order logic. In: Gadducci, F., Kehrer, T. (eds.) ICGT 2021. LNCS, vol. 12741, pp. 240–261. Springer, Cham (2021). https://doi.org/10.1007/978-3-030-78946-6_13

Application Domains

Formalisation, Abstraction and Refinement of Bond Graphs

Richard Banach[1]([⊠]) [iD] and John Baugh[2] [iD]

[1] Department of Computer Science, University of Manchester, Oxford Road,
Manchester M13 9PL, UK
richard.banach@manchester.ac.uk

[2] Department of Civil, Construction and Environmental Engineering, North Carolina State
University, Raleigh, NC 27695-7908, USA
jwb@ncsu.edu

Abstract. Bond graphs represent the structure and functionality of mechatronic systems from a power flow perspective. Unfortunately, presentations of bond graphs are replete with ambiguity, significantly impeding understanding. A formalisation of the essentials of bond graphs is given, together with a formalisation of bond graph transformation, which can directly express abstraction and refinement of bond graphs.

Keywords: Bond Graph · Formalisation · Abstraction · Refinement

1 Introduction

Bond graphs were introduced in the work of Paynter in 1959 [8,9]. These days the most authoritative presentation is [6]. From the large related literature we can cite [3,6,7].

Even the best presentations, though, are replete with ambiguity, often arising from a non-standard use of language, that leaves the reader who is more used to conventional parlance in engineering terminology, feeling insecure and confused. The topic is thus ripe for a reappraisal using the mathematical tools that bring precision to concepts in computer science. This paper introduces such a reformulation. For lack of space, only the essentials are covered, and the writing is rather terse. A fuller treatment is in [2].

The rest of the paper is as follows. Section 2 outlines the kind of physical theories that bond graphs are used for. Section 3 covers the essentials of bond graph structure. Section 4 shepherds the details in Sect. 3 to form the category $\mathcal{BGPatt}$, which we discuss briefly. Section 5 covers the essentials of bond graph transformation, relating this to $\mathcal{BGPatt}$. Section 6 shows how transformations can be used for abstraction and refinement of bond graphs. Section 7 concludes.

2 Classical Physical Theories for Classical Engineering

Bond graphs target the classical regime of physical theories well established at the end of the 19th century, and the engineering done using those theories. The usual remit of these includes mechanical, rotational, electrical, hydraulic, and thermal domains.

© The Author(s), under exclusive license to Springer Nature Switzerland AG 2023
M. Fernández and C. M. Poskitt (Eds.): ICGT 2023, LNCS 13961, pp. 145–162, 2023.
https://doi.org/10.1007/978-3-031-36709-0_8

Somewhat remarkably, the physical theories underpinning all these domains share a large degree of similarity, a fact exploited in the creation of the bond graph formalism. In this paper, for brevity, we briefly indicate the relevance to the mechanical and electrical domains. We axiomatise the common framework as follows.

[PT.1] A system consists of interconnected devices, and operates within an environment from which it is separated by a notional boundary. A system can input or output energy from the environment through specific devices. Aside from this, the system is assumed to be isolated from the environment.

[PT.2] The classical physics relevant to bond graphs is captured, in general, by a system of second order ordinary differential equations (ODE) of the form:

$$\Phi(q'', q', q) = e \tag{1}$$

Of most interest is the case where Φ is a linear constant coefficients (LCC) ODE:

$$L\frac{\mathrm{d}^2 q}{\mathrm{d}t^2} + R\frac{\mathrm{d}q}{\mathrm{d}t} + Kq = e \tag{2}$$

The system (1) or (2) concerns the behaviour of one (or more) generalised displacement(s), referred to as **gendis** with typical symbol q (*mech*: displacement; *elec*: charge). The gendis time derivative q' is called the **flow**, with typical symbol f (*mech*: velocity; *elec*: current). The gendis second time derivative q'' is called the generalised acceleration **genacc**, with typical symbol a (*mech*: acceleration; *elec*: induction). These all occur in the LHS of (1)–(2).

On the RHS of (1)–(2) is the **effort**, typical symbol e (*mech*: force; *elec*: voltage). The time integral (over a given time interval T) of the effort $\int_T e\,dt$ is called the generalised momentum **genmom** (accumulated over time T), with typical symbol p. The time derivative of the effort is normally of no interest.

[PT.3] Of particular importance among the variables mentioned is the product of effort and flow, because $e \times f$ is **power**, i.e., the rate at which energy is processed. The transfer and processing of power is crucial for the majority of engineered systems. According to **[PT.1]**, energy can only enter or exit a system through specific kinds of device. Therefore, all other devices conserve energy within the system.

[PT.4] Engineered systems are made by connecting relatively simple devices. We describe the most important ones now. Of particular utility are devices which are special cases of (2) that keep only one of the LHS terms.

> **Dissipator: R-device** (*mech*: dashpot; *elec*: resistor) $Rf = e$ (3)
>
> **Compliant: C-device** (*mech*: spring; *elec*: capacitor) $Kq = e$ (4)
>
> **Inertor: L-device** (*mech*: mass; *elec*: inductor) $La = e$ (5)

A dissipator is a device that can output energy into the environment in the form of heat. Compliants and inertors are devices that store energy. Specifically, the power they receive is accumulated within the device as stored energy, to be released back into the rest of the system later.

Sources input power to/from the system of interest in predefined ways.

> **Effort source: SE-device** (*mech*: force; *elec*: voltage) $e = \Phi_E(t)$ (6)
>
> **Flow source: SF-device** (*mech*: velocity; *elec*: current) $f = \Phi_F(t)$ (7)

Note that the power input and output to/from each of these cases is not determined by equations (6)–(7) alone (since the other variable is not specified), but by the behaviour of the rest of the system that they are connected to.

All the above devices are connected to the rest of the system via a single power connection, i.e., there is only one effort variable and one flow variable. Transformers and gyrators are devices that are connected to two power connections (two efforts and two flows), and allow non-trivial tradeoffs between the effort and the flow in the two connections.

Transformer: TR-device (*mech*: lever; *elec*: transformer)

$$e_1 = h\, e_2 \quad \text{and} \quad h\, f_1 = f_2 \tag{8}$$

Gyrator: GY-device (*mech*: gyroscope; *elec*: transducer)

$$e_1 = g\, f_2 \quad \text{and} \quad g\, f_1 = e_2 \tag{9}$$

Junctions are devices that distribute power among several power connections $1\ldots n$ (each with its own effort and flow), while neither storing nor dissipating energy. Aside from transformers and gyrators just discussed, the only remaining cases that arise are the common effort and common flow cases.

Common effort: E-device

(*mech*: common force; *elec*: common voltage, Kirchoff's Current Law)

$$e_1 = e_2 = \ldots = e_n \quad \text{and} \quad f_1 + f_2 + f_3 + \ldots + f_n = 0 \tag{10}$$

Common flow: F-device

(*mech*: common velocity; *elec*: common current, Kirchoff's Voltage Law)

$$e_1 + e_2 + e_3 + \ldots + e_n = 0 \quad \text{and} \quad f_1 = f_2 = \ldots = f_n \tag{11}$$

Noting that n is not fixed, **E** and **F** devices for different n are different devices.

[PT.5] From the bond graph perspective, the individual power connections to a device are conceptualised as power **port**s, through which power flows into or out of the device. Dissipators, compliants and inertors are therefore **one port** devices. Power sources are also one port devices. Transformers and gyrators are **two port** devices, while junctions are **three (or more) port** devices. For each category of device, all of its ports are individually labelled.

[PT.6] Since power is the product of an effort variable and a flow variable, each port is associated with an (effort, flow) variable pair whose values at any point in time define the power flowing through it.

[PT.7] All the variables involved in the description of a system are typed using a consistent system of dimensions and units. It is assumed that this typing is sufficiently finegrained that variables from different physical domains cannot have the same type. We do not have space to elaborate details, but since the only property of dimensions and units that we use is whether two instances are the same or not, it is sufficient to assume a set $\mathcal{DT} \times \mathcal{UT}$ of (dimension, unit) terms, that type the variables we need.

[PT.8] We refer to the elements of a system using a hierarchical naming convention. Thus, if Z-devices have ports p, then Z.p names the p ports of Z-devices. And if the effort variables of those ports are called e, then Z.$p.e$ names those effort variables.

Analogously, $Z.p.f$ would name the flow variables corresponding to $Z.p.e$. $Z.p.e.$DU names the dimensions and units of $Z.p.e$, while $Z.p.f.$DU names the dimensions and units of $Z.p.f$.

[PT.9] For every (effort, flow) variable pair in a system (belonging to a port p of device Z say), for example ($Z.p.e$, $Z.p.f$), there is a **directional indication** (determined by the physics of the device in question and the equations used to quantify its behaviour). This indicates whether the power given by the product $Z.p.e \times Z.p.f$ is flowing **in**to or **out**of the port when the value of the product is *positive*.

For the devices spoken of in **[PT.4]**, there is a standard assignment of **in/out** indicators to its ports. Thus, for **R, C, L** devices, the standard assignment to their single port is **in**. For **SE, SF** devices, the standard assignment to their single port is **out**. For **TR, GY** devices, the standard assignment is **in** for one port and **out** for the other, depicting positive power flowing through the device. For the **E** and **F** devices, we standardise on a symmetric **in** assignment to all the ports.

3   Bond Graph Basics

Bond graphs are graphs which codify the physical considerations listed above.

[UNDGR] An **undirected graph** is a pair (V, E) where V is a set of vertices, and E is a set of edges. There is a map ends : $E \rightarrow \mathbb{P}(V)$, where ($\forall edg \in E \bullet$ card(ends(edg)) $= 2$) holds, identifying the pair of *distinct* elements of V that any edge edg connects. When necessary, we identify the individual ends of an edge edg, where ends(edg) $= \{a, b\}$ using (a, edg) and (b, edg). If ends(edg) $= \{a, b\}$, then we say that edg is incident on a and b.

Our formulation of conventional power level bond graphs (DPLBGs, directed power level bond graphs) starts with PLBGs, which are undirected labelled graphs. It is important in the following to remember that the mathematical details are intended to follow, as closely as reasonable, the constraints that apply in the physical world. This prevents many appealing ways of formulating things mathematically from being applied, because they naturally force aspects of the model to be free, and decoupled from one another, whereas in the physical world, such freedom is not possible. PLBGs are assembled out of the following ingredients. Figure 1 illustrates the process.

[PLBG.1] There is an alphabet $\mathcal{VL} = \mathcal{BVL} \cup \mathcal{CVL}$ of vertex labels, with basic vertex labels $\mathcal{BVL} = \{\textbf{R, C, L, SE, SF, TR, GY, E, F}\}$, and user defined labels $\mathcal{CVL}$.

[PLBG.2] There is an alphabet $\mathcal{PL}$ of port labels and a map lab2pts : $\mathcal{VL} \rightarrow \mathbb{P}(\mathcal{PL})$, which maps each vertex element label to a set of port labels. (Below, we just say port, instead of port label, for brevity).

[PLBG.3] There are partial maps labpt2effDU , labpt2floDU : $\mathcal{VL} \times \mathcal{PL} \nrightarrow \mathcal{DT} \times \mathcal{UT}$ mapping each (vertex label, port) pair to the dimensions and units (not elaborated here) of the (forthcoming) effort and flow variables, and satisfying:

$$(lab, pt) \in \text{dom}(\text{labpt2effDU}) \Leftrightarrow pt \in \text{lab2pts}(lab) \tag{12}$$

$$(lab, pt) \in \text{dom}(\text{labpt2floDU}) \Leftrightarrow pt \in \text{lab2pts}(lab) \tag{13}$$

(a) **R**   (b) **R**.R   (c) **R**.R **in**

(d) *vr*.**R**   (e) *vr*.**R**.R   (f) *vr*.**R**.R **in**

(g) *vr*.**R**.R.*e*   (h) *vr*.**R**.R.*f*   (i) *vr*.**R**.R.*e* = *R* × *vr*.**R**.R.*f*

(j)

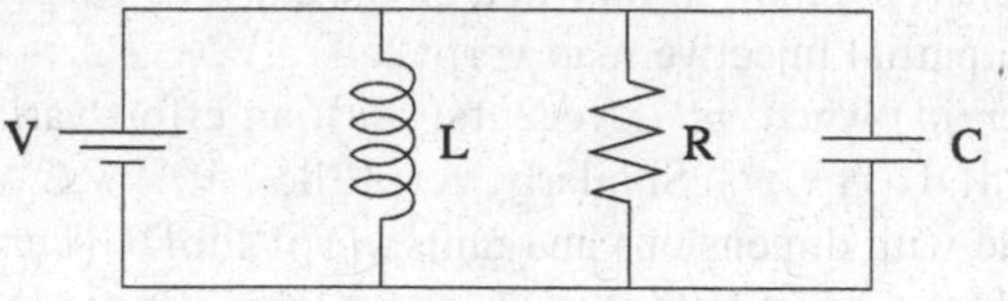

(k)

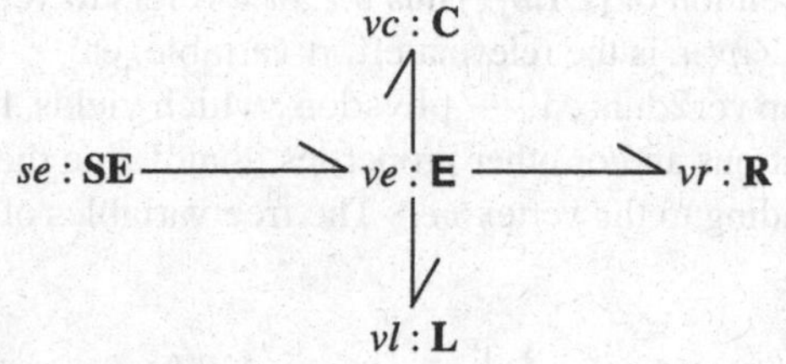

Fig. 1. Stages in bond graph construction: (a) a vertex label (for a dissipator); (b) adding a port; (c) adding a directional indicator; (d)–(f) assigning attributes (a)–(c) to a vertex *vr*; (g) *vr*'s effort variable; (h) *vr*'s flow variable; (i) *vr*'s constitutive equation; (j) a simple electrical circuit embodying a dissipator (among other components); (k) a bond graph of the circuit in (j). Dimensions and units are not shown.

[PLBG.4] There is an alphabet $\mathcal{IO} = \{$**in**, **out**$\}$ of standard directional indicators, and a partial map labpt2stdio : $\mathcal{VL} \times \mathcal{PL} \nrightarrow \mathcal{IO}$ satisfying:

$$(lab, pt) \in \mathrm{dom}(\mathrm{labpt2stdio}) \Leftrightarrow pt \in \mathrm{lab2pts}(lab) \tag{14}$$

The above clauses capture properties of PLBGs that are common to all vertices sharing the same label. Other properties are defined per vertex. PLBGs can now be constructed.

[PLBG.5] A power level bond graph PLBG is based on an undirected graph $BG = (V, E)$ as in **[UNDGR]**, together with additional machinery as follows.

[PLBG.6] There is a map ver2lab : $V \rightarrow \mathcal{VL}$, assigning each vertex a label.

When map ver2lab is composed with lab2pts, yielding map ver2pts = ver2lab $\S$ lab2pts : $V \rightarrow \mathbb{P}(\mathcal{PL})$, each vertex acquires a set of port labels.

When map ver2lab, with a choice of port, is composed with maps labpt2effDU and labpt2floDU, yielding maps verpt2effDU = ver2lab × **Id** $\S$ labpt2effDU : $V \times \mathcal{PL} \nrightarrow \mathcal{DT} \times \mathcal{UT}$ and verpt2floDU = ver2lab × **Id** $\S$ labpt2floDU : $V \times \mathcal{PL} \nrightarrow \mathcal{DT} \times \mathcal{UT}$, each (vertex, port) pair acquires dimensions and units for its effort and flow variables.

When map ver2lab, with a choice of port, is composed with map labpt2stdio, yielding partial map verpt2stdio = ver2lab × **Id** $\S$ labpt2io : $V \times \mathcal{PL} \nrightarrow \mathcal{IO}$, each (vertex, port) pair acquires its *standard* directional indicator.

[PLBG.7] In practice, and especially for **E**, **F** devices, directional indicators are often assigned per (vertex, port) pair rather than generically per (vertex label, port).

Thus there is a partial map verpt2io : $V \times \mathcal{PL} \rightarrowtail \mathcal{IO}$ satisfying

$$(ver, pt) \in \text{dom(verpt2io)} \Leftrightarrow pt \in \text{ver2pts}(ver) \tag{15}$$

and verpt2io(ver, pt) may, or may not, be the same as verpt2stdio(ver, pt).

There is a partial injective map verpt2eff : $V \times \mathcal{PL} \rightarrowtail \mathcal{PV}$ giving each (vertex, port) pair (ver, pt) where $pt \in \text{ver2pts}(ver)$, an effort variable with dimensions and units verpt2effDU(ver, pt). Similarly, verpt2flo : $V \times \mathcal{PL} \rightarrowtail \mathcal{PV}$ gives each (ver, pt) a flow variable with dimensions and units verpt2floDU(ver, pt). Also, we must have ran(verpt2eff) $\cap$ ran(verpt2flo) $= \varnothing$. These variables are referred to by extending the hierarchical convention of [PT.8]. Thus $v.Z.pt.e$ refers to vertex v, with label Z, having port pt, and so $v.Z.pt.e$ is the relevant effort variable, etc.

There is a map ver2defs : $V \rightarrow$ physdefs, which yields, for each vertex ver, a set of constitutive equations and/or other properties, that define the physical behaviour of the device corresponding to the vertex ver. The free variables of the properties in ver2defs satisfy:

$$\mathbf{FV}(\text{ver2defs}(ver)) \subseteq \bigcup_{pt \in \text{ver2pt}(ver)} (\text{verpt2eff}(ver, pt) \cup \text{verpt2flo}(ver, pt)) \tag{16}$$

Additionally, the properties in ver2defs(ver) can depend on generic parameters (from a set $\mathcal{PP}$ say), so there is a map ver2pars : $V \rightarrow \mathcal{PP}$ which satisfies:

$$\text{ver2pars}(ver) = \mathbf{Pars}(\text{ver2defs}(ver)) \tag{17}$$

Additionally, the properties in ver2defs(ver) can depend on some bound variables. When necessary, we refer to such variables using $\mathbf{BV}(\text{ver2defs}(ver))$.

[PLBG.8] There is a bijection Eend2verpt : $V \times E \rightarrowtail V \times \mathcal{PL}$, from edge ends in BG, to port occurrences:

$$(ver, edg) \in \text{dom(Eend2verpt)} \Leftrightarrow edg \text{ is incident on } ver \tag{18}$$

$$(ver, pt) \in \text{ran(Eend2verpt)} \Leftrightarrow pt \in \text{ver2pts}(ver) \tag{19}$$

$$(\forall ver \bullet ver \in \text{ends}(edg_1) \wedge ver \in \text{ends}(edg_2) \wedge edg_1 \neq edg_2 \Rightarrow$$
$$\text{Eend2verpt}(ver, edg_1) \neq \text{Eend2verpt}(ver, edg_2)) \tag{20}$$

For each edge $edg \in E$, where ends$(edg) = \{a, b\}$, the effort and flow variables at the ends of edg, have the same dimensions and units:

$$\text{verpt2effDU}(\text{Eend2verpt}(a, edg)) = \text{verpt2effDU}(\text{Eend2verpt}(b, edg)) \tag{21}$$

$$\text{verpt2floDU}(\text{Eend2verpt}(a, edg)) = \text{verpt2floDU}(\text{Eend2verpt}(b, edg)) \tag{22}$$

[PLBG.9] There is a map edge2dir : $E \rightarrow$ physdir, where physdir is a set of equalities and antiequalities between effort and flow variables, and such that for all edges $edg \in E$ (where ends$(edg) = \{a, b\}$):

$$\text{verpt2io}(\text{Eend2verpt}(a, edg)) \neq \text{verpt2io}(\text{Eend2verpt}(b, edg)) \Rightarrow$$
$$\text{edge2dir}(edg) = \{$$
$$\text{verpt2eff}(\text{Eend2verpt}(a, edg)) = \text{verpt2eff}(\text{Eend2verpt}(b, edg)),$$
$$\text{verpt2flo}(\text{Eend2verpt}(a, edg)) = \text{verpt2flo}(\text{Eend2verpt}(b, edg)) \} \qquad (23)$$

and

$$\text{verpt2io}(\text{Eend2verpt}(a, edg)) = \text{verpt2io}(\text{Eend2verpt}(b, edg)) \Rightarrow$$
$$\text{edge2dir}(edg) = \{$$
$$\text{verpt2eff}(\text{Eend2verpt}(a, edg)) = -\text{verpt2eff}(\text{Eend2verpt}(b, edg)),$$
$$\text{verpt2flo}(\text{Eend2verpt}(a, edg)) = \text{verpt2flo}(\text{Eend2verpt}(b, edg)) \} \qquad (24)$$

or, *the same with the minus sign between the flow variables*

The dynamics specified by a PLBG is the family of solutions to the collection of constraints specified by ver2defs (and ver2pars, edge2dir).

[PLBG.10] A PLBG is a DPLBG (directed PLBG, as in the literature) iff for each $edg \in E$, only case (23) is relevant. In such cases, edges become harpoons (half-arrows), showing the direction of positive power flow. In any case, the edges are called **bonds**.

A consequence of a unidirectional convention for variables along edges is that it permits the use of directed (rather than undirected) graphs as the underlying formalism. Although this makes the handling of edge ends a little easier, the impediments to bottom up bond graph construction that it imposes dissuaded us from following this approach.

4   The Category of Bond Graph Patterns $\mathcal{BGP}att$

The formulation of abstract bond graphs in Sect. 3 was extremely operational. In this section we show how these details may be shepherded into a structure within which we can discern a category, $\mathcal{BGP}att$, of bond graph patterns and morphisms. The extent to which this can be used as a basis for bond graph transformation will be discussed at the end of this section and in the next. We start with a familiar caveat on graph isomorphism.

[GRISO] Combinatorial graphs (e.g. bond graphs) have vertices and edges. The vertices and edges, in themselves, have no properties save their own identity, unless endowed with properties using, e.g., maps, such as appear in **[PLBG.6]-[PLBG.10]**. Starting with a graph G, and then manipulating it in different ways, may result in technically non-identical graphs, even when the intention is to arrive at 'the same' result. The different results will be isomorphic, though not identical. Writing the needed isomorphisms explicitly gets very tedious, so we will use the phrase 'the same' below, to indicate that we are suppressing these details. Similar observations apply to 'a copy' of part of the RHS of a transformation rule during rule application.

[BGP.1] A vertex label **Any** is introduced. It is a one port label, with an anonymous port '−'.

Any does not correspond to any physical device, but serves to label the vertex at the end of an edge at the periphery of a pattern (defined below) that is to be matched to a vertex (of a bond graph which is to be transformed).

Since **Any**-labelled vertices do not correspond to physical devices, they do not need all the attributes of normal vertices. They just need attributes for dimensions and units, and directional indication. These are given by extending the domains of the maps verpt2effDU, verpt2floDU, verpt2io, as needed, to the **Any**-labelled vertices.

[**BGP.2**] A **pattern** is a bond graph that contains zero or more **Any**-labelled vertices (along with their reduced set of attributes, as given in [**BGP.1**]).

Thus a bond graph is always a pattern, but a pattern is not a bond graph (i.e. a PLBG or a DPLBG) if it has one or more **Any**-labelled vertices.

[$\mathcal{BGP}att$.**OBJ**] The objects of the category $\mathcal{BGP}att$ are the patterns of [**BGP.2**].

Let P be a pattern. We define $Anys_P$ to be the set of all **Any**-labelled vertices of P, and $nonAnys_P$ to be the set of all vertices of P other than **Any**-labelled vertices. We define AAs_P to be the set of all edges of P between two $Anys_P$ vertices, NAs_P to be the set of all edges of P between a $nonAnys_P$ vertex and an $Anys_P$ vertex, and NNs_P to be the set of all edges of P between two $nonAnys_P$ vertices.

[**BGP.3**] Let A and B be patterns. We use A and B subscripts to label the individual technical ingredients from Sect. 3 belonging to A and B. A **matching** m of A to B, written $m : A \rightarrow B$ consists of: a map $m_V : V_A \rightarrow V_B$ on vertices; and an injective map $m_E : E_A \rightarrowtail E_B$ on edges. From m_E, a further injective map m_{ends} on edge ends is derived. These are all required to satisfy injective and homomorphic conditions:

$$m_E, m_{\text{ends}} \text{ are 1-1}, m_V \text{ need not be 1-1} \tag{25}$$

$$\text{ends}(edg) = \{a, b\} \Rightarrow \text{ends}(m_E(edg)) = \{m_V(a), m_V(b)\} \tag{26}$$

$$edg \text{ is incident on } a \Rightarrow m_{\text{ends}}(a, edg) = (m_V(a), m_E(edg)) \tag{27}$$

The injectivity on edges reflects the fact that physical devices have fixed numbers of connections, which each need to be connected to something for the device to function.

We further require that the bond graph properties of $nonAnys_A$ vertices and edges of A are mirrored in B (i.e., labels, definitions (which we assume to include identity of free variables and parameters), ports (and their effort and flow variables, their dimensions and units, and their directional indicators)):

$$\text{ver2lab}(a) \neq \textbf{Any} \Rightarrow$$

$$[\text{ver2lab}_A(a) = \text{ver2lab}_B(m_V(a))] \wedge \tag{28}$$

$$[\text{ver2defs}_A(a) = \text{ver2defs}_B(m_V(a)] \wedge \tag{29}$$

$$[\text{ver2pts}_A(a)) = \text{ver2pts}_B(m_V(a))] \wedge \tag{30}$$

$$[pt \in \text{ver2pts}_A(a) \Rightarrow$$

$$[\text{verpt2effDU}_A(a, pt) = \text{verpt2effDU}_B(m_V(a), pt)] \wedge \tag{31}$$

$$[\text{verpt2eff}_A(a, pt) = \text{verpt2eff}_B(m_V(a), pt)] \wedge \tag{32}$$

$$[\text{verpt2floDU}_A(a, pt) = \text{verpt2floDU}_B(m_V(a), pt)] \wedge \tag{33}$$

$$[\text{verpt2flo}_A(a, pt) = \text{verpt2flo}_B(m_V(a), pt)] \wedge \tag{34}$$

$$[\text{verpt2io}_A(a, pt) = \text{verpt2io}_B(m_V(a), pt)]] \tag{35}$$

We also require that the bond graph properties of $Anys_A$ vertices and edges of A are mirrored in B in line with their reduced attributes:

ver2lab$(a) = $ **Any** $\Rightarrow$

$[\,\text{verpt2effDU}_A(a, -) = \text{verpt2effDU}_B(m_V(a), pt)\,] \land$ (36)

$[\,\text{verpt2floDU}_A(a, -) = \text{verpt2floDU}_B(m_V(a), pt)\,] \land$ (37)

$[\,\text{verpt2io}_A(a, -) = \text{verpt2io}_B(m_V(a), pt)\,]$ (38)

where: edg is incident on a, and Eend2verpt$(m_V(a), m_E(edg)) = (m_V(a), pt)$

[$\mathcal{BGP}att$.MOR] The morphisms of the category $\mathcal{BGP}att$ are the matchings of **[BGP.3]**.

Theorem 1. $\mathcal{BGP}att$ *is a category, as claimed.*

Proof Sketch: The morphisms of $\mathcal{BGP}att$ are conventional homomorphisms of labelled graphs, but restricted in a number of ways. In particular, edges and edge ends are mapped injectively. This means that only $Anys$ vertices may map many-1 (provided their edge ends map to distinct edge ends of the target). The various labelling attributes of vertices map identically for $nonAnys$ vertices, and for $Anys$ vertices the much smaller set of labelling attributes that matter, also map identically. So morphisms are injective in all respects save the $Anys$ vertex map.

Isomorphisms can thus be bijective homomorphisms of labelled graphs that preserve all the attributes, with obvious identities. Since morphisms are constructed from functions (on vertices and edges (and their ends)) and equalities of attributes, associativity follows immediately. $\Box$

We observe that a large number of the conditions (25)–(38) are actually independent from one another. This creates scope for defining many minor variants on the notion of morphism, by removing one or more of these conditions, provided that none of the conditions which remain, are dependent on the removed ones. Every such variant gives rise to 'a different category, even if they all share the same objects.

We further observe that while $\mathcal{BGP}att$ is a category of concrete graphs, we can easily create an analogous category of abstract graphs by taking its objects as those of $\mathcal{BGP}att$, *up to isomorphism* (where the isomorphisms are understood to be those of $\mathcal{BGP}att$), with the analogous adaptation of the morphisms.

4.1   Commentary

Having covered the essentials of the $\mathcal{BGP}att$ category above, we reflect on the 'design choices' made, with an eye to using these insights in the construction of a notion of rule based bond graph transformation in the next section.

In particular, the high degree of injectivity demanded of the morphisms deserves comment. It arises from the strong coupling between a vertex's label and its permitted set of edges, via the functional dependence of the number and characteristics of the ports associated with that label. Such strong constraints are needed because the vertices represent actual physical devices and the edges represent actual physical connections.

Physical devices cannot acquire new physical connections or lose existing ones on a whim, which would be possible if there were no coupling. This coupling between a vertex and its permitted set of edges (via its label) is in stark contrast to the usual situation in graph transformation formalisms, in which the properties of the two are decoupled, this being exploited during transformation. We see the impact of this below.

5   Bond Graph Transformation

In practice, bond graphs are often simplified, transformed, or rewritten, in various ways. In the existing literature, this activity is always described informally. In this section, we address the transformation of bond graphs from a more formal perspective, benefiting from the categorical formulation of the previous section.

5.1   Rule Based Bond Graph Transformation

In Sect. 3, **[PLBG.1]**-**[PLBG.10]** provided the mechanics for constructing bond graphs. This was achieved using maps that took each vertex to its various components, e.g. label, ports, etc. In this section, bond graphs are transformed by applying transformation **rule**s, which are templates for how an actual bond graph may be changed in the region of a **matching of the rule** to a **match**, or **redex**. This draws on the machinery of $\mathcal{BGP}att$ from Sect. 4.

Rule based graph transformation has an extensive literature. The procedure we will describe is adapted from the so-called double pushout and single pushout constructions comprehensively presented in [4] and [5] and in work cited therein, as well as in more recent publications. However, our approach will be considerably simpler, so we will not need most of the machinery discussed there. Partly, this is because we can work at a lower level of abstract technically, and partly it is because there are fundamental obstacles to applying the standard approaches *verbatim* in this application domain. We axiomatise what we need as follows.

[BGTR.1] A bond graph transformation rule (L, p, R) (**rule** for short) is given, firstly, by a pair of patterns L and R, where L is called the left hand side (LHS) and R is called the right hand side (RHS). Secondly, there is a bijection $p : Anys_L \rightarrowtail\!\!\!\rightarrow Anys_R$, for which (36)–(38) hold (in which the matching m is p and the port pt is the anonymous port of $p_V(a)$, as needed).

Lemma 1. *Let (L, p, R) be a rule. Then:*

1. *p extends to a bijection between edge ends incident on Anys vertices.*
2. *For every AAs_L edge in L, either there is a unique AAs_R edge (where the respective ends correspond via p), or there are two NAs_R edges (where the $Anys_L$ and $Anys_R$ ends correspond via p).*
3. *For every AAs_R edge in R, either there is a unique AAs_L edge (where the respective ends correspond via p), or there are two NAs_L edges (where the $Anys_L$ and $Anys_R$ ends correspond via p).*
4. *There is a bijection between the $Anys_L$ vertices (and their incident edges/ends) that are not in the scope of either 1 or 2, and the $Anys_R$ vertices (and their incident edges/ends) that are not in the scope of either 1 or 2.*

Proof: This is a straightforward consequence of the bijective nature of p, and of the fact that $Anys$ vertices have a single edge end, extending p to a bijection between $Anys$ edge ends. The rest follows from the fact that each edge has exactly two ends.   □

[BGTR.2] Let (L, p, R) be a rule. Let G be a bond graph, and $m : L \rightarrow G$ be a matching. The **application** of the rule (L, p, R) at the matching m is the result of the following steps:

1. Remove from G, all $m_E(E_L)$ edges, and all $m_V(nonAnys_L)$ vertices. This creates D (which will not be a valid bond graph, since, in general, it will have ports (belonging to $m_V(Anys_L)$ vertices) that do not correspond to edge ends in D).
2. Add to D:
   (a) a copy[1] of the vertices in $nonAnys_R$;
   (b) a copy of the edges in R;
   (c) If $edgC$ is the copy of an edge edg of R, and if edg has edge end (v, edg) in R, let $(vC, edgC)$ be the edge end corresponding to (v, edg) where:
      (i) if $v \in nonAnys_R$ then vC is its copy;
      (ii) if $v \in Anys_R$ then $vC = m_V(p^{-1}(v))$.
   Call the resulting graph H.
3. Endow H with all the needed attributes to make it a PLBG (labels, ports, variables, definitions, directional indicators) by: (a) retaining the existing attributes inherited via $m(L)$ for D; (b) replicating the attributes from the edges and $nonAnys_R$ vertices to their copies in H.

Below, to shortcut the rather inconvenient language of 2.(c), we will say that 'the NAs_L edges have their $nonAnys_L$ end **redirected**' (to their destinations in R), and 'the NAs_L *matched* edges have their $nonAnys_L$ *matched* end **redirected**' (to their destinations in H), or similar language.

Note that, by Lemma 1, the 'other end' of an AAs_L edge is also an $Anys_L$ incident edge. Therefore, unless such an edge is mapped by p to a similar AAs_R edge, such an edge is redirected to two edges in R, which is reflected in H. Analogously, two NAs_L edges may both be redirected to the same AAs_R edge, similarly reflected in H.

Remark. We observe that the rule application described in **[BGTR.2]** definitely does not fall under the canonical framework of the double pushout approach (despite superficial appearances), since the entity D created in step 1 is not an object of $\mathcal{BGPatt}$. There are two relatively self-evident reactions to this state of affairs. (1) Simply accept things as they are, namely, accept that in this case, stepping out of the category $\mathcal{BGPatt}$ is necessary in order to define the required transformations. (2) Attempt to modify the definition of $\mathcal{BGPatt}$ so that it can accommodate the intermediate entities D, thereby placing the construction back in the legitimate double pushout approach.

Neither option is entirely satisfactory. (1) has the evident defect that it steps outside the clean categorical framework of the double pushout approach. Nevertheless, as

---

[1] When we say 'copy' of some graph theoretic concept, e.g., a set of vertices or a set of edges, we mean a distinct set of the same cardinality as the original, and endowed with attributes equivalent to the original, c.f. **[GRISO]**. In its turn, 'distinct' means having no element in common with any similar entity in the discourse.

argued in Sect. 4.1, the category it uses, $\mathcal{BGP}att$, captures the right properties for the given application domain. (2) has the merit of remaining within the double pushout approach. But, in breaking the link between a vertex label with its set of ports on the one hand, and the set of edges incident on a vertex carrying that label on the other hand —as must happen if D is to be a legitimate category object— we fatally undermine the suitability of such a category to capture the properties needed in the application domain.

Another possibility suggests itself: (3) Attempt to modify the actual definition of graph transformation, by involving another category besides $\mathcal{BGP}att$ (e.g. a modification of $\mathcal{BGP}att$ as envisaged in (2)), thus arriving at a more complex, and novel, kind of transformation formalism. This is certainly possible, e.g. an approach via opfibrations [1], but this is well beyond the scope of the present paper.

Proposition 1. *The rule application described in* **[BGTR.2]** *yields a legitimate PLBG* H.

Proof Sketch: By Lemma 1, p extends to a bijection between $Anys_L$ incident edge ends and $Anys_R$ incident edge ends. This can be composed at the L side with the injective m_{ends} and on the R side with the bijection between the corresponding $Anys_R$ incident edge ends and their copies in H. So there is a bijection between the images of the $Anys_L$ incident edge ends and the images of the $Anys_R$ incident edge ends. Since the rest of H is either pre-existing bond graph structure in D, or a copy of bond graph structure in R, the correctness of the combinatorial graph structure of H follows.

It remains to check that the dimensions, units, and directional indicators at the two ends of an edge in the PLBG H match suitably. They will do so in the part of D not affected by the removal of L. They will also do so in pattern R, a copy of which is added to D. This leaves the redirection mechanism to be checked. However, the constraints on matching, i.e. (36)–(38), and the analogues of those constraints demanded of p, ensure that like is replaced with like during the redirection, so that the needed properties hold.

The only remaining issue is the possible non-injectivity of m_V. However, all the properties we need are properties of edge ends, not of vertices, so the location of the edge ends is not germane, and any non-injectivity of m_V does not affect them.   □

Beyond the preceding, there is evidently a matching $m^{\mathrm{R}} : R \rightarrow H$, together with a bijection $q : m_V(Anys_L) \rightarrowtail m_V^{\mathrm{R}}(Anys_R)$, with similar properties to p.

The bijectivity of p implies that for every rule (L, p, R), there is an inverse rule (R, p^{-1}, L). And existence of the matching m^{R} implies that for every application of (L, p, R) to G using m to yield H, to which R can be matched using m^{R}, there is an application of (R, p^{-1}, L) to H using m^{R} to yield G, to which L can be matched using m.

[SOUND] The rule application described in **[BGTR.2]** for (L, p, R) and m is **sound** iff the family of solutions to the PLBG H, restricted to the variables at the ports of $m_V^{\mathrm{R}}(Anys_R)$, is contained in the family of solutions to the PLBG G, restricted to the variables at the ports of $m_V(Anys_L)$.

[COMP] The rule application described in **[BGTR.2]** for (L, p, R) and m is **complete** iff the family of solutions to the PLBG G, restricted to the variables at the ports of $m_V(Anys_L)$, is contained in the family of solutions to the PLBG H restricted to the variables at the ports of $m_V^{\mathrm{R}}(Anys_R)$.

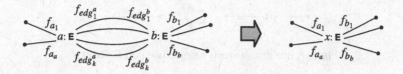

Fig. 2. A rule that shrinks a pattern whose *nonAnys* form a two vertex connected **E**-graph. The
• vertices are the **Any**-labelled vertices, with obvious bijection p.

[UPATH] A **path** in an undirected graph is a sequence of vertices, such that each
consecutive pair is the pair of edge ends of an edge. If needed, the edges in question can
be included in the path data.

[UCONN] A graph or pattern is **connected** iff for any two vertices, there is a path
between them.

[E-GR F-GR EF-GR] A pattern is an **E**-graph iff all its *nonAnys* vertices are **E**-
labelled. Similarly for an **F**-graph. A pattern is an **EF**-graph iff each *nonAnys* vertex
is either **E**-labelled, or **F**-labelled.

Theorem 2. *Let (L, p, R) be a rule in which L is a connected **E**-graph, and R is an*
***E**-graph with a single nonAnys vertex. Then any application of (L, p, R) is sound and*
complete. If L is not connected, then application is complete, but not necessarily sound.
*Similarly if L is an **F**-graph.*

Proof: We consider the special case in which L is an **E**-graph with two *nonAnys* vertices a, b, with one or more edges $\{edg_1 \ldots edg_k\}$ connecting them. See Fig. 2. Suppose
a is connected to **Any**-labelled vertices $\{a_1 \ldots a_a\}$ via suitable edges, and b is connected to **Any**-labelled vertices $\{b_1 \ldots b_b\}$ via suitable edges. Then a will have flow
variables $\{f_{edg_1^a} \ldots f_{edg_k^a}, f_{a_1} \ldots f_{a_a}\}$ at its ports connected to its edges, and b will
have flow variables $\{f_{edg_1^b} \ldots f_{edg_k^b}, f_{b_1} \ldots f_{b_b}\}$ at its ports connected to its edges.

Assuming the standard directional indicators for **E**-devices, i.e., **in**, the behaviours
at the two vertices matched to a, b in an application of (L, p, R) will be a copy of:

$$-f_{edg_1^a} - \ldots - f_{edg_k^a} = f_{a_1} + \ldots + f_{a_a} \tag{39}$$

$$-f_{edg_1^a} = f_{edg_1^b} \tag{40}$$

$$\ldots \ldots$$

$$-f_{edg_k^a} = f_{edg_k^b} \tag{41}$$

$$-f_{edg_1^b} - \ldots - f_{edg_k^b} = f_{b_1} + \ldots + f_{b_b} \tag{42}$$

Substituting (40)–(41) into (39), and adding the result to (42), yields:

$$0 = f_{a_1} + \ldots + f_{a_a} + f_{b_1} + \ldots + f_{b_b} \tag{43}$$

This, together with the constant effort condition, specifies the behaviour of G at the
ports matched to the **Any** ports of L. But the same conditions result from replacing the
vertices matched to a, b with a single vertex x say, and redirecting all edges incident on

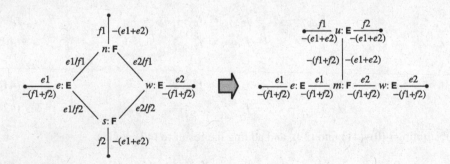

Fig. 3. A rule that shortcuts a single two port **E**-vertex.

the images of a, b to x. But this is what application of the rule (L, p, R) achieves. So application of (L, p, R) is complete.

But the procedure described can be reversed by considering the inverse rule (R, p^{-1}, L). If R is the LHS, its single *nonAnys* vertex imposes a condition like (43). The flow variables in the right of (43) can then be partitioned into two sets, corresponding to the two *nonAnys* vertices of L, and the incident edges can be redirected to these two vertices. Since the two vertices are connected by one or more edges in L, corresponding edges are created in the result PLBG, and a system of equations like (39)–(42) results, leading to soundness of application of (L, p, R).

If L is a connected **E**-graph with more than two *nonAnys* vertices, then the edges between them can be contracted one at a time following the above procedure, except that some of the $\{f_{a_1} \cdots f_{a_a}, f_{b_1} \cdots f_{b_b}\}$ flow variables will typically belong to ports to other *nonAnys* vertices, rather than to *Anys* vertices exclusively. This does not undermine the argument. The soundness and completeness properties of successive steps compose to give soundness and completeness for the application of (L, p, R) in its entirety.

If L is not connected, then following the above procedure will, at some point, fuse two vertices which are not connected by any edge. We will then have analogues of (39) and (42) (with 0 on the left), but none of (40)–(41). Absent $\{f_{edg_1^a} \cdots f_{edg_k^a}, f_{edg_1^b} \cdots f_{edg_k^b}\}$, (39) and (42) imply (43), but not *vice versa*, leading to the failure of soundness.   □

Corollary 1. *Let (L, p, R) be a rule where L has two **Any**-labelled vertices and a single **E**-labelled vertex, connected together by two edges, and where R has two **Any**-labelled vertices only, and a single edge connecting them, i.e. a single AAs_R edge.*

Fig. 4. A rule that optimises a shape that arises in bond graph construction from schematics.

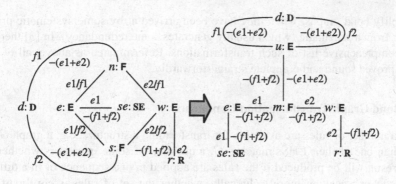

Fig. 5. An application of the rule in Fig. 4.

Then any application of (L, p, R) *is sound and complete. Similarly if L has a single* **F**-*labelled vertex.*

Proof: Figure 3 illustrates. This follows via a straightforward simplification of the previous arguments.

□

Figure 4 shows a rule (L, p, R) that simplifies a bond graph structure that can easily arise from the systematic construction of bond graphs from schematics. Only the values of the efforts and flows have been shown on the bonds, from which the bijection p can be easily inferred. For specific application, the rule needs to be completed with directional indicators, and appropriate (anti)equalities on the port effort and flow variables. Figure 5 shows an application of the rule to a bond graph containing a two port device **D**. Soundness and completeness follow from applying the distributive law in various ways to a formula of the shape $(a + b)(c + d)$. Figure 6 shows a more elaborate version of the same rule. These two examples have evident duals in which the roles of **E**-labelled vertices and **F**-labelled vertices are interchanged.

In various works in the bond graph literature there are *ad hoc* discussions of similar bond graph transformation rules (as we would term them) that can be safely applied

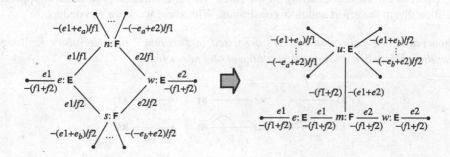

Fig. 6. A more elaborate version of the rule in Fig. 4.

to simplify bond graphs, when they have been arrived at by some systematic process starting from a schematic (which usually generates some redundancy). In [3] there is a more comprehensive list of such transformations. Reformulated as above, all of them can be proved sound and complete straightforwardly.

5.2   Bond Graph Transformation Confluence

When a number of rules are available for transforming a structure, e.g. a graph G, and more than one of their LHSs matches G, a question of some interest is whether 'the same' result will be produced if the rules are applied in one sequence or in a different sequence—or, speaking more technically, whether the set of rules is **confluent**. The following is an easy result guaranteeing confluence.

Theorem 3. *Let function* labels(X) *return the set of labels of the vertex set X. Let $\mathcal{R} = \{Rl_1 \equiv (L_1, p_1, R_1), \ldots, Rl_n \equiv (L_n, p_n, R_n)\}$ be a set of bond graph transformation rules, and let G be a bond graph. Then the application of $\mathcal{R}$ to G is confluent if:*

1. for all $1 \leq i \leq n$, an AAs edge does not occur as a subgraph of any L_i or R_i;
2. for every $1 \leq i < j \leq n$, we have that labels$(nonAnys(L_i)) \cap$ labels$(nonAnys(L_j)) = \varnothing.$

Figure 7 compactly shows two rules, $Rl1$ and $Rl2$, and a bond graph G. Rule $Rl1$'s LHS is shown, and the two dotted arrows indicate that vertex (labelled) **A1** is to be redirected to the lower **Any** vertex, and vertex **A2** is to be redirected to the upper **Any** vertex, so that the RHS of $Rl1$ is just a single AAs edge. Rule $Rl2$ is similar, but redirecting **B1** and **B2** to the **Any** vertices. If $Rl1$ is applied to G, bond graph H_1 results which is a single bond between **B1** and **B2**. There is a unique homomorphism from the LHS of $Rl2$ to H_1, but not a matching, because the edge map is not injective; both edges of the LHS of $Rl2$ have to map to the single edge in H_1. If we apply $Rl2$ first, we get H_2 which is a single bond between **A1** and **A2**, and we get no matching of $Rl1$'s LHS to it. Clearly there is no way of bringing H_1 and H_2 together with the rules we have.

The reasoning above was based on graph structure alone. This, unfortunately, is not enough to show the confluence of the system of rules discussed in Theorem 2 and Corollary 1 (including the analogues for **F**-labelled vertices). There is too much overlap between the labels occurring in the rules. To get confluence, we need to appeal, additionally, to the effort and flow constraints. With some work we can deduce:

Theorem 4. *The collection of rules discussed in Theorem 2 and Corollary 1, along with their **F**-labelled analogues, is confluent and normalising.*

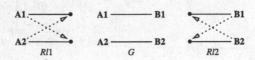

Fig. 7. An example showing the difficulties that arise when AAs edges are allowed to be either in, or created by, rule patterns.

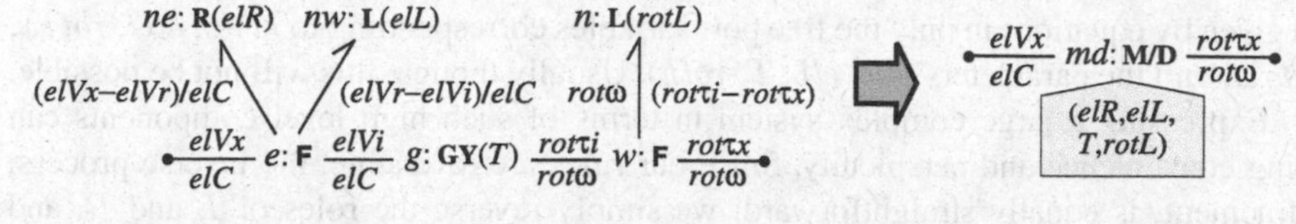

ne: $\mathbf{R}(elR)$ nw: $\mathbf{L}(elL)$ n: $\mathbf{L}(rotL)$

$(elVx-elVr)/elC$ $(elVr-elVi)/elC$ $rot\omega$ $(rot\tau i-rot\tau x)$

$\dfrac{elVx}{elC}$ e: $\mathbf{F}$ $\dfrac{elVi}{elC}$ g: $\mathbf{GY}(T)$ $\dfrac{rot\tau i}{rot\omega}$ w: $\mathbf{F}$ $\dfrac{rot\tau x}{rot\omega}$

$\dfrac{elVx}{elC}$ md: $\mathbf{M/D}$ $\dfrac{rot\tau x}{rot\omega}$

$(elR, elL, T, rotL)$

Fig. 8. An example abstraction, replacing the detail of an ideal but low level motor/dynamo with a high level abstract motor/dynamo device which conceals the details of its internal operation.

6 Bond Graph Abstraction and Refinement

The machinery we have developed for transforming bond graphs serves well for formally expressing hierarchy and abstraction. Parts of a large, low level bond graph can be formally abstracted by transformation to yield a smaller, more compact (and thus more perspicuous) one. Conversely, the reverse of the same process can formally expand a high level, abstract bond graph, refining it to a lower level of abstraction. The ingredient of our formalism, thus far unused, that enables us to do this effectively, is the user level device label.

We illustrate the process with a small idealised electric motor/generator example, adapted from [6]. Figure 8 shows the details of an abstraction rule (L, p, R). On the left, the upward pointing harpoons in L show the positive power flow into the dissipator (electrical resistor) ne, with value elR, the positive power flow into the inertor (electrical inductance) nw with value elL, and the positive power flow into the inertor (rotational flywheel) n with value $rotL$. When the horizontal bonds are turned into right pointing harpoons, the bond graph becomes that of an ideal DC electrical motor, taking in power from the left (via external voltage $elVx$ and external current elC), with the voltage drop $elVx - elVi$ accounted for by the power absorbed by ne and nw. The remaining internal power $elVi \times elC$ goes into the gyrator g, which has transduction coefficient T. This outputs rotational power $rot\tau i \times rot\tau \omega$, according to $elVi = T \times rot\tau \omega$ and $T \times elC = rot\tau i$. There is a further loss of power due to the drop in torque $rot\tau i - rot\tau x$ caused by the flywheel, which finally results in rotational power $rot\tau x \times rot\tau \omega$ being output via the driveshaft. When the direction of the horizontal harpoons is reversed, the pattern can be reinterpreted as an ideal electrical dynamo powered from the right via the driveshaft. Formally, in L, all the port variables corresponding to the bond values shown in L are free, as are the parameters elR, elL, T, $rotL$, and the equations that govern the behaviour of the devices e, ne, nw, g, n.

In the RHS of the rule R, we see a single $nonAnys$ vertex md labelled with device $\mathbf{M/D}$. The understanding would be that $\mathbf{M/D}$ is a user-introduced label for abstract ideal motors/dynamos in a given application context. The properties of $\mathbf{M/D}$ conceal and absorb the details of vertices e, ne, nw, g, n. This is done by aggregating the equations that govern the behaviour of the devices e, ne, nw, g, n, and existentially quantifying all their port variables, leaving the port variables corresponding to $elVx$, elC, $rot\tau x$, $rot\omega$ free. Note that the parameters elR, elL, T, $rotL$, now become parameters of $\mathbf{M/D}$.

If, inside the quantified formula for the behaviour of an abstract device like $\mathbf{M/D}$, the bound variables can be eliminated by some manipulation, then behaviour of $\mathbf{M/D}$ will be given by equations in only the free port variables corresponding to $elVx$, elC, $rot\tau x$, $rot\tau\omega$ (and the parameters elR, elL, T, $rotL$). Usually though, this will not be possible.

Expressing a large complex system in terms of such high level components can bring convenience and perspicuity. Since our rules are reversible, the inverse process, refinement, is equally straightforward: we simply reverse the roles of L and R, and replace a high level abstraction with a more detailed implementation.

7 Conclusions

In the preceding sections we have presented, rather tersely, the essential elements of a formalisation of bond graphs. We covered the ingredients of the physical theories that are in scope, and the graphical structures that capture their interrelationships in a precise manner. Given these basics, the category $\mathcal{BGP}att$ could be formulated, and given that, the long history of graph transformation frameworks provided ample inspiration for formulating a suitable rule based framework for bond graph transformation. This could be immediately leveraged to give a methodology for abstraction and refinement of bond graphs.

One topic not touched on above, is the 'causality' concept of bond graphs. More than the topics we covered, this suffers from an extraordinary level of imprecision and ambiguity in the conventional bond graph literature, overwhelmingly attributable to lack of precision in the use of language. It will be disentangled and formalised in [2].

References

1. Banach, R.: Term graph rewriting and garbage collection using opfibrations. Theor. Comput. Sci. **131**, 29–94 (1994)
2. Banach, R., Baugh, J.: Bond Graphs: An Abstract Formulation (2023, in preparation)
3. Borutzky, W.: Bond Graph Methodology. Springer, London (2010). https://doi.org/10.1007/978-1-84882-882-7
4. Corradini, A., Montanari, U., Rossi, F., Ehrig, H., Heckel, R., Lowe, M.: Alegbraic approaches to graph transformation, part i: basic concepts and double pushout approach. In: Rozenberg, G. (ed.) Handbook of Graph Grammars and Computing by Graph Transformation (3 vols.) vol. 1, pp. 163–245. World Scientific (1997)
5. Ehrig, H., et al.: Alegbraic approaches to graph transformation, Part II: single pushout approach and comparison with double pushout approach. In: Rozenberg, G. (ed.) Handbook of Graph Grammars and Computing by Graph Transformation (3 vols.) vol. 1, pp. 247–312. World Scientific (1997)
6. Karnopp, D., Margolis, D., Rosenberg, R.: System Dynamics: Modeling, Simulation, and Control of Mechatronic Systems, 5th edn. Wiley, Hoboken (2012)
7. Kypuros, J.: Dynamics and Control with Bond Graph Modeling. CRC (2013)
8. Paynter, H.: Analysis and Design of Engineering Systems. MIT Press, Cambridge (1961)
9. Paynter, H.: An Epistemic Prehistory of Bond Graphs. Elsevier, Amsterdam (1992)

A Rule-Based Procedure for Graph Query Solving

Dominique Duval, Rachid Echahed[(✉)], and Frédéric Prost

University Grenoble Alpes, Grenoble, France
{dominique.duval,rachid.echahed,frederic.prost}@imag.fr

Abstract. We consider a core language for graph queries. These queries, which may transform graphs to graphs, are seen as formulas to be solved with respect to graph databases. For this purpose, we first define a graph query algebra where some operations over graphs and sets of graph homomorphisms are specified. Then, the notion of pattern is introduced to represent a kind of recursively defined formula over graphs. The syntax and formal semantics of patterns are provided. Afterwards, we propose a new sound and complete procedure to solve patterns. This procedure, which is based on a set of rewriting rules, is terminating and develops only one needed derivation per pattern to be solved. Our procedure is generic in the sense that it can be adapted to different kinds of graph queries provided that the notions of graph and graph homomorphism are well defined.

Keywords: Rewriting systems · Graph Query Solving · Graph Databases

1 Introduction

Current developments in database theory show a clear shift from relational to graph databases [35]. Relational databases are now well mastered and have been largely investigated in the literature with an ISO standard language SQL [12,15]. On the other side, graphs are being widely used as a flexible data model for numerous database applications [35]. So that various graph query languages such as SPARQL [37], Cypher [23] or G-CORE [2] to quote a few, as well as an ongoing ISO project of a standard language, called GQL[1], have emerged for graph databases.

Representing data graphically is quite legible. However, there is always a dilemma in choosing the right notion of graphs when modeling applications. This issue is already present in some well investigated domains such as modeling languages [8] or graph transformation [34]. Graph data representation does not escape from such dilemma. We can quote for example RDF graphs [38] on which SPARQL is based or variants of Property Graphs [23] currently used in several languages such as Cypher, G-CORE or in GQL.

This work has been partly funded by the project VERIGRAPH : ANR-21-CE48-0015.
[1] https://www.gqlstandards.org/.

M. Fernández and C. M. Poskitt (Eds.): ICGT 2023, LNCS 13961, pp. 163–183, 2023.
https://doi.org/10.1007/978-3-031-36709-0_9

In addition to the possibility of using different graph representations for data, graph database languages feature new kinds of queries such as graph-to-graph queries, cf. CONSTRUCT queries in SPARQL or G-CORE, besides the classical graph-to-relation (table) queries such as SELECT or MATCH queries in SPARQL or Cypher. The former constitute a class of queries which transform a graph database into another graph database. The latter transform a graph into a multiset of solutions represented in general by means of a table just as in the classical relational framework.

In general, graph query processing integrates features shared with graph transformation techniques and goal solving or logic programming (variable assignments). Our main aim in this paper is to define an operational semantics, based on rewriting techniques, for graph queries. We propose a generic rule-based calculus, called *gq-narrowing* which is parameterized by the actual interpretations of graphs and their matches (homomorphisms). That is to say, the obtained calculus can be adapted to different definitions of graphs and the corresponding notion of match. The proposed calculus consists of a dedicated rewriting system and a narrowing-like [4,25,25] procedure which follows closely the formal semantics of patterns or queries, the same way as (SLD-)Resolution calculus is related to formal models underlying Horn or Datalog [27] clauses. The use of rewriting techniques in defining the proposed operational semantics paves the way to syntactic analysis and automated verification techniques for the proposed core language.

In order to define a sound and complete calculus, we first propose a uniform formal semantics for queries. For practical reasons, we were inspired by existing graph query languages and consider graph-to-graph queries and graph-to-table queries as two facets of one same syntactic object that we call *pattern*. The proposed patterns can be nested at will as in declarative functional terms and may include aggregation operators as well as graph construction primitives. The semantics of a pattern is defined as a set of matches, that is to say, a set of graph homomorphisms and not only a set of variable assignments as proposed in [3,22,23]. From such a set of matches, one can easily display either the table by considering the images of the variables as defined by the matches or the graph, target of the matches, or even both the table and the graph. The proposed semantics for patterns allows also to write nested patterns in a natural way, that is, new data graphs can be constructed on the fly before being queried.

The paper is organized as follows: The next section introduces a graph query algebra featuring some key operations needed to express the proposed calculus. Section 3 defines the syntax of patterns and queries as well as their formal semantics. In Sect. 4, a sound and complete calculus is given. First we introduce a rewriting system describing how query results are found. Then, we define gq-narrowing, which is associated with the proposed rules. Concluding remarks and related work are given in Sect. 5. Due to lack of space, omitted proofs as well as a new query form combining CONSTRUCT and SELECT query forms can be found in [20].

2   A Graph Query Algebra

During graph query processing, different intermediate results can be computed and composed. In this section, we introduce a Graph Query Algebra $\mathcal{GQ}$ which consists of a family of operations over graphs, matches (graph homomorphisms) and expressions. These different items are used later on to define the semantics of queries in Sects. 3 and 4.

The algebra $\mathcal{GQ}$ is defined over a signature Σ_{gq}. The main sorts of Σ_{gq} are Gr, Som, Exp and Var to be interpreted as graphs, sets of matches, expressions and variables, respectively. The sort Var is a subsort of Exp. The main operators of Σ_{gq} are:

$$\text{Match} : \text{Gr}, \text{Gr} \rightarrow \text{Som} \qquad\qquad \text{Join} : \text{Som}, \text{Som} \rightarrow \text{Som}$$
$$\text{Bind} : \text{Som}, \text{Exp}, \text{Var} \rightarrow \text{Som} \qquad \text{Filter} : \text{Som}, \text{Exp} \rightarrow \text{Som}$$
$$\text{Build} : \text{Som}, \text{Gr} \rightarrow \text{Som} \qquad\qquad \text{Union} : \text{Som}, \text{Som} \rightarrow \text{Som}$$

The above sorts and operators are given as an indication while being inspired by concrete languages. They may be modified or tuned according to actual graph query languages. Various interpretations of sorts Gr and Som can be given. In order to provide concrete examples, we have to fix an actual interpretation of these sorts. For all the examples given in the paper, we have chosen to interpret the sort Gr as generalized RDF graphs [38]. This choice is not a limitation, we might have chosen other notions of graphs such as property graphs [23]. Our choice here is motivated by the simplicity of the RDF graph definition (set of triples). Below, we define generalized RDF graphs. They are the usual RDF graphs with the ability to contain isolated nodes. Let $\mathcal{L}$ be a set, called the set of *labels*, made of the union of two disjoint sets $\mathcal{C}$ and $\mathcal{V}$, called respectively the set of *constants* and the set of *variables*.

Definition 1 (graph). *Every element $t = (s, p, o)$ of $\mathcal{L}^3$ is called a triple and its members s, p and o are called respectively the* subject, *the* predicate *and the* object *of t. A graph G is a pair $G = (G_N, G_T)$ made of a finite subset G_N of $\mathcal{L}$ called the set of* nodes *of G and a finite subset G_T of $\mathcal{L}^3$ called the set of* triples *of G, such that the subject and the object of each triple of G are nodes of G. The nodes of G which are neither a subject nor an object are called the* isolated nodes *of G. The set of* labels *of a graph G is the subset $\mathcal{L}(G)$ of $\mathcal{L}$ made of the nodes and predicates of G, then $\mathcal{C}(G) = \mathcal{C} \cap \mathcal{L}(G)$ and $\mathcal{V}(G) = \mathcal{V} \cap \mathcal{L}(G)$. The graph with an empty set of nodes and an empty set of triples is called the* empty graph *and is denoted by $\emptyset$. Given two graphs G_1 and G_2, the graph G_1 is a* subgraph *of G_2, written $G_1 \subseteq G_2$, if $(G_1)_N \subseteq (G_2)_N$ and $(G_1)_T \subseteq (G_2)_T$, thus $\mathcal{L}(G_1) \subseteq \mathcal{L}(G_2)$. The* union $G_1 \cup G_2$ *is the graph defined by $(G_1 \cup G_2)_N = (G_1)_N \cup (G_2)_N$ and $(G_1 \cup G_2)_T = (G_1)_T \cup (G_2)_T$, then $\mathcal{L}(G_1 \cup G_2) = \mathcal{L}(G_1) \cup \mathcal{L}(G_2)$.*

In the rest of the paper we write graphs as sets of triples and nodes: for example $G = \{(s_1, o_1, p_1), n_1, n_2\}$ is the graph with four nodes n_1, n_2, s_1, p_1 and one triple (s_1, o_1, p_1).

Example 1. We define a toy database which is used as a running example throughout the paper. The database consists of *persons* who *are* either *professors* or *students*, with *topics* such that each professor (resp. student) *teaches* (resp. *studies*) some topics. The graph G_{ex} is described below by its triples (on the left) and by a diagram (on the right), in which plain arrows represent *is* relation, dashed lines represent *teaches* relation and dotted lines represent *studies* relation.

(Alice, is, Professor)	(Alice, teaches, Math)
(Bob, is, Professor)	(Bob, teaches, CS)
(Charly, is, Student)	(Charly, studies, Math)
(David, is, Student)	(David, studies, Math)
(Eva, is, Student)	(Eva, studies, CS)

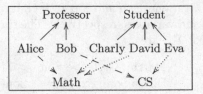

Definition 2 (match). *A* graph homomorphism *from a graph L to a graph G, denoted $m : L \to G$, is a function from $\mathcal{L}(L)$ to $\mathcal{L}(G)$ which* preserves nodes *and* preserves triples, *in the sense that $m(L_N) \subseteq G_N$ and $m^3(L_T) \subseteq G_T$. A* match *is a graph homomorphism $m : L \to G$ which* fixes $\mathcal{C}$, *in the sense that $m(c) = c$ for each c in $\mathcal{C}(L)$.*

A match $m : L \to G$ determines two functions $m_N : L_N \to G_N$ and $m_T : L_T \to G_T$, restrictions of m and m^3 respectively. A match $m : L \to G$ is invertible if and only if both functions m_N and m_T are bijections. This means that a function m from $\mathcal{L}(L)$ to $\mathcal{L}(G)$ is an invertible match if and only if $\mathcal{C}(L) = \mathcal{C}(G)$ with $m(c) = c$ for each $c \in \mathcal{C}(L)$ and m is a bijection from $\mathcal{V}(L)$ to $\mathcal{V}(G)$: thus, L is the same as G up to variable renaming. It follows that the symbol used for naming a variable does not matter as long as graphs are considered only up to invertible matches.

Notice that RDF graphs [38] are graphs according to Definition 1 but without isolated nodes, and where constants are either IRIs (Internationalized Resource Identifiers) or literals and where all predicates are IRIs and only objects can be literals. Blank nodes in RDF graphs are the same as variable nodes in our graphs. An isomorphism of RDF graphs, as defined in [38], is an invertible match.

Below we introduce some useful definitions on matches. Notice that we do not consider a match m as a simple variable assignment but rather as a graph homomorphism with clear source and target graphs. This nuance in the definition of matches is key in the rest of the paper since it allows us to define the notion of nested patterns in a straightforward manner.

Definition 3 (compatible matches). *Two matches $m_1 : L_1 \to G_1$ and $m_2 : L_2 \to G_2$ are compatible, written as $m_1 \sim m_2$, if $m_1(x) = m_2(x)$ for each $x \in \mathcal{V}(L_1) \cap \mathcal{V}(L_2)$. Given two compatible matches $m_1 : L_1 \to G_1$ and $m_2 : L_2 \to G_2$, let $m_1 \bowtie m_2 : L_1 \cup L_2 \to G_1 \cup G_2$ denote the unique match such that $m_1 \bowtie m_2 \sim m_1$ and $m_1 \bowtie m_2 \sim m_2$ (which means that $m_1 \bowtie m_2$ coincides with m_1 on L_1 and with m_2 on L_2).*

Definition 4 (building a match). *Let* $m : L \rightarrow G$ *be a match and* R *a graph. The match Build-match* $(m, R) : R \rightarrow G \cup H_{m,R}$ *is the unique match (up to variable renaming) such that for each variable* x *in* R:

$$Build\text{-}match\,(m, R)(x) = \begin{cases} m(x) \text{ if } x \in \mathcal{V}(R) \cap \mathcal{V}(L), \\ \text{some fresh variable } var(m, x) \text{ if } x \in \mathcal{V}(R) - \mathcal{V}(L) \end{cases}$$

and $H_{m,R}$ *is the image of* R *by Build-match* (m, R).

Definition 5 (set of matches, assignment table). *Let* L *and* G *be graphs. A set* $\underline{m}$ *of matches, all of them from* L *to* G, *is denoted* $\underline{m} : L \Rightarrow G$ *and called a* homogeneous set of matches, *or simply a* set of matches, *with* source L *and* target G. *The* image *of* L *by* $\underline{m}$ *is the subgraph* $\underline{m}(L) = \cup_{m \in \underline{m}}(m(L))$ *of* G. *We denote* $Match(L, G) : L \Rightarrow G$ *the set of all matches from* L *to* G. *When* L *is the empty graph this set has one element which is the inclusion* $\emptyset_G : \emptyset \rightarrow G$. *We denote* $\underline{i}_G = Match(\emptyset, G) : \emptyset \Rightarrow G$ *this singleton and* $\varnothing_G : \emptyset \Rightarrow G$ *its empty subset. The* assignment table $Tab(\underline{m})$ *of* $\underline{m}$ *is the two-dimensional table with the elements of* $\mathcal{V}(L)$ *in its first row, then one row for each* m *in* $\underline{m}$, *and the entry in row* m *and column* x *equals to* $m(x)$.

Thus, the assignment table $Tab(\underline{m})$ describes the set of functions $\underline{m}|_{\mathcal{V}(L)} : \mathcal{V}(L) \Rightarrow \mathcal{L}$, made of the functions $m|_{\mathcal{V}(L)} : \mathcal{V}(L) \rightarrow \mathcal{L}$ for all $m \in \underline{m}$. A set of matches $\underline{m} : L \Rightarrow G$ is determined by the graphs L and G and the assignment table $Tab(\underline{m})$.

Example 2. In order to determine whether professor $?p$ teaches topic $?t$ which is studied by student $?s$, we may consider the following graph L_{ex}, where $?p$, $?t$ and $?s$ are variables. In all examples, variables are preceded by a "?".

There are three matches from L_{ex} to the graph G_{ex} given in Example 1. The set $\underline{m}_{ex}$ of all these matches is $\underline{m}_{ex} = \{m_1, m_2, m_3\} : L_{ex} \Rightarrow G_{ex}$:

$$\underline{m}_{ex} : L_{ex} \Rightarrow G_{ex} \text{ with } Tab(\underline{m}_{ex}) =$$

$\underline{m}_{ex}$	$?p$	$?t$	$?s$
m_1	Alice	Maths	Charly
m_2	Alice	Maths	David
m_3	Bob	CS	Eva

Query languages usually provide a term algebra dedicated to express operations over integers, booleans and so forth. We do not care here about the way basic operations are chosen but we want to deal with aggregation operations as in most database query languages. Thus, one can think of any kind of term algebra with operators which are classified as either basic operators (unary or binary) and aggregation operators (always unary). For defining the syntax and semantics of aggregation functions we follow [19]. We consider that all expressions are well typed. Typically, and not exclusively, the sets Op_1, Op_2 and Agg of *basic unary* operators, *basic binary* operators and *aggregation* operators can be:

$$Op_1 = \{-, \text{not}\},$$
$$Op_2 = \{+, -, \times, /, =, >, <, \text{and}, \text{or}\},$$
$$Agg = \{\text{max}, \text{min}, \text{sum}, \text{avg}, \text{count}\}.$$

Definition 6 (syntax of expressions). Expressions e *and their sets of* in-scope *variables* $\mathcal{V}(e)$ *are defined recursively as follows, with* $c \in \mathcal{C}$, $x \in \mathcal{V}$, $op_1 \in Op_1$, $op_2 \in Op_2$, $agg \in Agg$ *and* g *a set of expressions:*

$$e ::= c \mid x \mid op_1 \, e \mid e \; op_2 \, e \mid agg(e) \mid agg(e \text{ by } g),$$
$$\mathcal{V}(c) = \varnothing, \; \mathcal{V}(x) = \{x\}, \; \mathcal{V}(op_1 \, e) = \mathcal{V}(e), \; \mathcal{V}(e \; op_2 \, e') = \mathcal{V}(e) \cup \mathcal{V}(e'),$$
$$\mathcal{V}(agg(e)) = \mathcal{V}(e) \text{ and } \mathcal{V}(agg(e \text{ by } g)) = \mathcal{V}(e)$$

(variables in g *must be distinct from those in* e*).*

The *value* of an expression e with respect to a set of matches $\underline{m}$, as stated in Definition 7, is a family of constants $\underline{m}(e) = (m(e)_{\underline{m}})_{m \in \underline{m}}$ indexed by the set $\underline{m}$. In general $m(e)_{\underline{m}}$ depends on e and m and it may also depend on other matches in $\underline{m}$ when e involves aggregation operators. Whenever e is free from any aggregation operator then $m(e)_{\underline{m}}$ does not depend on the matches different from m in $\underline{m}$, so that it can be written simply $m(e)$. To each basic operator op is associated a function $[[op]]$ (or simply op) from constants to constants if op is unary and from pairs of constants to constants if op is binary. To each aggregation operator agg is associated a function $[[agg]]$ (or simply agg) from *multisets* of constants to constants. Note that each family of constants $\underline{c} = (c_m)_{m \in \underline{m}}$ determines a multiset of constants $\{\!\{c_m \mid m \in \underline{m}\}\!\}$, which is also denoted $\underline{c}$ when there is no ambiguity.

Definition 7 (evaluation of expressions). *Let* L, G *be graphs,* e *an expression such that* $\mathcal{V}(e) \subseteq \mathcal{V}(L)$ *and* $\underline{m} : L \Rightarrow G$ *a set of matches. The* value *of* e *with respect to* $\underline{m}$ *is the family*

$$\underline{m}(e) = (m(e)_{\underline{m}})_{m \in \underline{m}}$$

defined recursively as follows. It is assumed that each $m(e)_{\underline{m}}$ *in this definition is a constant.*

$m(c)_{\underline{m}} = c$, $m(x)_{\underline{m}} = m(x)$, $m(op_1 \, e)_{\underline{m}} = [[op_1]] \, m(e)_{\underline{m}}$,
$m(e \; op_2 \, e')_{\underline{m}} = m(e)_{\underline{m}} \, [[op_2]] \, m(e')_{\underline{m}}$, $m(agg(e))_{\underline{m}} = [[agg]](\underline{m}(e))$,
$m(agg(e \text{ by } g))_{\underline{m}} = [[agg]](\underline{m}|_{g,m}(e))$ *where* $\underline{m}|_{g,m}$ *is*

the *group of* m *in* $\underline{m}$ *with respect to* g, *i.e., the subset of* $\underline{m}$ *made of the matches* m' *in* $\underline{m}$ *such that* $m'(e')_{\underline{m}} = m(e')_{\underline{m}}$ *for every expression* e' *in* g.

Note that $m(agg(e))_{\underline{m}}$ *is the same for all* m *in* $\underline{m}$, *while* $m(agg(e\ by\ g))_{\underline{m}}$ *is the same for all* m *in* $\underline{m}$ *which are in a common group with respect to* g.

Example 3. Consider $\underline{m}_{ex} = \{m_1, m_2, m_3\} : L_{ex} \Rightarrow G_{ex}$ as in Example 2, denoted simply $\underline{m}$ for readability. Let us evaluate the expressions count($?s$) and count($?s$ by $?p$).

The evaluation of count($?s$) with respect to $\underline{m}$ runs as follows:

$$\begin{aligned}
\underline{m}(\text{count}(?s)) &= (\,m_i(\text{count}(?s))_{\underline{m}}\,)_{i=1,2,3}\\
m_i(\text{count}(?s))_{\underline{m}} &= \text{count}(\underline{m}(?s))\\
\underline{m}(?s) &= (\,m_i(?s)_{\underline{m}}\,)_{i=1,2,3}\\
m_i(?s)_{\underline{m}} &= m_i(?s) \quad \text{for } i = 1,2,3
\end{aligned}$$

Since $m_1(?s) = $ Charly, $m_2(?s) = $ David and $m_3(?s) = $ Eva we get:

$$\begin{aligned}
\underline{m}(?s) &= (\,\text{Charly, David, Eva}\,)\\
\text{count}(\underline{m}(?s)) &= 3\\
m_i(\text{count}(?s))_{\underline{m}} &= 3 \quad \text{for } i = 1,2,3\\
\underline{m}(\text{count}(?s)) &= (\,3,3,3\,)
\end{aligned}$$

The evaluation of count($?s$ by $?p$) with respect to $\underline{m}$ runs as follows:

$$\begin{aligned}
\underline{m}(\text{count}(?s\ by\ ?p)) &= (\,m_i(\text{count}(?s\ by\ ?p))_{\underline{m}}\,)_{i=1,2,3}\\
m_i(\text{count}(?s\ by\ ?p))_{\underline{m}} &= \text{count}(\underline{m}|_{\{?p\},m_i}(?s))
\end{aligned}$$

Since $m_1(?p) = m_2(?p) = $ Alice and $m_3(?p) = $ Bob we get

$$\underline{m}|_{\{?p\},m_1} = \underline{m}|_{\{?p\},m_2} = \{m_1, m_2\} \text{ and } \underline{m}|_{\{?p\},m_3} = \{m_3\}$$

Then count($\underline{m}|_{\{?p\},m_i}(?s)$) = 2 for $i = 1, 2$ and count($\underline{m}|_{\{?p\},m_3}(?s)$) = 1 and finally $\underline{m}(\text{count}(?s\ by\ ?p)) = (\,2,2,1\,)$.

We conclude this section with the definition of the algebra $\mathcal{GQ}$ over Σ_{gq}. Whenever needed, we extend the target of matches: for every graph H and every match $m : L \to G$ where G is a subgraph of H we write $m : L \to H$ when m is considered as a match from L to H.

Definition 8 ($\mathcal{GQ}$ algebra). *The algebra* $\mathcal{GQ}$ *is the algebra over the signature* Σ_{gq} *where the sorts* Gr, Som, Exp *and* Var *are interpreted respectively as the set of graphs (Definition 1), the set of sets of matches (Definition 5), the set of expressions (Definition 6) and its subset of variables, and where the operators are interpreted by the operations with the same name, as follows.*

– *For all graphs* L *and* G:
  Match$(L,G) : L \Rightarrow G$ *is the set of all matches from* L *to* G.
– *For all sets of matches* $\underline{m} : L \Rightarrow G$ *and* $\underline{p} : R \Rightarrow H$:
  Join$(\underline{m}, \underline{p}) = \{m \bowtie p \mid m \in \underline{m} \wedge p \in \underline{p} \wedge m \sim p\} : L \cup R \Rightarrow G \cup H$.
– *For each set of matches* $\underline{m} : L \Rightarrow G$ *and each expression* e, *let* $H_m = \{m(e)_{\underline{m}} \mid m \in \underline{m}\}$ *and for each match* m *in* $\underline{m}$ *let* $p_m : \{x\} \Rightarrow H_m$ *be the match such that* $p_m(x) = m(e)_{\underline{m}}$. *Then for each fresh variable* $x \notin \mathcal{V}(L)$:
  Bind$(\underline{m}, e, x) = \{m \bowtie p_m \mid m \in \underline{m}\} : L \cup \{x\} \Rightarrow G \cup H_m$.

- For each set of matches $\underline{m} : L \Rightarrow G$ and each expression e:
  $Filter(\underline{m}, e) = \{m \mid m \in \underline{m} \wedge m(e)_{\underline{m}} = true\} : L \Rightarrow G.$
- For each set of matches $\underline{m} : L \Rightarrow G$ and each graph R:
  $Build(\underline{m}, R) = \{Build\text{-}match\,(m, R) \mid m \in \underline{m}\} : R \Rightarrow G \cup Build(\underline{m}, R)(R)$
  where $Build(\underline{m}, R)(R) = \cup_{m \in \underline{m}} Build\text{-}match\,(m, R)(R).$
- For all sets of matches $\underline{m} : L \Rightarrow G$ and $\underline{p} : L \Rightarrow H$:
  $Union(\underline{m}, \underline{p}) = (\underline{m} : L \Rightarrow G \cup H) \cup (\underline{p} : L \Rightarrow G \cup H) : L \Rightarrow G \cup H.$

Note that we could handle other kinds of graphs in the same way. Here, the *Bind* operation illustrates the interest of accepting isolated nodes in graphs as done in Definition 1 contrary to RDF graphs.

Example 4. As in Example 2 consider the set of matches $\underline{m}_{ex} : L_{ex} \Rightarrow G_{ex}$. We know from Example 3 that the value of count($?s$) with respect to $\underline{m}_{ex}$ is $(3,3,3)$ and that the value of count($?s$ by $?p$) with respect to $\underline{m}_{ex}$ is $(2,2,1)$.
Thus, $Bind\,(\underline{m}_{ex}, \text{count}(?s), ?n) = \underline{m}'_{ex} : L_{ex} \cup \{?n\} \Rightarrow G_{ex} \cup \{3\}$ and $Build\,(\underline{m}'_{ex}, \{?n\}) = \underline{m}''_{ex} : \{?n\} \Rightarrow G_{ex} \cup \{3\}$ with assignment tables:

$$Tab(\underline{m}'_{ex}) = \begin{array}{|c|c|c|c|} \hline ?p & ?s & ?c & ?n \\ \hline \text{Alice} & \text{Charly} & \text{Maths} & 3 \\ \text{Alice} & \text{David} & \text{Maths} & 3 \\ \text{Bob} & \text{Eva} & \text{CS} & 3 \\ \hline \end{array} \qquad Tab(\underline{m}''_{ex}) = \begin{array}{|c|} \hline ?n \\ \hline 3 \\ \hline \end{array}$$

$Bind\,(\underline{m}_{ex}, \text{count}(?s \text{ by } ?p), ?n) = \underline{n}'_{ex} : L_{ex} \cup \{?n\} \Rightarrow G_{ex} \cup \{2,1\}$ with $Tab(\underline{n}'_{ex})$ below. Now let $R'_{ex} = \{(?p, \text{supervises}, ?n)\}$, then $Build\,(\underline{n}'_{ex}, R'_{ex}) = \underline{n}''_{ex} : R'_{ex} \Rightarrow G_{ex} \cup \{(\text{Alice}, \text{supervises}, 2), (\text{Bob}, \text{supervises}, 1)\}$ with assignment tables:

$$Tab(\underline{n}'_{ex}) = \begin{array}{|c|c|c|c|} \hline ?p & ?s & ?c & ?n \\ \hline \text{Alice} & \text{Charly} & \text{Maths} & 2 \\ \text{Alice} & \text{David} & \text{Maths} & 2 \\ \text{Bob} & \text{Eva} & \text{CS} & 1 \\ \hline \end{array} \qquad Tab(\underline{n}''_{ex}) = \begin{array}{|c|c|} \hline ?p & ?n \\ \hline \text{Alice} & 2 \\ \text{Bob} & 1 \\ \hline \end{array}$$

3 Patterns and Queries

The syntax of graph databases is still evolving. We do not consider all technical syntactic details of a real-world language nor all possible constraints on matches. We focus on a core language. Its syntax reflects significant aspects of graph queries. Conditions on graph paths, which can be seen as constraints on matches, are omitted in this paper in order not to make the syntax too cumbersome. We consider mainly two syntactic categories: *patterns* and *queries*, in addition to *expressions* already mentioned. Queries are either SELECT queries, as in most query languages or CONSTRUCT queries, as in SPARQL and G-CORE. A SELECT query applied to a graph returns a *table* which describes a multiset of *solutions* or variable bindings, while a CONSTRUCT query applied to a graph returns a graph. Besides that, a pattern applied to a graph returns a set of matches. Patterns are the basic blocks for building queries. They are defined in Sect. 3.1 together with their semantics. Queries are defined in Sect. 3.2 and their semantics is easily derived from the semantics of patterns. In this Section, as in Sect. 2, the set of *labels* $\mathcal{L}$ is the union of the disjoint sets $\mathcal{C}$ and $\mathcal{V}$, of *constants* and *variables* respectively. We assume that the set $\mathcal{C}$ of constants contains the numbers and strings and the boolean values *true* and *false*.

3.1   Patterns

In Definition 9, the signature for patterns is built by extending the signature Σ_{gq} with a sort Pat for patterns and several operators involving patterns. For instance the operator BASIC in the term BASIC (L) turns a graph L into a pattern. Other operators such as JOIN, BIND or FILTER are rather classical and were inspired by existing database query languages. Operator BUILD is specific to graph-to-graph queries. The following definition of patterns can be enriched by more specific operators if needed. The formal semantics of patterns is given by an evaluation function in Definition 10.

Definition 9 (syntax of patterns). *The signature Σ_{gq} is extended with a sort* Pat *for patterns and the following operators:*

- *If P is a pattern then $[P]$ is a graph, called the* scope graph *of P.*
- *The symbol $\square$ is a pattern, called the* empty pattern,
- *If L is a graph then $P = $ BASIC (L) is a pattern, called a* basic pattern.
- *If P_1 and P_2 are patterns then $P = P_1$ JOIN P_2 is a pattern.*
- *If P_1 is a pattern, e an expression such that $\mathcal{V}(e) \subseteq \mathcal{V}([P_1])$ and x a variable such that $x \notin \mathcal{V}([P_1])$ then $P = P_1$ BIND e AS x is a pattern.*
- *If P_1 is a pattern and e an expression such that $\mathcal{V}(e) \subseteq \mathcal{V}([P_1])$ then $P = P_1$ FILTER e is a pattern.*
- *If P_1 is a pattern and R a graph then $P = P_1$ BUILD R is a pattern.*
- *If P_1 and P_2 are patterns such that $[P_1] = [P_2]$ then $P = P_1$ UNION P_2 is a pattern.*

The semantics of a pattern P over a graph G is a set of matches $[[P]]_G : [P] \Rightarrow G^{(P)}$ where the source of matches is $[P]$, the so-called scope graph of P and the target graph is $G^{(P)}$. The target graph is obtained by transforming the initial graph G according to the shape of pattern P. These notions are made precise in the following Definition 10. The different pattern operations defined in Definition 9 could be seen as elementary actions that may be used to transform graph databases while computing sets of matches. The induced graph transformation by patterns is similar to traditional algebraic graph transformation processes like DPO [14], AGREE [13] etc. in the following sense. An algebraic graph transformation process does not only transform a graph G into a graph H, but it also transforms an instance of a left-hand side graph L (of a rule) in G into an instance of a right-hand side graph R in H. Such instances are similar to matches. Moreover, patterns are interpreted as sets of matches, so that the induced graph transformation can be seen as a kind of conflict-free simultaneous "parallel" graph transformation process [16].

Definition 10 (evaluation of patterns, set of solutions). *The* set of
solutions *or the* value *of a pattern P over a graph G is a set of matches*
$[[P]]_G : [P] \Rightarrow G^{(P)}$ *from the scope graph $[P]$ of P to a graph $G^{(P)}$ that contains*
G. *This value* $[[P]]_G : [P] \Rightarrow G^{(P)}$ *is defined inductively as follows:*

- $[[\square]]_G = \varnothing_G : \emptyset \Rightarrow G$.
- $[[\text{BASIC}\,(L)]]_G = Match\,(L, G) : L \Rightarrow G$
- $[[P_1 \, \text{JOIN} \, P_2]]_G = Join\,([[P_1]]_G, [[P_2]]_{G^{(P_1)}}) : [P_1] \cup [P_2] \Rightarrow G^{(P_1)(P_2)}$
- $[[P_1 \, \text{BIND} \, e \, \text{AS} \, x]]_G = Bind\,([[P_1]]_G, e, x) : [P_1] \cup \{x\} \Rightarrow G^{(P_1)} \cup \{m(e)_{\underline{m}} \mid m \in \underline{m}\}$
  where $\underline{m} = [[P_1]]_G$
- $[[P_1 \, \text{FILTER} \, e]]_G = Filter\,([[P_1]]_G, e) : [P_1] \Rightarrow G^{(P_1)}$
- $[[P_1 \, \text{BUILD} \, R]]_G = Build\,([[P_1]]_G, R) : R \Rightarrow G^{(P_1)} \cup [[P_1]]_G(R)$
- $[[P_1 \, \text{UNION} \, P_2]]_G = Union\,([[P_1]]_G, [[P_2]]_{G^{(P_1)}}) : [P_1] \Rightarrow G^{(P_1)(P_2)}$.

Remark 1. In all cases, the graph $G^{(P)}$ is built by adding to G "whatever is
required" for the evaluation. When P is the empty pattern, the value of P over
G is the empty subset $\varnothing_G$ of $Match\,(\emptyset, G)$. When P is a BIND, isolated nodes
have to be added to G, justifying the use of isolated nodes in graphs.

Syntactically, each operator OP builds a pattern P from a pattern P_1 and
a parameter *param*, which is either a pattern P_2 (for JOIN and UNION), a
pair (e, x) made of an expression and a variable (for BIND), an expression e
(for FILTER) or a graph R (for BUILD). Semantically, for every pattern $P =
P_1 \, \text{OP} \, param$, let us denote $\underline{m}_1 : X_1 \Rightarrow G_1$ for $[[P_1]]_G : [P_1] \Rightarrow G^{(P_1)}$ and
$\underline{m} : X \Rightarrow G'$ for $[[P]]_G : [P] \Rightarrow G^{(P)}$. In every case it is necessary to evaluate
$\underline{m}_1$ before evaluating *param*: for JOIN and UNION this is because pattern P_2
is evaluated on G_1, for BIND and FILTER because expression e is evaluated
with respect to $\underline{m}_1$, and for BUILD because of the definition of $Build$. Note that
the semantics of $P_1 \, \text{JOIN} \, P_2$ and $P_1 \, \text{UNION} \, P_2$ is not symmetric in P_1 and P_2
in general, unless $G^{(P_1)} = G$ and $G^{(P_2)} = G$, which occurs when P_1 and P_2 are
basic patterns. Given a pattern $P = P_1 \, \text{OP} \, param$, the pattern P_1 is called a
subpattern of P, as well as P_2 when $P = P_1 \, \text{JOIN} \, P_2$ or $P = P_1 \, \text{UNION} \, P_2$. The
semantics of patterns is defined in terms of the semantics of its subpatterns (and
the semantics of its other arguments, if any).

Definition 11. *For every pattern P, the set $\mathcal{V}(P)$ of in-scope variables of P
is the set $\mathcal{V}([P])$ of variables of the scope graph $[P]$. An expression e is over a
pattern P if $\mathcal{V}(e) \subseteq \mathcal{V}(P)$.*

Example 5. Let R_{ex} be the graph $R_{ex} = \{(?p, \text{teaches}, ?z), (?s, \text{studies}, ?z)\}$,
where $?p$, $?z$ and $?s$ are variables. Note that R_{ex} is the same as L_{ex} (in Example 2), except for the name of one variable. In order to determine when professor
$?p$ teaches some topic which is studied by student $?s$, whatever the topic, we
consider the following pattern P_{ex}.

$$P_{ex} = \text{BASIC}\,(L_{ex})\,\text{BUILD}\,R_{ex}$$
$$= \text{BASIC}\,(\,\{(?p, \text{teaches}, ?t), (?s, \text{studies}, ?t)\}\,)$$
$$\text{BUILD}\,\{(?p, \text{teaches}, ?z), (?s, \text{studies}, ?z)\}$$

Note that the variable $?z$ in R_{ex} does not appear in L_{ex}. Since there are three matches from L_{ex} to G_{ex} (Example 2), the value of P_{ex} over G_{ex} is:

$$\underline{p}_{ex} : R_{ex} \Rightarrow G'_{ex} \text{ with } Tab(\underline{p}_{ex}) = $$

$?p$	$?z$	$?s$
Alice	$?z_1$	Charly
Alice	$?z_2$	David
Bob	$?z_3$	Eva

where $?z_1$, $?z_2$ and $?z_3$ are three fresh variables and G'_{ex} is the union of G_{ex} with the following graph H_{ex}:

(Alice, teaches, $?z_1$)	(Charly, studies, $?z_1$)
(Alice, teaches, $?z_2$)	(David, studies, $?z_2$)
(Bob, teaches, $?z_3$)	(Eva, studies, $?z_3$)

3.2 Queries

We consider two kinds of queries: CONSTRUCT queries, specific to graph database languages, as one may find in SPARQL or G-CORE and SELECT queries, rather classical, close to SQL language, also called MATCH queries in some languages such as Cypher. The semantics of queries is defined from the semantics of patterns. According to Definition 10, all patterns have a graph-to-set-of-matches semantics. In contrast, CONSTRUCT queries have a graph-to-graph semantics and SELECT queries have a graph-to-multiset-of-solutions or graph-to-table semantics.

Definition 12 (syntax of queries). *Let S be a set of variables, R a graph and P a pattern. A query Q has one of the following shapes:*

> *either* CONSTRUCT R WHERE P *or* SELECT S WHERE P

Definition 13 (result of CONSTRUCT queries). *Given a pattern P_1 and a graph R, consider the query $Q =$ CONSTRUCT R WHERE P_1 and the pattern $P = P_1$ BUILD R. The result of the query Q over a graph G, denoted $Result_C(Q, G)$, is the subgraph of $G^{(P)}$ image of R by the set of matches $[[P]]_G$.*

Thus, the result of a CONSTRUCT query Q over a graph G is the graph $Result_C(Q, G) = [[P]]_G(R)$ built by "gluing" the graphs $m(R)$ for all matches m in $[[P]]_G$, where $m(R)$ is a copy of R with each variable $x \in \mathcal{V}(R) - \mathcal{V}(P)$ replaced by a fresh variable (which means, fresh for each m and each x).

Example 6. Consider the query:

$$\begin{aligned} Q_{C,ex} = \ & \text{CONSTRUCT } R_{ex} \text{ WHERE } \text{BASIC } (L_{ex}) \\ = \ & \text{CONSTRUCT } \{(?p, \text{teaches}, ?z), (?s, \text{studies}, ?z)\} \\ & \text{WHERE } \text{BASIC } (\{(?p, \text{teaches}, ?t), (?s, \text{studies}, ?t)\}) \end{aligned}$$

The corresponding pattern P_{ex} and the value $\underline{p}_{ex} : R_{ex} \Rightarrow G'_{ex}$ of P_{ex} over G_{ex} are as in Example 5. It follows that the result of the query $Q_{C,ex}$ over G_{ex} is the subgraph of G'_{ex} image of R_{ex} by $\underline{p}_{ex}$: it is the graph H_{ex} from Example 5.

Remark 2. CONSTRUCT queries in SPARQL are similar to CONSTRUCT queries considered in this paper: the variables in $\mathcal{V}(R) - \mathcal{V}(P_1)$ play the same role as the blank nodes in SPARQL. By considering BUILD patterns, thanks to the functional orientation of the definition of patterns, our language allows BUILD subpatterns: this is new and specific to the present study.

For SELECT queries we proceed as for CONSTRUCT queries: we define a transformation from each SELECT query Q to a BUILD pattern P and a transformation from the result of pattern P to the result of query Q. Definition 14 below would deserve more explanations. However this is not the subject of this paper, see [18] for details about how turning a table into a graph (reification).

Definition 14 (result of SELECT queries). *For every set of variables $S = \{s_1, ..., s_n\}$, let $Gr(S)$ denote the graph made of the triples (r, c_j, s_j) for $j \in \{1, ..., n\}$ where r is a fresh variable and c_j is a fresh constant string for each j. Given a pattern P_1 and a set of variables $S = \{s_1, ..., s_n\}$ consider the query $Q = \text{SELECT } S \text{ WHERE } P_1$ and the pattern $P = P_1 \text{ BUILD } Gr(S)$. The value of P over a graph G is a set of matches $[[P]]_G$ whose assignment table has $n + 1$ columns, corresponding to the variables $r, s_1, ..., s_n$. The result of the query Q over a graph G, denoted $Result_S(Q, G)$, is the multiset of solutions made of the rows of the assignment table of $[[P]]_G$ after dropping the column ?r.*

Example 7. Consider the query:

$$Q_{S,ex} = \text{SELECT } \{?p, ?s\} \text{ WHERE BASIC } (L_{ex})$$

Let $R_{S,ex} = Gr(\{?p, ?s\}) = \{(?r, A_p, ?p), (?r, A_s, ?s)\}$ where $?r$ is a fresh variable and A_p, A_s are fresh distinct strings. Then the corresponding pattern is:

$$P_{S,ex} = \text{BASIC } (L_{ex}) \text{ BUILD } R_{S,ex}$$

The value of $P_{S,ex}$ over G_{ex} is:

$$\underline{p}_{S,ex} : R_{S,ex} \Rightarrow G'_{S,ex} \quad \text{with} \quad Tab(\underline{p}_{S,ex}) =$$

?r	?p	?s
$?r_1$	Alice	Charly
$?r_2$	Alice	David
$?r_3$	Bob	Eva

where $?r_1$, $?r_2$ and $?r_3$ are three fresh variables and $G'_{S,ex}$ is the union of G_{ex} with the following graph $H_{S,ex}$:

$(?r_1, A_p, \text{Alice})$ $(?r_1, A_s, \text{Charly})$
$(?r_2, A_p, \text{Alice})$ $(?r_2, A_s, \text{David})$
$(?r_3, A_p, \text{Bob})$ $(?r_3, A_s, \text{Eva})$

It follows that:

$$Result_S(Q_{S,ex}, G_{ex}) =$$

?p	?s
Alice	Charly
Alice	David
Bob	Eva

4 A Sound and Complete Calculus

In this section we propose a calculus for *solving* patterns and queries based on a relation over patterns called *gq-narrowing*. It computes sets of solutions of patterns (Definition 10) and results of queries (Definitions 13 and 14) over any graph. This calculus is sound and complete with respect to the set-theoretic semantics given in Sect. 3. It is based on the notion of *configuration* (Definition 15) and a function *Solve* for transforming configurations, which is defined by a rewriting system $\mathcal{R}_{gq}$ (Fig. 1).

In logic-oriented programming languages, narrowing [4] or resolution [30] derivations are used to solve goals and may have the following shape :

$$g_0 \rightsquigarrow_{[\sigma_0]} g_1 \rightsquigarrow_{[\sigma_1]} g_2 \cdots g_n \rightsquigarrow_{[\sigma_n]} g_{n+1}$$

where g_0 is the initial goal to solve (e.g., conjunction of atoms, equations or a (boolean) term), g_{n+1} is a "terminal" goal such as the empty clause, unifiable equations or the constant *true* and g_{i+1} is deduced from g_i by using a clause or a rule in the considered program via subtitution σ_i. From such a derivation, a solution is obtained by simple composition of local substitutions $\sigma_n \circ \ldots \sigma_1 \circ \sigma_0$ with restriction to variables of the initial goal g_0.

In this paper, g_0 is a configuration and the underlying program is a set of rewriting rules augmented by the graph of a considered database. This rewriting system $\mathcal{R}_{gq}$ defines the behavior of the function *Solve* for rewriting configurations. Then three functions are easily derived from *Solve*: *Solve*$_P$ for solving patterns, *Solve*$_C$ and *Solve*$_S$ for solving CONSTRUCT and SELECT queries, respectively.

An important difference between the setting developed in this paper and classical logic-oriented languages comes from the use of functional composition "$\circ$" in $\sigma_n \circ \ldots \sigma_1 \circ \sigma_0$. Depending on the shape of the considered patterns, solutions can be obtained by using additional composition operators such as Join (Definitions 8 and 10) which composes only compatible substitutions computed by different parts of a derivation (e.g., $Join(\sigma_k \circ \ldots \sigma_0, \sigma_n \circ \ldots \sigma_{k+1})$).

Remember from the previous sections that $\emptyset$ is the empty graph, $i_G = Match(\emptyset, G) = \{\emptyset_G : \emptyset \to G\}$ and $\square$ is the empty pattern.

Definition 15 (configuration). *A* configuration $[P, \underline{m}]$ *is made of a pattern P and a set of matches* $\underline{m} : L \Rightarrow G$*. Let Config denote the set of configurations. An* initial *configuration is of the form* $[P, i_G : \emptyset \Rightarrow G]$ *for some graph G and a* terminal *configuration is of the form* $[\square, \underline{m}]$*.*

Roughly speaking, a configuration $[P, \underline{m} : L \Rightarrow G]$ represents a state where the considered pattern is P and the current graph database is G, which is the target of the current set of matches $\underline{m}$. In this section we define a function:
 Solve : *Config* $\to$ *Config* by a rewriting system $\mathcal{R}_{gq}$.
This function gives rise to three functions *Solve*$_P$, *Solve*$_C$ and *Solve*$_S$ such that:
 Solve$_P(P, G) = [[P]]_G$ for every pattern P and graph G,
 Solve$_C(Q, G) = Result_C(Q, G)$ for every CONSTRUCT query Q and graph G,

and $Solve_S(Q, G) = Result_S(Q, G)$ for every SELECT query Q and graph G. For patterns this runs as follows: in order to find the set of solutions $[[P]]_G$ of a pattern P over a graph G we start from the initial configuration $[P, \underline{i}_G]$ and we apply the rewriting system $\mathcal{R}_{gq}$ to $Solve([P, \underline{i}_G])$ until we reach a terminal configuration $[\square, \underline{m} : L \Rightarrow G']$, then $\underline{m}$ is the value $[[P]]_G$ of P over G. Notice that graph G' contains G but it is not necessarily equal to G.

In Fig. 1, we provide the rewriting system $\mathcal{R}_{gq}$ which defines the function $Solve$. This function is defined by structural induction on the first component of configurations, i.e., on the patterns. The second argument of configurations, i.e., the sets of matches, in the left-hand sides of the rules, is always a variable of the form $\underline{m} : L \Rightarrow G$ or simply $\underline{m}$. This variable can be handled easily in the pattern-matching process of the left-hand sides of the rules: there is no need to use higher-order pattern-matching nor unification. In the rules of $\mathcal{R}_{gq}$, the letters P, P_1 and P_2 are variables ranging over *patterns* (sort Pat) while variables L, G and R are ranging over *graphs* (sort Gr) and $\emptyset$ is the constant denoting the empty graph. Symbol e is a variable of sort Exp and x is a variable of subsort Var while $\underline{m}$, $\underline{m}'$ and $\underline{p}$ are variables ranging over sets of matches (sort Som). The rules of Fig. 1 are not dedicated to the graphs used in this paper (Definition 1) but are rather parameterized by the kind of graphs and their corresponding homomorphisms. Indeed, rule r_1 needs the computation of possible matches between two graphs (cf. $Match(L, G)$). The nature of graphs is not specified. They can be RDF graphs, Property graphs, attributed oriented graphs, attributed hypergraphs, constrained graphs etc. The other rules use some operations already introduced in Definition 8, such as *Match, Join, Bind, Filter, Build* and *Union*. These operations are to be straightforwardly tuned according to the considered definitions of graphs and graph homomorphisms.

$$
\begin{array}{rl}
r_0 : & Solve([\square, \underline{m} : L \Rightarrow G]) \rightarrow [\square, \varnothing_G : \emptyset \Rightarrow G] \\
r_1 : & Solve([\,BASIC(L), \underline{m} : L \Rightarrow G]) \rightarrow [\square, \underline{p}] \text{ where } \underline{p} = Match(L, G) \\
r_2 : & Solve([P_1 \text{ JOIN } P_2, \underline{m}]) \rightarrow Solve_{JL}(Solve([P_1, \underline{m}]), P_2) \\
r_3 : & Solve_{JL}([\square, \underline{m}], P) \rightarrow Solve_{JR}(\underline{m}, Solve([P, \underline{m}])) \\
r_4 : & Solve_{JR}(\underline{m}, [\square, \underline{m}']) \rightarrow [\square, \underline{p}] \text{ where } \underline{p} = Join(\underline{m}, \underline{m}') \\
r_5 : & Solve([P \text{ BIND } e \text{ AS } x, \underline{m}]) \rightarrow Solve_{BI}(Solve([P, \underline{m}]), e, x) \\
r_6 : & Solve_{BI}([\square, \underline{m}], e, x) \rightarrow [\square, \underline{p}] \text{ where } \underline{p} = Bind(\underline{m}, e, x) \\
r_7 : & Solve([P \text{ FILTER } e, \underline{m}]) \rightarrow Solve_{FR}(Solve([P, \underline{m}]), e) \\
r_8 : & Solve_{FR}([\square, \underline{m}], e) \rightarrow [\square, \underline{p}] \text{ where } \underline{p} = Filter(\underline{m}, e) \\
r_9 : & Solve([P \text{ BUILD } R, \underline{m}]) \rightarrow Solve_{BU}(Solve([P, \underline{m}]), R) \\
r_{10} : & Solve_{BU}([\square, \underline{m}], R) \rightarrow [\square, \underline{p}] \text{ where } \underline{p} = Build(\underline{m}, R) \\
r_{11} : & Solve([P_1 \text{ UNION } P_2, \underline{m}]) \rightarrow Solve_{UL}(Solve([P_1, \underline{m}]), P_2) \\
r_{12} : & Solve_{UL}([\square, \underline{m}], P) \rightarrow Solve_{UR}(\underline{m}, Solve([P, \underline{m}])) \\
r_{13} : & Solve_{UR}(\underline{m}, [\square, \underline{m}']) \rightarrow [\square, \underline{p}] \text{ where } \underline{p} = Union(\underline{m}, \underline{m}')
\end{array}
$$

Fig. 1. $\mathcal{R}_{gq}$: Rewriting rules for patterns

Rule r_0 considers the degenerated case when one looks for solutions of the empty pattern $\square$. In this case there is no solution and the empty set of matches $\varnothing_G$ is computed. Rule r_1 is key in this calculus because it considers basic patterns of the form $\text{BASIC}(L)$ where L is a graph which may contain variables. In this case $Solve([\text{BASIC}(L), \underline{m}])$ consists in finding all matches from L to G. These matches can instantiate variables in L. Thus, the constraint $\underline{p} = Match(L, G)$ of rule r_1 instantiates variables occurring in graph L. This variable instantiation process is close to the narrowing or the resolution-based calculi.

In the context of functional-logic programming languages, several strategies of narrowing-based procedures have been developed to solve goals including even a needed strategy [4]. In this paper, we do not need all the power of narrowing procedures because manipulated data are mostly flat (mainly constants and variables). Thus the unification process used at every step in the narrowing relation is beyond our needs while simple pattern-matching as in classical rewriting is not enough since variables in patterns P cannot be instantiated by simply rewriting the initial term $Solve([P, \underline{i}_G : \emptyset \Rightarrow G])$. Consequently, we propose hereafter a new relation induced by the above rewriting system that we call gq-narrowing. Before the definition of this relation, we recall briefly some notations about first-order terms. Readers not familiar with such notations may consult, e.g., [5].

Definition 16 (position, subterm replacement, substitution, $t\downarrow_{gq}$). *A position is a sequence of positive integers identifying a subterm in a term. For a term t, the empty sequence, denoted Λ, identifies t itself. When t is of the form $g(t_1, \ldots, t_n)$, the position $i.p$ of t with $1 \leq i \leq n$ and p a position in t_i, identifies the subterm of t_i at position p. The subterm of t at position p is denoted $t|p$ and the result of replacing the subterm of t at position p with term s is written $t[s]_p$. We write $t\downarrow_{gq}$ for the term obtained from t where all expressions of $\mathcal{GQ}$-algebra (i.e., operations such as Join, Bind, Filter, Match, etc.) have been evaluated. A substitution σ is a mapping from variables to terms. When $\sigma(x) = u$ with $u \neq x$, we say that x is in the domain of σ. We write $\sigma(t)$ to denote the extension of the application of σ to a term t which is defined inductively as $\sigma(c) = c$ if c is a constant or c is a variable outside the domain of σ. Otherwise $\sigma(f(t_1, \ldots, t_n)) = f(\sigma(t_1), \ldots, \sigma(t_n))$.*

We write $\mathcal{P}_{gq}(\mathcal{V})$ for the term algebra over the set of variables $\mathcal{V}$ generated by the operations occurring in the rewriting system $\mathcal{R}_{gq}$.

Definition 17 (gq-narrowing $\rightsquigarrow$). *The rewriting system $\mathcal{R}_{gq}$ defines a binary relation $\rightsquigarrow$ over terms in $\mathcal{P}_{gq}(\mathcal{V})$ that we call gq-narrowing relation. We write $t \rightsquigarrow_{[u,lhs\rightarrow rhs,\sigma]} t'$ or simply $t \rightsquigarrow t'$ and say that t is gq-narrowable to t' iff there exists a rule $lhs \rightarrow rhs$ in the rewriting system $\mathcal{R}_{gq}$, a position u in t and a substitution σ such that $\sigma(lhs) = t|_u$ and $t' = t[\sigma(rhs)\downarrow_{gq}]_u$. Then $\rightsquigarrow^*$ denotes the reflexive and transitive closure of the relation $\rightsquigarrow$.*

Notice that in the definition of term $t' = t[\sigma(rhs)\downarrow_{gq}]_u$ above, the substitution σ is not applied to t as in narrowing ($\sigma(t[rhs]_u \downarrow_{gq})$ but only to the right-hand side ($\sigma(rhs)$). This is mainly due to the possible use of additional function composition such as *Join* operation. If we consider again rule r_1 in Fig. 1, t' would be of the following shape $t' = t[\Box, (Match(\sigma(L), G) : \sigma(L) \Rightarrow G)\downarrow_{gq}]_u$. In this case, the evaluation of *Match* operation instantiates possible variables occurring in the pattern $\text{BASIC}(\sigma(L))$ just like classical narrowing procedures.

The first nice property of the proposed calculus is termination (Proposition 1). In addition, all gq-narrowing derivation steps for solving patterns are needed since at each step only one position is candidate to a gq-narrowing step (Proposition 2). Last but not least, the proposed calculus is sound and complete with respect to the formal semantics given in Sect. 3 (Theorems 1 and 2).

Proposition 1 (termination). *The relation $\rightsquigarrow$ is terminating.*

Proposition 2 (determinism). *Let $t_0 \rightsquigarrow t_1 \rightsquigarrow \ldots \rightsquigarrow t_n$ be a gq-narrowing derivation with $t_0 = Solve([P, \underline{i}_G])$. For all $i \in [0..n-1]$, there exists at most one position u_i in t_i such that t_i is gq-narrowable.*

Theorem 1 (soundness). *Let G be a graph, P a pattern and $\underline{m}$ a set of matches such that $Solve([P, \underline{i}_G]) \rightsquigarrow^* [\Box, \underline{m}]$. Then for all morphisms m in $\underline{m}$, there exists a morphism m' equals to m up to renaming of variables such that m' is in $[[P]]_G$.*

Theorem 2 (completeness). *Let G_1, G_2 and X be graphs, P a pattern and $h : X \Rightarrow G_2$ a match in $[[P]]_{G_1}$. Then there exist graphs G'_2 and X', a set of matches $\underline{m} : X' \Rightarrow G'_2$, a derivation $Solve([P, \underline{i}_{G_1}] \rightsquigarrow^* [\Box, \underline{m}]$ and a match $m : X' \Rightarrow G'_2$ in $\underline{m}$ such that m and h are equal up to variable renaming.*

Now we use the *Solve* function for tackling patterns and queries as in Sect. 3.2. The following three corollaries are obvious consequences of the above results. Remember that the *value* of a pattern P over a graph G is a set of matches $[[P]]_G$, while the *result* of a query Q over a graph G is a graph $Result_C(Q, G)$ when Q is a CONSTRUCT query and a table $Result_S(Q, G)$ when Q is a SELECT query. For SELECT queries we use the graph $Gr(S)$ associated to the set of variables S as in Definition 14 (see [18] for details).

Corollary 1 (solving patterns). *Let P be a pattern. For every graph G there is exactly one gq-narrowing derivation of the form:*

$$Solve([P, \underline{i}_G : \emptyset \Rightarrow G]) \rightsquigarrow^* [\Box, \underline{m}]$$

and then $\underline{m}$ is the value $Solve_P(P, G) = [[P]]_G$ of P over G.

Example 8. As in Example 5 we consider the pattern:

$$
\begin{aligned}
P_{ex} &= \text{BASIC } (L_{ex}) \text{ BUILD } R_{ex} \\
&= \text{BASIC } (\{(?p, \text{teaches}, ?t), (?s, \text{studies}, ?t)\}) \\
&\quad \text{BUILD } \{(?p, \text{teaches}, ?z), (?s, \text{studies}, ?z)\}
\end{aligned}
$$

The gq-narrowing derivation is as follows:

$$
\begin{aligned}
Solve([P_{ex}, \underline{i}_{G_{ex}}]) &\rightsquigarrow_{r_9} Solve_{BU}(Solve([\text{ BASIC } (L_{ex}), \underline{i}_{G_{ex}}], R_{ex})) \\
&\rightsquigarrow_{r_1} Solve_{BU}([\Box, Match(L_{ex}, G_{ex})], R_{ex}) \\
&\rightsquigarrow_{r_{10}} [\Box, Build(Match(L_{ex}, G_{ex}), R_{ex})]
\end{aligned}
$$

We know from Example 5 that $Build(Match(L_{ex}, G_{ex}), R_{ex}) = \underline{p}_{ex} : R_{ex} \Rightarrow G'_{ex}$, thus we get the required result.

Corollary 2 (solving CONSTRUCT queries). *Let R be a graph and P a pattern. Consider the query $Q = $ CONSTRUCT R WHERE P. For every graph G there is exactly one gq-narrowing derivation of the form:*

$$Solve([P \text{ BUILD } R, \underline{i}_G]) \leadsto^* [\square, \underline{m}]$$

and then the graph image of R by $\underline{m}$ is the result $Solve_C(Q, G) = Result_C(Q, G)$ of Q over G.

Corollary 3 (solving SELECT queries). *Let S be a set of variables and P a pattern. Consider the query $Q = $ SELECT S WHERE P. For every graph G there is exactly one gq-narrowing derivation of the form:*

$$Solve([P \text{ BUILD } Gr(S), \underline{i}_G]) \leadsto^* [\square, \underline{m}]$$

and then the table obtained by dropping the first column from $Tab(\underline{m})$, as in Definition 14, is the result $Solve_S(Q, G) = Result_S(Q, G)$ of Q over G.

5 Conclusion and Related Work

We propose a sound and compete rule-based calculus for a core graph query language (cf. Figure 1 and Corollaries 2 and 3). The calculus is generic. We illustrate it on RDF graphs to keep examples concise but it can easily be extended and adapted to various data graphs, e.g. Property Graphs, provided that the notion of matches is well defined (cf. rule r_1 in Fig. 1). The syntax of queries was inspired by current implemented languages such as SPARQL, Cypher or preliminary papers about GQL. We were particularly keen to tackle graph-to-graph queries in addition to classical graph-to-relation queries.

Due to the different outcomes of SELECT and CONSTRUCT queries, the composition of graph queries is not straightforward, which contrasts with the situation in the context of relational databases [15,28]. A particular query nesting, namely EXISTS subqueries, has been implemented in some languages such as Cypher [23] or PGQL [33]. In this paper, we propose the notion of patterns as the main syntactic means to formulate queries. Patterns are defined as terms (trees) on purpose, in order to make it easier to nest patterns at will. Classical composition of (graph) homomorphisms ensures for free the composition of the semantics of nested patterns. It is only at the end of the resolution of a pattern that one chooses to act as a SELECT query and return a table or to act as a CONSTRUCT query and return a graph. One may also choose to act as both kinds of queries in a novel query form we call CONSELECT in [20] by returning at the same time a table and a graph when a pattern is solved.

The results of this paper can be extended to actual graph query languages. For instance, path variables may be added to the syntax and matches between two graphs L and G, as in rule r_1 of Fig. 1, can be constrained by positive, negative or path constraints and written $Match(L, G, \Phi)$ where Φ represents constraints in a given logic (e.g., [31]).

To our knowledge, the proposed calculus is the first sound and complete rule-based calculus dedicated to graph query languages featuring graph-to-graph queries and aggregation operators. The proposed procedure is terminating and does not develop unnecessary derivations. The reader familiar with rewriting systems would notice that a naive and straightforward operationalization of the formal semantics would lead to a rewriting system with fewer rules than those proposed in Fig. 1 but which is not confluent and not all of its derivations yield sound answers.

Among related work, we quote first the use of declarative (functional and logic) languages in the context of relational databases (see, e.g. [1,9,26]). In these works, the considered databases follow the relational paradigm which differs from the graph-oriented one that we are tackling in this paper. Our aim here is not to make connections between graph query languages and functional logic ones. We are rather interested in investigating formally graph query languages, and particularly in using dedicated rewriting techniques for such languages.

The notion of *pattern* present in this paper is close to the syntactic notions of *clauses* in [23] or *graph patterns* in [3]. For such syntactic notions, some authors associate as semantics sets of variables bindings (tables) as in [22,23,32] or simply graphs as in [2]. In our case, we associate both variable bindings and graphs since we associate sets of graph homomorphisms to patterns. This semantics is borrowed from a previous work on formal semantics of graph queries based on category theory [17]. Our semantics allows composition of patterns in a natural way. Such composition of patterns is not easy to catch if the semantics is based only on variable bindings but can be recovered when queries have graph outcomes as in G-CORE [2].

Last but not least, the patterns and queries considered in this paper may be seen as formulas of a logic having graphs as interpretations or models. In [21, 24], a graph logic, called *Nested Graph Conditions* (NGC) has been introduced and used to express conditions on graphs. NGC Formulas can be nested and have graph homomorphisms as semantics just like the patterns considered in the present paper. NGC allows one to state conditions on the shape of graphs. In [36], an extension of NGC to attributed graphs has been proposed. NGC formulas can be used as graph queries but do not provide some of desirable query features such as aggregation operators nor do they allow graph transformations as in CONSTRUCT queries or patterns with BUILD operator. Actually, a pattern with a BUILD operator acts as a formula of a dynamic logic (see, e.g. [6]). An operationalization of NGC graph queries has been proposed in [7] where the authors define a set of rewrite rules based on the PO (pushout) approach to compute query solutions. With [7], we share the same abstract definition of queries [7, Definition 2] in the sense that any query is characterized by a so-called *request graph* which corresponds to the notion of *scope graph* of patterns. However, the notion of a query answer according to [7, Definition 3] is defined as a graph homomorphism having the queried graph G as co-domain. This is a particular case of our definition of a query answer. Indeed, the co-domain of a query answer, as we define in the present paper, is the graph G' equals to the queried graph G augmented by the different actions underlying the BUILD and BIND operators involved in the considered pattern or query. So, the proposed rewriting system in [7] departs from ours. Their rules are well adapted to NGC conditions and thus fail to add new items (nodes or edges) to queried graphs and do not take into account aggregations.

Future work includes an implementation of the proposed calculus as well as the investigation of validation techniques for graph database languages, including verification methods e.g., [10,11] or test techniques as proposed for example in [29].

References

1. Almendros-Jiménez, J.M., Becerra-Terón, A.: A safe relational calculus for functional logic deductive databases. Electron. Notes Theor. Comput. Sci. **86**(3), 168–204 (2003)
2. Angles, R., et al.: G-CORE: A core for future graph query languages. In: Das, G., Jermaine, C.M., Bernstein, P.A. (eds.), Proceedings of the 2018 International Conference on Management of Data, SIGMOD Conference 2018, Houston, TX, USA, June 10–15, 2018, pp. 1421–1432. ACM (2018)
3. Angles, R., Arenas, M., Barceló, P., Hogan, A., Reutter, J.L., Vrgoc, D.: Foundations of modern query languages for graph databases. ACM Comput. Surv. **50**(5), 68:1–68:40 (2017)
4. Antoy, S., Echahed, R., Hanus, M.: A needed narrowing strategy. J. ACM **47**(4), 776–822 (2000)
5. Baader, F., Nipkow, T.: Term Rewriting and All That. Cambridge University Press (1998)
6. Balbiani, P., Echahed, R., Herzig, A.: A dynamic logic for termgraph rewriting. In: Ehrig, H., Rensink, A., Rozenberg, G., Schürr, A. (eds.) ICGT 2010. LNCS, vol. 6372, pp. 59–74. Springer, Heidelberg (2010). https://doi.org/10.1007/978-3-642-15928-2_5
7. Beyhl, T., Blouin, D., Giese, H., Lambers, L.: On the operationalization of graph queries with generalized discrimination networks. In: Echahed, R., Minas, M. (eds.) ICGT 2016. LNCS, vol. 9761, pp. 170–186. Springer, Cham (2016). https://doi.org/10.1007/978-3-319-40530-8_11
8. Bork, D., Karagiannis, D., Pittl, B.: A survey of modeling language specification techniques. Inf. Syst. **87** (2020)
9. Braßel, B., Hanus, M., Müller, M.: High-level database programming in curry. In: Hudak, P., Warren, D.S. (eds.) PADL 2008. LNCS, vol. 4902, pp. 316–332. Springer, Heidelberg (2007). https://doi.org/10.1007/978-3-540-77442-6_21
10. Brenas, J.H., Echahed, R., Strecker, M.: Verifying graph transformation systems with description logics. In: Lambers, L., Weber, J. (eds.) ICGT 2018. LNCS, vol. 10887, pp. 155–170. Springer, Cham (2018). https://doi.org/10.1007/978-3-319-92991-0_10
11. Brenas, J.H., Echahed, R., Strecker, M.: Reasoning formally about database queries and updates. In: ter Beek, M.H., McIver, A., Oliveira, J.N. (eds.) FM 2019. LNCS, vol. 11800, pp. 556–572. Springer, Cham (2019). https://doi.org/10.1007/978-3-030-30942-8_33

12. Chamberlin, D.D., Boyce, R.F.: SEQUEL: a structured English query language. In: Rustin, R. (ed.) FIDET 1974: Data models: data-structure-set versus relational: Workshop on Data Description, Access, and Control, May 1–3, 1974, pp. 249–264. Michigan, Ann Arbor (1974)

13. Corradini, A., Duval, D., Echahed, R., Prost, F., Ribeiro, L.: AGREE – algebraic graph rewriting with controlled embedding. In: Parisi-Presicce, F., Westfechtel, B. (eds.) ICGT 2015. LNCS, vol. 9151, pp. 35–51. Springer, Cham (2015). https://doi.org/10.1007/978-3-319-21145-9_3

14. Corradini, A., Montanari, U., Rossi, F., Ehrig, H., Heckel, R., Löwe, M.: Algebraic approaches to graph transformation - part I: basic concepts and double pushout approach. In: Handbook of Graph Grammars and Computing by Graph Transformations, Volume 1: Foundations, pp. 163–246 (1997)

15. Date, C.J.: A guide to the SQL standard: a user's guide to the standard relational language SQL (1987)

16. de la Tour, T.B., Echahed, R.: Parallel rewriting of attributed graphs. Theor. Comput. Sci. **848**, 106–132 (2020)

17. Duval, D., Echahed, R., Prost, F.: An algebraic graph transformation approach for RDF and SPARQL. In: Hoffmann, B., Minas, M. (eds.) Proceedings of the Eleventh International Workshop on Graph Computation Models, Online-Workshop, 24th June 2020, vol. 330. EPTCS, pp. 55–70 (2020)

18. Duval, D., Echahed, R., Prost, F.: All you need is CONSTRUCT. CoRR, abs/2010.00843 (2020)

19. Duval, D., Echahed, R., Prost, F.: Querying RDF databases with sub-constructs. In: Kutsia, T. (ed.), Proceedings of the 9th International Symposium on Symbolic Computation in Software Science, SCSS 2021, Hagenberg, Austria, September 8–10, 2021, vol. 342, EPTCS, pp. 49–64 (2021)

20. Duval, D., Echahed, R., Prost, F. A rule-based operational semantics of graph query languages. CoRR, abs/2202.10142 (2022)

21. Ehrig, H., Ehrig, K., Habel, A., Pennemann, K.-H.: Constraints and application conditions: from graphs to high-level structures. In: Ehrig, H., Engels, G., Parisi-Presicce, F., Rozenberg, G. (eds.) ICGT 2004. LNCS, vol. 3256, pp. 287–303. Springer, Heidelberg (2004). https://doi.org/10.1007/978-3-540-30203-2_21

22. Francis, N., et al.: Formal semantics of the language cypher. CoRR, abs/1802.09984 (2018)

23. Francis, N., et al.: Cypher: An evolving query language for property graphs. In: SIGMOD Conference, pp. 1433–1445. ACM (2018)

24. Habel, A., Pennemann, K.: Correctness of high-level transformation systems relative to nested conditions. Math. Struct. Comput. Sci. **19**(2), 245–296 (2009)

25. Habel, A., Plump, D.: Complete strategies for term graph narrowing. In: Fiadeiro, J.L. (ed.) WADT 1998. LNCS, vol. 1589, pp. 152–167. Springer, Heidelberg (1999). https://doi.org/10.1007/3-540-48483-3_11

26. Hanus, M.: Dynamic predicates in functional logic programs, vol. 2004. EAPLS (2004)

27. Huang, S.S., Green, T.J., Loo, B.T.: Datalog and emerging applications: an interactive tutorial. In: Sellis, T.K., Miller, R.J., Kementsietsidis, A., Velegrakis, Y. (eds.) Proceedings of the ACM SIGMOD International Conference on Management of Data, SIGMOD 2011, Athens, Greece, June 12–16, 2011, pp. 1213–1216. ACM (2011)

28. Kim, W.: On optimizing an SQL-like nested query. ACM Trans. Database Syst. **7**(3), 443–469 (1982)

29. Lambers, L., Schneider, S., Weisgut, M.: Model-based testing of read only graph queries. In: 13th IEEE International Conference on Software Testing, Verification and Validation Workshops, ICSTW 2020, Porto, Portugal, October 24–28, 2020, pp. 24–34. IEEE (2020)
30. Lloyd, J.W.: Foundations of Logic Programming, 2nd edn. Springer, Heidelberg (1987). https://doi.org/10.1007/978-3-642-96826-6
31. Navarro, M., F. Orejas, E. Pino, and L. Lambers. A navigational logic for reasoning about graph properties. J. Log. Algebraic Methods Program. 118 (2021)
32. Pérez, J., Arenas, M., Gutiérrez, C.: Semantics and complexity of SPARQL. ACM Trans. Database Syst. 34(3), 16:1–16:45 (2009)
33. PGQL 1.5 Specification. https://pgql-lang.org/spec/1.5/, August 2022
34. Rozenberg, G. (ed.) Handbook of Graph Grammars and Computing by Graph Transformations, Volume 1: Foundations. World Scientific (1997)
35. Sakr, S., et al.: The future is big graphs: a community view on graph processing systems. Commun. ACM 64(9), 62–71 (2021)
36. Schneider, S., Lambers, L., Orejas, F.: Automated reasoning for attributed graph properties. Int. J. Softw. Tools Technol. Transfer 20(6), 705–737 (2018). https://doi.org/10.1007/s10009-018-0496-3
37. SPARQL 1.1 Query Language. W3C Recommendation, March 2013
38. RDF 1.1 Concepts and Abstract Syntax. W3C Recommendation, February 2014

Advanced Consistency Restoration
with Higher-Order Short-Cut Rules

Lars Fritsche[1]($\boxtimes$)(iD), Jens Kosiol[2](iD), Adrian Möller[1](iD), and Andy Schürr[1](iD)

[1] Technical University Darmstadt, Darmstadt, Germany
{lars.fritsche,andy.schuerr}@es.tu-darmstadt.de,
adrian.moeller@stud.tudarmstadt.de
[2] Philipps-Universität Marburg, Marburg, Germany
kosiolje@mathematik.uni-marburg.de

Abstract. Sequential model synchronisation is the task of propagating changes from one model to another correlated one to restore consistency. It is challenging to perform this propagation in a least-changing way that avoids unnecessary deletions (which might cause information loss). From a theoretical point of view, so-called *short-cut (SC) rules* have been developed that enable provably correct propagation of changes while avoiding information loss. However, to be able to react to every possible change, an infinite set of such rules might be necessary. Practically, only small sets of pre-computed *basic SC rules* have been used, severely restricting the kind of changes that can be propagated without loss of information. In this work, we close that gap by developing an approach to compute more complex required SC rules on-the-fly during synchronisation. These *higher-order SC rules* allow us to cope with more complex scenarios when multiple changes must be handled in one step. We implemented our approach in the model transformation tool eMoflon. An evaluation shows that the overhead of computing *higher-order SC rules* on-the-fly is tolerable and at times even improves the overall performance. Above that, completely new scenarios can be dealt with without the loss of information.

Keywords: Model synchronisation · Triple-graph grammars · Least-change synchronisation · Consistency restoration

1 Introduction

Model-Driven Engineering (MDE) [6] provides the necessary means to tackle the challenges of modern software systems, which become more and more complex and distributed. Using MDE, a system can be described by various models that describe different aspects and provide specific views onto the system itself, where each view may overlap to some degree with other views of the same system.

This work was partially funded by the German Research Foundation (DFG), project "Triple Graph Grammars (TGG) 3.0".

M. Fernández and C. M. Poskitt (Eds.): ICGT 2023, LNCS 13961, pp. 184–203, 2023.
https://doi.org/10.1007/978-3-031-36709-0_10

Consequently, if one view changes, we have to propagate these changes to other views that share the same information. This is necessary to ensure that all models together consistently represent the overall system state. The propagation process is often called *model synchronisation*, where we distinguish between *sequential* and *concurrent* model synchronisation. In this paper, we will focus on the former, where only one model is changed at a time, while the latter case would also incorporate changes to multiple models at once.

As our methodology of choice, we employ *Triple Graph Grammars* (TGGs) [25], which are a declarative way to specify a consistency relationship between possibly heterogeneous models by means of a set of graph grammar rules. By transforming these rules, we can automatically derive different *consistency restoring operators* that served as central ingredients for various model management processes, such as batch translators [7,16,25], consistency checkers [21] or sequential [12,14,20,22] and concurrent synchronisers [10,24,27]. In general, synchronisers are required to avoid information loss. Most TGG-based approaches, in fact, do not meet this requirement: They propagate changes by deleting (parts of) the model that is to be synchronised and then retranslating (the missing parts) from the updated model [14,17,20–22]. This procedure is both inefficient and prone to loss of information that is private to the synchronised model. For this purpose, we introduced *Short-Cut Rules* [11,12,19], which implement consistency and information-preserving modifications. Before, an inconsistency was resolved by revoking previous translation steps and retranslating these elements from anew. A *Short-Cut Rule* does the same but in one step, which lets us identify and preserve information that is deleted just to be recreated. Even more, we can provide conditions under which the application of a *Short-Cut Rule* reestablishes or at least improves consistency [18,19]. The translation and, thus, revocation of former translation steps is based on basic TGG rules that describe atomic consistency-preserving model changes. The same holds for our first works on *Short-Cut Rules* [11,12], where we created them by combining pairs of basic TGG rules. While the resulting *Short-Cut Rules* describe useful repairs in many situations, we found that there are cases where we have to repair multiple inconsistencies at once to yield a better result. Hence, we propose to concatenate TGG rules that describe a sequence of atomic model changes and derive *Short-Cut Rules* from these to repair multiple TGG rule applications at once. Since there are usually infinitely many possibilities of concatenating TGG rules and thus infinitely many *Short-Cut Rules*, we propose an analysis that determines at runtime which TGG rules to concatenate to construct a *Short-Cut Rule* for a particular scenario.

This paper is structured as follows: We introduce TGGs and *Short-Cut Rules* together with our running example in Sect. 2. In Sect. 3, we will present our novel analysis to determine non-trivial consistency restoring operators in the form of *Short-Cut Rules* at runtime. Section 4 presents our evaluation investigating how much information is preserved and whether it comes with additional costs. Finally, in Sect. 5, we discuss related works and summarise open challenges in Sect. 6.

2   Fundamentals

In our running example throughout this paper, we will define consistency
between terrace house and construction planning models to which we will refer
in the following as source and target, respectively. Due to the complexity of
the derived repair operations later in this paper, we had to choose a rather
small example. Yet, TGGs have been successfully applied in industrial applications [2,3,13]. As a first step, we introduce a third correspondence model between
source and target, which connects elements from both sides and thus makes corresponding information traceable. Figure 1 depicts the three metamodels with
the source metamodel on the left, the target metamodel on the right and the
correspondence metamodel in between. The source metamodel consists solely of
the House class and HouseType enum. Each House has a reference to the next
(neighbour) House and contains information about its HouseType, which will
determine the architecture of it on the target side. Additionally, each building contains information about its architect, which has no representation in the
target model. On the opposite side, there are six classes. There is a (construction) Plan for every row of houses that contains the corresponding Constructions.
Each Construction contains the name of its assigned (construction) company and
a sequence of Construction Steps consisting of Cellar, Basement and Saddle Roof
elements that have to be processed in the given order. Finally, we connect both
metamodels using the correspondence type depicted as a hexagon, which maps
Buildings to their corresponding Construction.

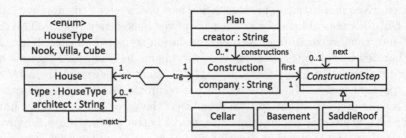

Fig. 1. Source, Target and Correspondence Metamodel

To define a consistency relationship between source and target models, we
employ Triple Graph Grammars (TGGs) [25], which are a declarative and rule-
based way to achieve this. As the name indicates, the rules of a TGG span the
language of all consistent triple graphs, where our source, target and correspondence model are interpreted as graph-like structures. Figure 2 shows the TGG
ruleset of our running example consisting of two rules on the left. Before a rule
can be applied, its precondition must be met, meaning that all context elements
in black must exist and all attribute conditions hold. When the rule is applied,
all elements depicted in green and annotated with ++ are created. The first
rule is Nook Rule, which can be applied without any precondition. It creates a

Nook House together with a corresponding Construction on the other side and a connecting correspondence link. Since this will create the first Building in the row, we also create a Plan on the target side. Finally, each Nook House must have a Basement but no Cellar or Saddle Roof. The other rules Cube Rule and Villa Rule are very similar in that they require a Building in the row that must already exist together with a corresponding Construction and Plan. Given that, they create a House with the Cube or Villa HouseType as the next House in line and a corresponding Construction, where a Cube has a Cellar and Basement, while a Villa will have a Basement and Saddle Roof.

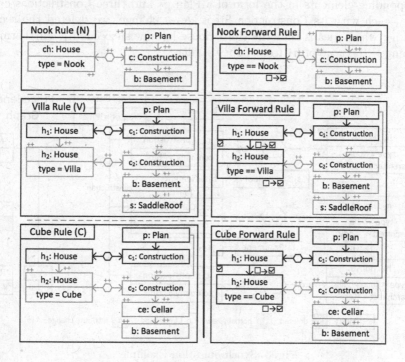

Fig. 2. TGG Rules

Using these rules, we can create consistent models from scratch. More interestingly, we can transform these rules to obtain forward translation rules as is depicted on the right in Fig. 2. The main difference to our original rules is that their created source elements have now become part of the precondition. Assuming that we want to translate a source to a target model, this makes sense as a source model must already exist beforehand. To avoid translating elements more than once, we introduce the annotations $\square \rightarrow \boxtimes$ and $\boxtimes$.[1] Source elements that were created in the original TGG rule are annotated with $\square \rightarrow \boxtimes$ in the derived forward rule. This indicates that they mark elements upon translation

---

[1] These markings are not needed for the original TGG rules.

and only if the element has not been marked yet. Other source elements that were already black are annotated with ☑, which means that these elements must have already been marked; different formalisations of such a marking mechanism are available [16,18,22]. Also, attribute assignments now turn to attribute constraints.

We want to incrementally synchronise changes from one model to another. Figure 3 depicts a small model, which was consistent w.r.t. our TGG but then was changed on the source side. On the left, we see a Nook House h_1, followed by a Villa h_2 and a Cube h_3, while on the right side there are the expected corresponding elements in the form of a Plan p_1 and three Constructions c_1, c_2 and c_3, each with its Construction Steps. As a change, we deleted the second House h_2, which means that the third House h_3 now succeeds h_1. Furthermore, we changed the type of h_3 from Cube to Villa.

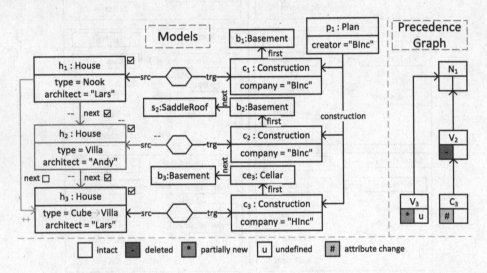

Fig. 3. Synchronisation Example

On the right side, we can see a so-called *precedence graph*, which is of particular interest for the rest of this paper. Some of its nodes correspond to the rule applications that created the original triple graph; others represent rule applications that could have possibly been used to create the newly added elements on the source side. Edges denote sequential dependence between rule applications. Each *precedence node* contains the initials of the corresponding TGG rule together with a small index, which is the same as the created elements of this rule application. Apart from that, each node has two boxes where the left and right depict the consistency state of the source and the target side, respectively. Blank boxes indicate that the elements on this side are still intact w.r.t. the corresponding TGG rule, which means that they have not been tampered with or that the changes had no consistency violating effect, e.g., by changing the architect. Green boxes containing a "+" indicate that new elements have been detected, which can be translated by a TGG rule, while "*" would mean that there are new elements but they cannot be translated because some elements that would need to be marked together with them are already marked, i.e., they are still part of another (possibly inconsistent) rule application. In both cases, the right box for the target side is annotated with "u", expressing our lack of information on whether the corresponding target elements already exist. In contrast, red boxes containing a "-" indicate that all translatable elements on this side were deleted, meaning that created elements on the opposite side should be removed as well. A red box with "/" states that some but not all translatable elements were deleted, which means that the remaining elements must be translated differently to before. Finally, an orange box with "#" means that an attribute was changed such that a rule application has become invalid, e.g., the HouseType. A formalisation of precedence graphs via (partial) comatchs can be found in [18].

Regarding our example, we have an intact Nook Rule application N_1 on top. The Villa Rule application V_2 depends on N_1 as it provides the previous House, Construction and Plan. V_2 itself is no longer intact as indicated by "-" on the source side due to the deletion of House h_2. Cube Rule application C_3, which depends on V_2 is also no longer intact due to an attribute change within House h_3 from HouseType Cube to Villa, which is denoted as "#". Due to this attribute change and the new edge between House h_1 and h_3, we could retranslate h_3 using Villa Rule as represented via the precedence node V_3. However, since h_3 is still part of the former rule application C_3, we find an "*" annotation on the source side because we need to revoke this rule application before retranslating h_3.

Those annotations and the actual precedence graph are constructed using the results of an incremental graph pattern matcher (IGPM) that tracks all rules' pre- and postconditions. If a postcondition is violated then the IGPM engine will notify us that the postcondition can no longer be matched and we can analyse which model change caused this. For preconditions, we also track whether the source (or target) side are matched for forward (or backward) rules, which we refer to as source (or target) matches. This gives us all possible translation steps even if the necessary context has not yet been created on the target side that would enable the translation to be executed. Note, however, that some steps may be mutually exclusive either because they would translate the same elements or because they rely on other rules to create necessary context on the opposite side.

One sees that the above-described change of moving the third house and changing an attribute should only trigger a change of Construction Steps while some information such as the architect should persist. To achieve this, we use so-called *Short-Cut Rules* [12], which describe consistency-preserving operations, e.g., moving a house in the row without losing the target side information. Basically, a *Short-Cut Rule* revokes a rule application and applies another one instead, while preserving those elements that would be deleted just to be recreated. For our example, a *Short-Cut Rule* could exchange the HouseType while adding the missing Construction Steps and removing the now superfluous Construction Steps. A TGG *Short-Cut Rule* is created by overlapping an inversed TGG rule (the *replaced* rule that revokes a former rule application) with another TGG rule (the *replacing* rule). Deleted elements from the replaced rule that are overlapped with created elements from the replacing rule are preserved and become context as they would be unnecessarily deleted and then recreated. Consequently, deleted elements from the replaced rule that are not in the overlap must be removed and, analogously, created elements from the replacing rule that are not part of the overlap must be created. Finally, only the attribute conditions from the replacing rule must hold after applying the *Short-Cut Rule* as we want to revoke the former replaced rule application.

The precedence graph tells us which TGG rules to overlap with each other. There is an invalid Villa Rule application, while there is a new Cube Rule application that could be applied instead to make the now necessary transformation steps for Cube. We, thus, overlap Villa Rule with Cube Rule such that the parent House is not overlapped as it has changed, while the created Houses are assumed to be the same as well as their corresponding Constructions, the Plans and Basements. Note that generally, there are many ways to overlap rules, which means that there may be many possible *Short-Cut Rules* and that this process is completely automated. Figure 4 depicts the resulting *Short-Cut Rule* on the left. It tells us that we can move a House and change its HouseType from Villa to Cube at the same time if, on the opposite side, we remove the superfluous Cellar and add a new Saddle Roof.

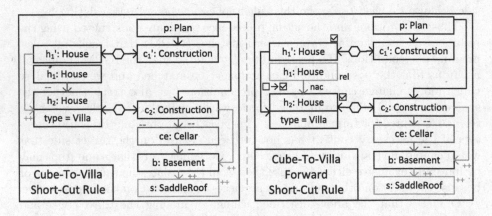

Fig. 4. Cube-To-Villa *Short-Cut Rule*

As with TGG rules, *Short-Cut Rules* can be transformed to yield forward and backward operationalised versions. On the right of Fig. 4, we can see its forward operationalisation. Intuitively, we have to make sure that a user made the same changes as the *Short-Cut Rule* on the source side so that applying the forward operationalisation will propagate these changes correctly. Similar to before, formerly created elements on the source side must be marked because they are new, while black elements remain marked. Deleted elements on the source side must have been deleted by a user change. Hence, we must ensure that these elements do no longer exist, which is expressed as a *negative application condition* depicted in blue and annotated with "nac". Also note that some context elements are omitted from the rule that would stem from the replaced rule, e.g., the context construction. We can leave some of them out, if they are not needed to perform the Short-Cut Rule. This is the case for the House h1, which is needed to check whether the edge to h2 has been deleted. The "rel" annotation on the right side indicates that h1 is relaxed. Relaxed means that this element does not necessarily need to exist, because if it was deleted together with its adjacent edges, then the "nac" is satisfied.

Using Cube-To-Villa *forward Short-Cut Rule*, we can now resolve the situation from before by first processing the deleted second House and deleting its corresponding parts and then repairing the target side of the third House by using this rule. While the result looks very similar to the one of translating the whole source model from scratch, we still have the information about each Constructions' architect, which means that this information is no longer lost during synchronisation. Another advantage is that moving a House in a long row of terrace Houses can be very expensive as all succeeding rule applications would have to be revoked. In many cases, *Short-Cut Rules* can also help with this issue by preserving the consistency of subsequent steps and thereby boost the performance.

3 Higher-Order Short-Cut Rules

Formally, a short-cut rule is sequentially composed from a rule that only deletes structure and another one that only creates structure [11]. (We have generalised that kind of sequential composition to arbitrary rules in [19].) In practical applications so far, we made use of a static, finite set of short-cut rules; we composed each inverse of a rule from the given TGG with every rule from the TGG [9,10,12]. Intuitively, such a *Short-Cut Rule* replaces the action of the inverse rule by the one of its second input rule (while preserving common elements). The *Short-Cut Rule* from Sect. 2 is created in that way. The set of *Short-Cut Rules* that we obtained in this way enables repairs in many situations. But certain complex model changes are not supported yet, namely situations in which the common effect of several TGG rules should be replaced by the effect of another set of TGG rules. For this, one needs *Short-Cut Rules* that are not just computed from the TGG rules (and their inverses) but from (arbitrary long) *concurrent rules* [8], i.e., sequential compositions, of the TGG rules (and their inverses). Yet, in contrast to before, we cannot precalculate this set as there are usually infinitely many ways to concatenate an arbitrarily large set of TGG rules and, thus, infinitely many *Short-Cut Rules*. Hence, we must investigate each inconsistency and deduce what TGG rules to concatenate at runtime to create a helpful *higher-order Short-Cut Rule* that repairs multiple rule applications at once. In this chapter, we will explain the process of deriving these new rules using our running example and conclude with a discussion on its correctness.

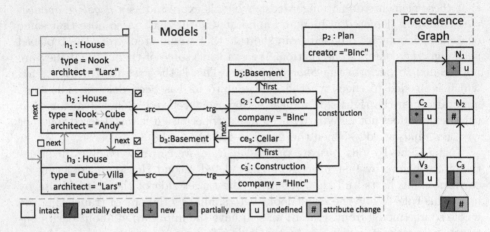

Fig. 5. Synchronisation Example

3.1   Exemplifying the Need for Higher-Order Short-Cut Rules

Figure 5 shows another example model together with its precedence graph. We can see two Houses, where one was of HouseType Nook and the other of HouseType Cube together with their corresponding Constructions. Then, the first House's HouseType was changed from Nook to Cube and another House of House-Type Nook was added at the beginning of this row. Similar to Fig. 3, this results in an inconsistency shown in the precedence graph on the right. There, we see that the Nook Rule application has become invalid due to an attribute constraint violation. Intuitively, we would expect that a new Construction is created for the new Nook House, which is then added to the already existing Plan. However, our consistency specification makes this rather hard to achieve as the Nook Rule creates this Plan together with the first House in a row. By handling one rule application at a time, we have to translate the new Nook House, create a corresponding Construction *and* another Plan. Then, we would either have to revoke the former Nook Rule application and retranslate the Cube House or use a *Short-Cut Rule* to transform a Nook Rule to a Cube Rule application. Since Cube Rule does not create a Plan but requires one, we would either way have to delete the old Plan and then try to connect the newly created or preserved Constructions with the new Plan. While this restores consistency, this procedure has two disadvantages. First, any information stored in the original Plan would be lost and, second, we would have to fix all succeeding rule applications to connect them with the new Plan. Hence, we want to create a *Short-Cut Rule* that preserves the original Plan by translating the new Nook House and repairing the Cube House in one step. Since also the order of the two original Houses was changed, this repair step necessarily needs to involve reacting to that and performing the required repair for the former Cube House that now is a Villa. Hence, a higher-order *Short-Cut Rule* is needed.

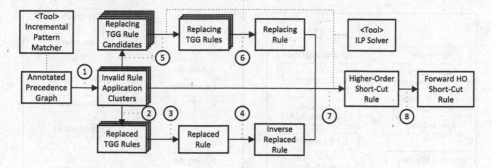

Fig. 6. Construction Process of Higher-Order Short-Cut Rules

3.2 Deriving Higher-Order Short-Cut Rules On-the-Fly

Figure 6 shows our proposed process on how to obtain the necessary *higher-order Short-Cut Rule* using the information from an annotated precedence graph. Remember, to create a *Short-Cut Rule*, we need two ingredients: a *higher-order replaced* and *higher-order replacing* rule; i.e., one concurrent rule, built from inverses of the TGG rules, and one built from the TGG rules. The first is trivial to obtain as we can directly derive it from neighboured inconsistencies in our precedence graph (②) by forming sets of inconsistent precedence graph nodes that are connected (①). Such a set is found by picking an inconsistent precedence node and exploring its adjacent neighbours. Every neighbour that is also inconsistent is added to the set and we explore its neighbours recursively. Note that here, and also for the synthesis of the replacing rule, the order in which we compose more than two rules is irrelevant since sequential rule composition is associative [4]. In contrast, nodes that are still intact or depict alternative rule applications are ignored and not explored further. We, thereby, identify all inconsistencies that should be repaired in one step.

In our case, we only have one set consisting of the precedence graph nodes N_2 and C_3. These nodes represent formerly valid TGG rule applications for which we know what elements were created and which were used as precondition. We also know what changes invalidated them. Thus, we can use this information to concatenate Nook Rule and Cube Rule ((③)) because we know that the Plan, House and Construction required by C_3 were created by N_2. In general, the resulting *replaced* rule can have context elements that stem from rule applications that are still considered intact. The resulting inverse *replaced* rule (④) is depicted on the left of Fig. 7 and can delete all elements created by the corresponding rules.

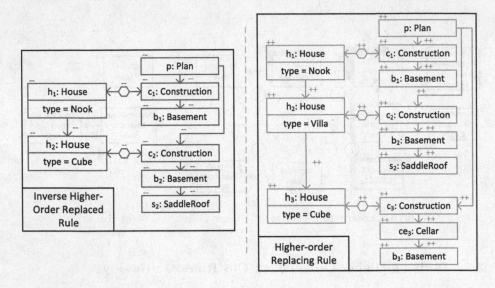

Fig. 7. Inverse *Higher-Order Replaced* Rule & *Higher-Order Replacing* Rule

The much more challenging task is to synthesise the *replacing* rule. While we used some inconsistencies to deduce our inverse *replaced* rule, we now need information that tells us how to restore consistency (⑤). We, thus, need to use the precedence graph nodes with "+" annotations, which indicate new translation options using entirely new elements and those with "*" that describe translation alternatives for elements that are still part of an older rule application. Yet, not all of these nodes are relevant as some translations could just be carried out without using *Short-Cut Rules* or if they are needed for fixing another inconsistency. We can identify the relevant nodes as those that (partially) overlap with nodes from our set of connected inconsistencies. Intuitively, these overlapped nodes are the ones, we would like to reuse and, thus, make part of a new rule application that is again consistent. Naturally, "+" annotated nodes cannot overlap with any inconsistencies as their source elements are entirely new. However, at times it is necessary to repair elements and, simultaneously, translate new elements. We, thus, also add nodes annotated with "+" if they are connected with a "*" or "+" annotated node. Having a set of possible rule applications, we now have to construct our *replacing* rule (⑥). However, we only have a partial knowledge on how to concatenate the identified rules because these changes have not yet been propagated. Regarding our example, we know that we have to concatenate Nook Rule, Villa Rule and Cube Rule, but only on the source side we know exactly how.

In general, there are more challenges that we have to overcome. First, we have to make sure that after the repair, all elements on the source side are again part of an intact rule application. This is a hard task because we may find that there are multiple ways to translate a specific element, e.g., by different rules that compete with each other. Then, selection of an appropriate set of rule applications is required. In some cases, this choice can influence how much information on the target side is preserved. Furthermore, we must select a set of rule applications such that the precondition of every selected rule application holds – through other rules selected for this replacing rule or through original rule applications that are still intact. Note that for space reasons, our example is rather small and does not show such a scenario. Finally, we must ensure that the newly repaired rule applications will not introduce a cyclic dependency, i.e., a situation where rule applications would mutually guarantee their preconditions; in our example, for instance, a cyclic row of houses.

To achieve these conditions, we encode the selection of a suitable subset of to be concatenated rules and the calculation of their overlap into an ILP problem automatically. A solution to such a problem will essentially describe an applicable sequence of rule applications that do not lead to a dead-end. We encode competing rule applications as mutually exclusive binary variables and define constraints ensuring that a rule is only chosen if (one or in combination many) other rules create the elements needed by its precondition. Hence, we have to ensure via constraints that context elements are mapped to created elements of compatible type (the same or a sub-type) and that edges can only be mapped if their corresponding source and target nodes are also mapped accordingly. We also make sure that all remaining source elements that are part of the inconsistency must be handled by the replacing rule application. This encoding is inspired by former approaches that present the finding of applicable TGG rule sequences as an ILP problem, e.g., [21,26]. Regarding our example, there are no competing rule applications but considering the precondition of Villa Rule and Cube Rule, we know that the Plan must be the one created by the Nook Rule application. The Construction could be taken from the Nook Rule or in case of Cube Rule from Villa Rule. Due to our knowledge that the needed House of Cube Rule stems from a Villa Rule application, the constraints will make it infeasible for Cube Rule to take the Nook Rule's Construction as there is no correspondence node between them. The identified rules are concatenated ((6)) and form the *replacing* rule as depicted on the right of Fig. 7.

Using both the inverse *replaced* and *replacing* rule as well as the information of the actual inconsistency ((7)), we can now construct the *Short-Cut Rule*. As with the search for a *replacing* rule, we know what elements to overlap on the source side. This means that we only have to calculate the overlap for the correspondence and target side. While there are still many possible overlaps, e.g., between all Constructions of both rules, we usually want to find the one(s) that preserve more elements. To support least surprise, for target elements that have a corresponding source element, we prefer solutions that only identify these target elements if their corresponding source elements are also identified. We also encode the overlapping as an ILP problem to find an optimal solution w.r.t. preserving elements. Currently, we do not differentiate between elements but the optimisation function can be customised to favour elements that, e.g., contain attributes, which have no representation on the other side and should be prioritised. In our computation of overlaps, we also support node type inheritance.

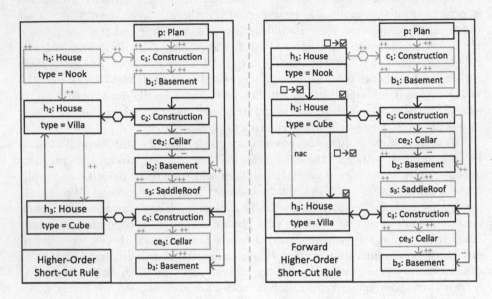

Fig. 8. Higher-Order Short-Cut Rule and Forward Higher-Order Short-Cut Rule

The resulting higher-order *Short-Cut Rule* is depicted in Fig. 8 on the left. It results from overlapping the Houses h_1 and h_2 from the replaced rule with h_2 and h_3 from the replacing rule. We also overlap p, c_1, b_1, c_2 and b_2 from the replaced rule's target side with p, c_2, b_2, c_3 and b_3 from the replacing rule. Finally, we also overlap correspondence links and edges if their corresponding source and target nodes are also mapped onto another. By applying this rule, we can preserve the two Houses on the left together with their corresponding Constructions and Basements. We also preserve the Plan and create a new Nook House with corresponding target structures. Since h_2 was a Nook House, we now assign the new HouseType Cube, which means that we have to add a new Cellar on the target side and make it the first Construction Step. Similarly, h_3 was a Cube House but is now transformed to a Villa House, which means that the former Cellar is now superfluous and thus removed. Instead, we add a new Saddle Roof as a final Construction Step. As before, we can operationalise this rule ((8)) to obtain a forward (and likewise backward) variant as depicted in Fig. 8 on the right. Applying it to our example from Fig. 5, we can now translate the new Nook, while simultaneously repairing the other two Houses and preserving the Plan with its content.

3.3   Discussion

From a technical point of view, there are some aspects worth mentioning. As a result of the construction process, we do not only get a higher-order *Short-Cut Rule* but also (implicitly) its match w.r.t. the analysed inconsistency. Hence, we can directly apply the rule and repair it. Besides that, however, we do not store the generated rule for applying it at other locations, which has two reasons. First, a higher-order *Short-Cut Rule* is quite specific to a certain inconsistency that is characterized by a set of related broken rule applications and how exactly each of them has been invalidated. To efficiently determine whether an existing rule can be applied to another set of inconsistencies can, thus, be very challenging and iterating and trying out all stored rules will probably become more expensive than simply constructing the rule from scratch. Second, as a higher-order *Short-Cut Rule* targets multiple inconsistencies at once, the space of related inconsistencies is generally very large. Hence, we argue that encountering the exact same constellation of inconsistencies is less likely. Yet, this has to be investigated further for real-world scenarios.

In the following, we shortly discuss applicability and correctness of the individual process steps. In ①, we identify clusters of related inconsistencies by analysing the precedence graph. This information is then used to create the *replaced* and *replacing* rule. Choosing the cluster too small, e.g., to save performance, might lead to a *Short-Cut Rule* that is not applicable if necessary context has been altered and this change was not considered. In that case, we can still fall back to restoring consistency by revoking the rule applications and applying alternative translation steps instead, although, this has the risk of losing some information. In contrast, if the set is chosen bigger than necessary, then the resulting ILP problems also tend to become harder to solve as the search space increases. In our implementation, these clusters contain all related inconsistencies that are (transitively) connected with each other without any consistent steps in between. We, thereby, ensure that the cluster is not chosen too small as inconsistencies that are (remotely) related are repaired together. However, at the moment, we cannot guarantee that our chosen clusters are minimal because some inconsistencies may be related but do not need to be handled together, i.e., could be repaired by standard *Short-Cut Rules* or smaller *higher-order Short-Cut Rules*. In the future, we would like to investigate ways to identify situations where these clusters could be broken up to yield less complex ones, which has the potential of improving the performance of our approach. Both ②, ③ and ④ are trivial as we use information about formerly valid rule applications for which we rely on a pattern matcher to identify these locations.

Regarding ⑤, we may encounter multiple alternative translation options for one element, which may lead to an exponentially increasing number of possible translation sequences and, thus, performance issues. However, under certain technical (and not too strict) circumstances, any choice can be part of a synchronisation process that finally restores consistency [18, Theorem 6.17] (also compare [21, Theorem 4]). Yet, as discussed for ①, different choices may lead to different amounts of preserved model elements. Creating the inverse of the resulting rule in ⑥ is again trivial. Regarding the underlying theory of (*higher-order*) *Short-Cut Rules*, we refer to the literature for details [11, 18, 19]. Importantly, for our purposes in ⑦ and ⑧, both *replaced* and *replacing rule* are composed from the rules of the given TGG, through our computation of *Short-Cut Rules* and the information provided by the IGPM, we ensure consistent matching of the replaced and the replacing rule, and we prevent the introduction of cyclic dependencies. This means that we meet criteria that guarantee language-preserving applications of *Short-Cut Rules* (cf. [18, Theorem 4.16] or [19, Theorem 17]). The according application of their operationalised versions during synchronisation then incrementally improves consistency (compare [18, Proposition 6.15]).

4   Evaluation

We implemented our approach and integrated it into the state-of-the-art graph transformation tool eMoflon[2]. In our evaluation, we pose the following research questions: **(RQ1)** Does the new runtime analysis introduce any additional cost in comparison with using a precalculated set of *Short-Cut Rules*? **(RQ2)** Does our new approach indeed preserve more information in certain cases? To answer these questions, we investigated four different scenarios of our running example with three different resolution techniques: The *legacy* algorithm revokes invalid rule applications and retranslates parts of the model, while *SC* stands for the original *Short-Cut Rule* framework with precalculcated repair rules and *HO SC* stands for our new *higher-order Short-Cut Rules*. We compare both *SC* and *HO SC* with *legacy* to show the impact of repairing models instead of retranslating them (partially). Also note that in case no *Short-Cut Rule* was found, our algorithm falls back to using the *legacy* algorithm.

In the first and second scenario, we have a fixed number of 2 000 rows of Houses with around 26 000 nodes on both source and target side and we increase the number of applied changes in steps of 50 from 50 to 250. In contrast, the third and fourth scenario comprise two long rows of Houses that we iteratively increase by adding new Houses in steps of 20 from 20 to 200, while we apply only one change. This change, however, will be more expensive to resolve as the rows grow. In the first and third scenario, we add new Nook Houses at the start of rows. *HO SC* can synthesise a *higher-order Short-Cut Rule* to react to this change, while *legacy* has to re-translate each affected row and *SC* has to separately translate the newly created Nook Houses and then move the affected Constructions to the newly translated Plan (using precalculated *Short-Cut Rules*). In the second and fourth scenario, we relocate subrows. This is a worst-case scenario for *HO SC*. In this case, every moved House's Construction must be repaired and moved to the new Plan. While *legacy* has to perform this via retranslation, both *SC* and *HO SC* can use the same repair rules – only *SC* does not have the offset of creating

---

[2] www.emoflon.org/.

them on-the-fly. We measure the time it takes to resolve the inconsistencies as well as how many elements are deleted in the process. Every measurement was repeated five times and we show the average runtime while the number of deleted elements did not fluctuate. The evaluation was executed on a system with an AMD Ryzen 9 3900x with 64 GB RAM. It can be replicated using our prepared VM[3], which comes with a detailed explanation on how to reproduce the results.

Figure 9 shows the measured times for all scenarios. For the first and third scenario, where *legacy* and *SC* have to transitively retranslate/repair, creating the needed higher-order *Short-Cut Rules* on-the-fly even has performance benefits. For the other two cases, the results indicate that for the worst-case scenario where *SC* will have the same effect as *HO SC*, constructing higher-order *Short-Cut Rules* and applying it instead of performing many smaller repair steps only introduces a small cost **(RQ1)**. Figure 10 shows the largest number of deleted elements (nodes and edges) for all four scenarios. As can be seen, using *HO SC* has the potential of preserving more information than *SC* and *legacy* in the right cases **(RQ2)**. In the others, both *SC* and *HO SC* outperform the *legacy* algorithm; this cannot preserve any information at all.

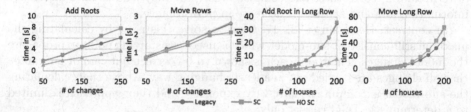

Fig. 9. Time Comparison

Threats to Validity. There are several points worth mentioning. Our approach still has to be evaluated on a real-world example with non-synthetic data and user changes. Another point is the TGG of our running example, which currently does not include competing rule applications, i.e., cases where there are multiple options to translate an element. Since our construction approach uses ILP solving, incorporating these alternative translations can greatly increase the search space and thus increase the overhead of creating a higher-order *Short-Cut Rule*. Note, however, that we still need and use the ILP solver to find out how to concatenate the replacing rule candidates such that more elements can be preserved on the opposite side of the change. Also, we use it to find and maximise the overlap between replaced and replacing rules.

---

[3] www.zenodo.org/record/7728966.

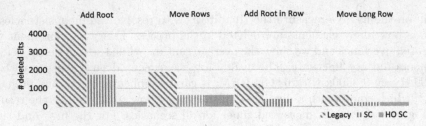

Fig. 10. Comparison of deleted elements

5   Related Work

In this section, we relate our approach to other incremental model synchronisation approaches. Due to space limitations, we will focus on TGG-based approaches. There have been several works that implement model synchronisation by transitively revoking invalid rule applications and retranslating possibly large parts of a model [14,20,22]. While these approaches come with proofs of termination, correctness and completeness, they may lead to an unnecessary information loss and at times a decrease in performance.

In contrast, Hermann et al. [17] tried to solve this issue by calculating the maximal still-consistent sub-model, which is used as a starting point to propagate the model changes. Their approach is shown to be correct but cannot guarantee that all changes are carried out in a least-changing way. Above that, calculating the sub-model is computationally very expensive and their approach is limited to a deterministic TGG ruleset only.

A very similar approach to ours is given by Giese and Hildebrandt [13] where they propose repairing rules that closely resemble *Short-Cut Rules*. These also save nodes instead of unnecessarily deleting them and are able to propagate relocations of model elements. However, neither a construction scheme to obtain these rules is provided nor is the preservation of information proven to be correct. Yet, using our approach from [12], we are indeed able to construct the same rules as mentioned in their work. Similarly, Blouin et al. [5] also propose to use custom repair rules.

Another approach was initially proposed by Greenyer et al. [15] and later formalised by Orejas and Pino [23]. They propose that elements are never directly deleted but rather marked. Thereby, these elements are still free to be re-used during synchronisation, e.g., by letting a forward rule transformation take elements from the set of these elements instead of creating new ones and, thus, prevent them from being deleted. Yet, to the best of our knowledge, their approach was never implemented.

Finally, Anjorin et al. [1] discussed design guidelines for TGGs that lead to less transitive dependencies between rule applications and can, thus, be synchronised without triggering a large cascade of retranslation steps. Yet in many cases, remodelling a TGG to comply with these guidelines means to change the defined language and may for that reason not be desired.

In summary, there are several TGG-based approaches that acknowledged the limitations of propagating changes by retranslating possibly large parts of a model. Some approaches follow a similar approach as ours, yet without a formal foundation or construction concept. Others have, to the best of our knowledge, never been implemented. While we provide both, we extend our previous works [11,12] by broadening the practical applicability and preserving information in more complex scenarios than before.

6 Conclusion

In this paper, we presented a novel approach to construct non-trivial repair rules in the form of higher-order *Short-Cut Rules* from a given consistency specification. These repair rules are constructed on demand by analysing a precedence graph that is annotated based on user changes. Our approach was fully implemented into eMoflon, a state-of-the-art graph transformation tool, which forms the base of our evaluation. In the evaluation, we showed that higher-order *Short-Cut Rules* can preserve information in more complex cases, while introducing only a small overhead to the runtime and at times even outperforming other strategies in eMoflon. Next, we would like to investigate whether we can identify situations statically where a higher-order *Short-Cut Rule* is needed to improve performance when smaller *Short-Cut Rules* suffice. For concurrent synchronisation scenarios, these rules could also be used to check whether sequences of user changes to both models correspond to or contradict each other. For the latter, we could highlight the differences and resolve such a conflict.

References

1. Anjorin, A., Leblebici, E., Kluge, R., Schürr, A., Stevens, P.: A systematic approach and guidelines to developing a triple graph grammar. In: Proceedings of the 4th International Workshop on Bidirectional Transformations. CEUR Workshop Proceedings, vol. 1396, pp. 81–95 (2015). http://tubiblio.ulb.tu-darmstadt.de/76241/
2. Anjorin, A., Weidmann, N., Oppermann, R., Fritsche, L., Schürr, A.: Automating test schedule generation with domain-specific languages: a configurable, model-driven approach. In: MoDELS 2020: ACM/IEEE 23rd International Conference on Model Driven Engineering Languages and Systems, Virtual Event, Canada, 18–23 October 2020, pp. 320–331. ACM (2020). https://doi.org/10.1145/3365438.3410991
3. Becker, S.M., Westfechtel, B.: Incremental integration tools for chemical engineering: an industrial application of triple graph grammars. In: Bodlaender, H.L. (ed.) WG 2003. LNCS, vol. 2880, pp. 46–57. Springer, Heidelberg (2003). https://doi.org/10.1007/978-3-540-39890-5_5
4. Behr, N., Krivine, J.: Compositionality of rewriting rules with conditions. Compositionality 3 (2021). https://doi.org/10.32408/compositionality-3-2
5. Blouin, D., Plantec, A., Dissaux, P., Singhoff, F., Diguet, J.-P.: Synchronization of models of rich languages with triple graph grammars: an experience report. In: Di Ruscio, D., Varró, D. (eds.) ICMT 2014. LNCS, vol. 8568, pp. 106–121. Springer, Cham (2014). https://doi.org/10.1007/978-3-319-08789-4_8

6. Brambilla, M., Cabot, J., Wimmer, M.: Model-Driven Software Engineering in Practice, 2nd edn. Synthesis Lectures on Software Engineering. Morgan & Claypool Publishers (2017). https://doi.org/10.2200/S00751ED2V01Y201701SWE004

7. Ehrig, H., Ehrig, K., Ermel, C., Hermann, F., Taentzer, G.: Information preserving bidirectional model transformations. In: Dwyer, M.B., Lopes, A. (eds.) FASE 2007. LNCS, vol. 4422, pp. 72–86. Springer, Heidelberg (2007). https://doi.org/10.1007/978-3-540-71289-3_7

8. Ehrig, H., Ehrig, K., Prange, U., Taentzer, G.: Fundamentals of Algebraic Graph Transformation. Monographs in Theoretical Computer Science. An EATCS Series. Springer, Heidelberg (2006). https://doi.org/10.1007/3-540-31188-2

9. Fritsche, L.: Local consistency restoration methods for triple graph grammars. Ph.D. thesis, Technical University of Darmstadt, Germany (2022). http://tuprints.ulb.tu-darmstadt.de/21443/

10. Fritsche, L., Kosiol, J., Möller, A., Schürr, A., Taentzer, G.: A precedence-driven approach for concurrent model synchronization scenarios using triple graph grammars. In: Lämmel, R., Tratt, L., de Lara, J. (eds.) Proceedings of the 13th ACM SIGPLAN International Conference on Software Language Engineering, SLE 2020, Virtual Event, USA, 16–17 November 2020, pp. 39–55. ACM (2020). https://doi.org/10.1145/3426425.3426931

11. Fritsche, L., Kosiol, J., Schürr, A., Taentzer, G.: Short-cut rules –sequential composition of rules avoiding unnecessary deletions. In: Mazzara, M., Ober, I., Salaün, G. (eds.) STAF 2018. LNCS, vol. 11176, pp. 415–430. Springer, Cham (2018). https://doi.org/10.1007/978-3-030-04771-9_30

12. Fritsche, L., Kosiol, J., Schürr, A., Taentzer, G.: Avoiding unnecessary information loss: correct and efficient model synchronization based on triple graph grammars. Int. J. Softw. Tools Technol. Transfer **23**(3), 335–368 (2020). https://doi.org/10.1007/s10009-020-00588-7

13. Giese, H., Hildebrandt, S., Neumann, S., Wätzoldt, S.: Industrial case study on the integration of SysML and AUTOSAR with triple graph grammars. Technical report, 57 (2012)

14. Giese, H., Wagner, R.: From model transformation to incremental bidirectional model synchronization. Softw. Syst. Model. **8**(1), 21–43 (2009). https://doi.org/10.1007/s10270-008-0089-9

15. Greenyer, J., Pook, S., Rieke, J.: Preventing information loss in incremental model synchronization by reusing elements. In: France, R.B., Kuester, J.M., Bordbar, B., Paige, R.F. (eds.) ECMFA 2011. LNCS, vol. 6698, pp. 144–159. Springer, Heidelberg (2011). https://doi.org/10.1007/978-3-642-21470-7_11

16. Hermann, F., Ehrig, H., Golas, U., Orejas, F.: Formal analysis of model transformations based on triple graph grammars. Math. Struct. Comput. Sci. **24**(4), 240408 (2014). https://doi.org/10.1017/S0960129512000370

17. Hermann, F., et al.: Model synchronization based on triple graph grammars: correctness, completeness and invertibility. Softw. Syst. Model. **14**(1), 241–269 (2013). https://doi.org/10.1007/s10270-012-0309-1

18. Kosiol, J.: Formal foundations for information-preserving model synchronization processes based on triple graph grammars. Ph.D. thesis, University of Marburg, Germany (2022). https://archiv.ub.uni-marburg.de/diss/z2022/0224

19. Kosiol, J., Taentzer, G.: A generalized concurrent rule construction for double-pushout rewriting: generalized concurrency theorem and language-preserving rule applications. J. Log. Algebraic Methods Program. **130**, 100820 (2023). https://doi.org/10.1016/j.jlamp.2022.100820

20. Lauder, M., Anjorin, A., Varró, G., Schürr, A.: Efficient model synchronization with precedence triple graph grammars. In: Ehrig, H., Engels, G., Kreowski, H.-J., Rozenberg, G. (eds.) ICGT 2012. LNCS, vol. 7562, pp. 401–415. Springer, Heidelberg (2012). https://doi.org/10.1007/978-3-642-33654-6_27

21. Leblebici, E.: Inter-model consistency checking and restoration with triple graph grammars. Ph.D. thesis, Darmstadt University of Technology, Germany (2018). http://tuprints.ulb.tu-darmstadt.de/7426/

22. Leblebici, E., Anjorin, A., Fritsche, L., Varró, G., Schürr, A.: Leveraging incremental pattern matching techniques for model synchronisation. In: de Lara, J., Plump, D. (eds.) ICGT 2017. LNCS, vol. 10373, pp. 179–195. Springer, Cham (2017). https://doi.org/10.1007/978-3-319-61470-0_11

23. Orejas, F., Pino, E.: Correctness of incremental model synchronization with triple graph grammars. In: Di Ruscio, D., Varró, D. (eds.) ICMT 2014. LNCS, vol. 8568, pp. 74–90. Springer, Cham (2014). https://doi.org/10.1007/978-3-319-08789-4_6

24. Orejas, F., Pino, E., Navarro, M.: Incremental concurrent model synchronization using triple graph grammars. In: FASE 2020. LNCS, vol. 12076, pp. 273–293. Springer, Cham (2020). https://doi.org/10.1007/978-3-030-45234-6_14

25. Schürr, A.: Specification of graph translators with triple graph grammars. In: Mayr, E.W., Schmidt, G., Tinhofer, G. (eds.) WG 1994. LNCS, vol. 903, pp. 151–163. Springer, Heidelberg (1995). https://doi.org/10.1007/3-540-59071-4_45

26. Weidmann, N., Anjorin, A.: Schema compliant consistency management via triple graph grammars and integer linear programming. In: FASE 2020. LNCS, vol. 12076, pp. 315–334. Springer, Cham (2020). https://doi.org/10.1007/978-3-030-45234-6_16

27. Weidmann, N., Fritsche, L., Anjorin, A.: A search-based and fault-tolerant approach to concurrent model synchronisation. In: Lämmel, R., Tratt, L., de Lara, J. (eds.) Proceedings of the 13th ACM SIGPLAN International Conference on Software Language Engineering, SLE 2020, Virtual Event, USA, 16–17 November 2020, pp. 56–71. ACM (2020). https://doi.org/10.1145/3426425.3426932

Formalization and Analysis of BPMN Using Graph Transformation Systems

Tim Kräuter[1]([✉]) [iD], Adrian Rutle[1] [iD], Harald König[1,2] [iD], and Yngve Lamo[1] [iD]

[1] Western Norway University of Applied Sciences, Bergen, Norway
{tkra,aru,yla}@hvl.no
[2] University of Applied Sciences, FHDW, Hanover, Germany
harald.koenig@fhdw.de

Abstract. The Business Process Modeling Notation (BPMN) is a widely used standard notation for defining intra- and inter-organizational workflows. However, the informal description of the BPMN execution semantics leads to different interpretations of BPMN elements and difficulties in checking behavioral properties. In this paper, we propose a formalization of the execution semantics of BPMN that, compared to existing approaches, covers more BPMN elements while facilitating property checking. Our approach is based on a higher-order transformation from BPMN models to graph transformation systems. As proof of concept, we have implemented our approach in an open-source web-based tool.

Keywords: BPMN · Higher-order model transformation · Graph transformation · Model checking · Formalization

1 Introduction

In today's fast-paced business environment, organizations with complex workflows require a powerful means to accurately map, analyze, and optimize their processes. Business Process Modeling Notation (BPMN) [18] is a widely used standard to define these workflows. However, the informal description of the BPMN execution semantics leads to different interpretations of BPMN models and difficulties in checking behavioral properties [4]. Formalizing BPMN would reduce the cost of business process automation drastically by facilitating the detection of errors and optimization potentials in process models already during design time. To this end, we propose a formalization that covers most of the BPMN elements used in practice and supports checking behavioral properties. General behavioral properties such as *Safeness* and *Soundness* were adapted to BPMN [6], which can uncover potential flaws in BPMN models leading to deadlocks or other undesirable execution states.

In this paper, we consider two fundamental concepts when formalizing the execution semantics of BPMN. First, *state structure*, i.e., how models are represented during execution. The state structure corresponds to the type graph in Graph Transformation (GT) systems. Second, *state-changing elements*, i.e.,

M. Fernández and C. M. Poskitt (Eds.): ICGT 2023, LNCS 13961, pp. 204–222, 2023.
https://doi.org/10.1007/978-3-031-36709-0_11

which elements in a model encode state changes. These elements are implemented using GT rules. In our approach, we automatically generate GT rules based on a Higher-Order model Transformation (HOT) for each specific BPMN model, as shown in Fig. 1.

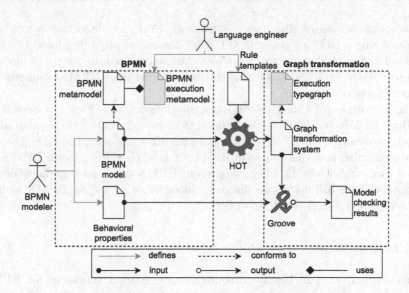

Fig. 1. Overview of the approach

To begin the BPMN modeling process, a modeler first defines the BPMN model and its corresponding behavioral properties for evaluation. This model must adhere to the BPMN metamodel as outlined in the BPMN specification by the Object Management Group [18]. To create the state structure for BPMN, the BPMN execution metamodel is defined by language engineers, utilizing the BPMN metamodel as a foundation. Typically, an execution metamodel is created by extending the languages metamodel.

Furthermore, we define a HOT from BPMN models to GT systems. We call the transformation *higher-order* since the resulting graph-transformation systems represent model-transformations themselves [23]. The HOT creates a GT system, i.e., GT rules and a start graph for a given BPMN model. It is defined using rule generation templates, which describe how GT rules should be generated for each state-changing element in BPMN (see Sect. 3). The obtained GT system conforms to the execution type graph, which corresponds to the BPMN execution metamodel. In the figure, we have colored both artifacts blue to visualize that they contain the same information. Ultimately, we use Groove as an execution engine for the GT system and check the behavioral properties defined earlier.

Our approach has been implemented in a user-friendly, open-source web-based tool, the *BPMN Analyzer*, which can be used online without needing installation. The BPMN Analyzer was validated using a comprehensive test suite. Additionally, our approach is versatile as it can be applied to formalize other behavioral languages, such as activity diagrams, state charts, and more. To define the execution semantics of an alternate behavioral language, one simply needs to establish a new execution metamodel and HOT (see the language engineer in Fig. 1).

The contribution of this paper is twofold. First, we introduce a new approach utilizing a HOT to generate GT rules instead of providing fixed GT rules to formalize the semantics of a behavioral language. Second, we apply our approach to BPMN, resulting in a formalization covering most BPMN elements that supports property checking.

The remainder of this paper is structured as follows. First, in Sect. 2, we introduce BPMN and point out the theoretical background of this contribution. Second, we describe the BPMN semantics formalization using the HOT (Sect. 3) before explaining how this can be utilized for model checking general BPMN and custom properties (Sect. 4). Then, we present BPMN Analyzer implementing our approach in Sect. 5. Finally, we discuss related work regarding BPMN element coverage in Sect. 6 and conclude in Sect. 7.

2   Preliminaries

In this section, we will briefly introduce the execution semantics of BPMN, and readers are encouraged to consult [11] or the BPMN specification [18] for more in-depth information. Furthermore, our application of GTs to formalize the execution semantics of BPMN will be outlined in addition to a brief overview of the theoretical principles that underlie our use of GTs.

2.1   BPMN

Figure 2 depicts the structure of BPMN models with the corresponding concrete syntax BPMN symbols contained in clouds.

A BPMN model is represented by a Collaboration that has participants and messageFlows between InteractionNodes. Each participant is a Process containing flowNodes connected by SequenceFlows. A FlowNode is either an Activity, Gateway, or Event. Many types of Activities, Gateways, and Events exist. Activities represent certain tasks to be carried out during a process, while events may happen during the execution of these tasks. Furthermore, gateways model conditions, parallelizations, and synchronizations [11].

The BPMN execution semantics is described using the concept of *tokens* [18], which can be located at sequence flows and specific flow nodes. Tokens are consumed and created by flow nodes according to the connected sequence flows. The FlowNode is colored purple in Fig. 2 since it represents the *state-changing elements* of BPMN, as described in Sect. 3.

A BPMN process is triggered by one of its start events, leading to a token at each outgoing flow of the triggered start event. Activities can start when at least

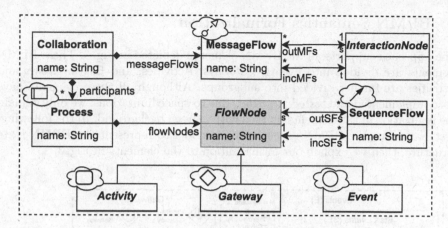

Fig. 2. Excerpt of the BPMN metamodel [18]

one token is on an incoming sequence flow. The start of an activity will move the incoming token to the activity. When an activity terminates, it deletes its token and adds one at each outgoing sequence flow. Furthermore, different gateway types exist, such as parallel and exclusive gateways. Parallel gateways represent forks and joins, meaning they delete one token for each incoming sequence flow and add one token for each outgoing sequence flow. Exclusive gateways represent an XOR by deleting a token from one incoming sequence flow and adding a token to exactly one of the outgoing sequence flows. Events delete and add tokens similar to activities but have additional semantics depending on their type. For example, message events will add or delete messages.

2.2   Theoretical Background

We use typed attributed graphs for the formalization of the BPMN execution semantics. Each state, i.e., token distribution during the execution of a BPMN model, is represented as an attributed graph typed by the BPMN execution type graph, which we introduce in Sect. 3.

Regarding GT, we utilize the single-pushout (SPO) approach with negative application conditions (NAC) [9], as implemented in Groove [20]. In addition, we utilize *nested rules* with quantification to make parts of a rule repeatedly applicable or optional [21,22]. Moreover, we utilize the NACs to implement more intricate parts in the BPMN execution semantics, such as the termination of processes.

Formal definitions of SPO rules, their application, and the corresponding extensions of the theory (NACs, nested rules) are well-known, see [9,21]. We do not repeat them to instead focus on our practical contribution.

3   BPMN Semantics Formalization

The approach supports all the BPMN elements depicted in Fig. 3. These BPMN elements are divided into Events, Gateways, Activities, and Edges. Events and Activities are further divided into subgroups. Although all these elements have been implemented and tested (see [16]), due to space limitations, we only explain the realization of the elements marked with a green background. In the following, first, we define the BPMN execution metamodel to represent the BPMN state structure, then we explain our formalization of the elements in Fig. 3.

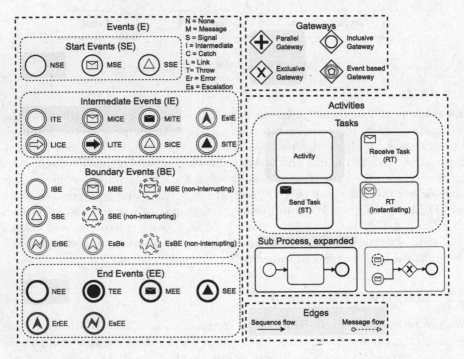

Fig. 3. Overview of the supported BPMN elements (structure adapted from [14])

3.1   BPMN Execution Metamodel

In our formalization of BPMN, we utilize a token-based representation of the execution semantics, similar to the approach used in the informal description of the BPMN specification [18]. To describe processes holding tokens during execution, we define the execution metamodel shown in Fig. 4, depicted as a UML class diagram. In the context of our approach, we fulfill the role of the language engineer by defining the execution metamodel (see Fig. 1).

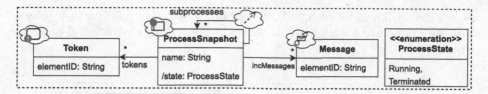

Fig. 4. BPMN execution metamodel

We use ProcessSnapshot to denote a running BPMN process with a specific token distribution that describes one state in the history of the process execution. Every ProcessSnapshot has a set of tokens, incoming messages, and subprocesses. A ProcessSnapshot has the state Terminated if it has no tokens or subprocesses. Otherwise, it has the state Running. A Token has an elementID, which points to the BPMN Activity or the SequenceFlow at which it is located. A Message has an elementID pointing to a MessageFlow. To concisely depict graphs conforming to this type graph, we introduce a concrete syntax in the clouds attached to the elements. Our concrete syntax extends the BPMN syntax by adding process snapshots, subprocess relations, tokens, and messages. Tokens are represented as colored circles drawn at their specified positions in a model. In addition, we use colored circles at the top left of the bounding box, representing instances of the BPMN Process; these circles represent process snapshots. The token's color must match the color of the process snapshot holding the token. The concrete syntax was inspired by the bpmn-js-token-simulation [2].

Our BPMN execution metamodel was not created by extending the BPMN metamodel and adding missing concepts such as tokens and messages. We chose to create a minimal execution model only containing concepts needed during execution. This is only possible since our rules are generated by the HOT for each specific BPMN model such that the structure of each model is already implicitly encoded in the rules. This design choice leads to smaller states in the graph transformation system when compared to an execution metamodel that extends the BPMN metamodel.

The execution metamodel is a UML class diagram without operations, which can be seen as an attributed type graph [12]. We keep the execution metamodel and the execution type graph separate (see Fig. 1) because the execution metamodel should be independent of the formalism used to define the execution semantics. One can reuse the execution metamodel when changing the formalism or concrete tool implementing the formalism (in our case, Groove) by adjusting how the execution metamodel is transformed. Using the execution metamodel as the type graph, we can now define how the start graph and GT rules for the different BPMN elements are created.

Since our approach is based on a HOT from BPMN to GT systems, we generate a *start graph* and *GT rules* for each given BPMN model (see Fig. 1). Generating the start graph for a BPMN model is straightforward. First, for each process in the BPMN model, we generate a process snapshot if the process contains a none start event (NSE). An NSE describes a start event without a trigger (none). Then, for each NSE, we add one token to each outgoing sequence flow. An example of a start graph is shown in Fig. 5 using abstract and concrete syntax.

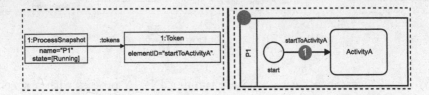

Fig. 5. Example start graph in abstract (left) and concrete syntax (right)

The HOT generates one or more GT rules for each FlowNode, i.e., state-changing element in a BPMN model. In order to provide a better understanding of the transformation process, we will begin by presenting two example results, namely the generated rules for an activity (as shown in Fig. 3). Following this, we will delve into an explanation of how our HOT creates these rules as well as others.

Figure 6 depicts an example GT rule ($L \rightarrow R$) to start an activity in abstract syntax. The rule is straightforward, moving a token from the incoming sequence flow to the activity itself.

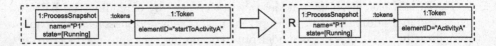

Fig. 6. Example GT rule to start an activity (abstract syntax)

For the rest of the paper, we will depict all rules in the concrete syntax introduced earlier. The rule from Fig. 6 depicted in concrete syntax is shown on the left in Fig. 7. The rule on the right in Fig. 7 implements the termination of an activity, which will move one token from the activity to the outgoing sequence flow.

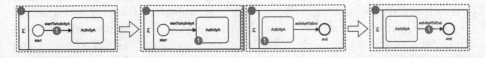

Fig. 7. Example GT rule to start (left) and terminate (right) an activity

To summarize, we described two example rules and introduced a concrete syntax to depict them concisely and understandably. In the following subsections, we use this concrete syntax to define how these rules and rules for other flow nodes are generated by our HOT. Elements of the HOT are depicted using rule generation templates that show how specific rules are created for various flow nodes. Defining the rule generation templates and thus the HOT from BPMN to GT systems is the second task of the language engineer in our approach (see Fig. 1).

Our HOT defines a formal execution semantics of BPMN, similar to other approaches that formalize BPMN by mapping to Petri Nets or other formalisms systems [7].

3.2   Process Instantiation and Termination

Start events do not need GT rules since the generated start graph of the GT system will contain a token for each outgoing sequence flow of an NSE. Other types of start events are triggered in corresponding throw event rules.

Figure 8 depicts the rule generation template for end events (NEE in Fig. 3). All rule generation templates show a state-changing element (FlowNode) with surrounding flows in the left column and the applicable rule generation in the right column. The left column shows instances of the BPMN metamodel (Fig. 2), and the right column shows the generated rules typed by the BPMN execution metamodel (see Fig. 4). If more than one rule is generated from a FlowNode, an expression defines how each rule is generated. For example, the expression $\forall sf \in$ E.incSFs for the rule generation template of end events (see Fig. 8) generates one rule for each incoming sequence flow sf of the end event E. We use "." in expressions to navigate along the associations of the BPMN metamodel shown in Fig. 2. In the example, E.incSFs means following all incSFs links for a FlowNode object, resulting in a set of SequenceFlow objects.

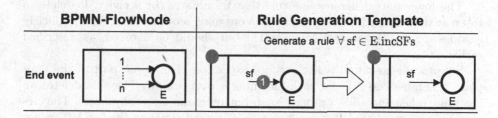

Fig. 8. Rule generation templates for start and end events

The generated end event rules delete tokens one by one for each incoming sequence flow. However, they do not terminate processes. Process termination is implemented with a generic rule—independent of the input BPMN model— which is applicable to all process snapshots. The termination rule in Fig. 9 is automatically generated once during the HOT. The rule changes the state of the process snapshot from running to terminated if it has neither tokens nor subprocesses.

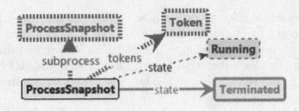

Fig. 9. Termination rule in Groove

The Groove syntax is the following. The thin black elements in Fig. 9 need to be present and will be preserved during transformation, while the dashed blue elements need to be present but will be removed. Furthermore, the fat green elements will be created and the dashed fat red elements represent the NACs, whose presence prevent the rule from being applied.

3.3   Activities and Subprocesses

Figure 10 depicts the rule generation templates for activities and subprocesses (see Fig. 3). Activity execution is divided into two steps implemented in two parts in the first rule template. The upper part generates one rule for each incoming sequence flow to start the activity. An activity can be started using a token positioned at any of its incoming sequence flows. This part generates the sample rule on the right of Fig. 7. Having multiple incoming or outgoing sequence flows for a flow node is considered bad practice since the implicitly encoded gateways should be explicit to avoid confusion. Our formalization still supports those models not to force modelers to rewrite them, but we recommend using static analyzers to avoid such models [3].

The lower part generates one rule that terminates the activity. It deletes a token at the activity and adds one at each outgoing sequence flow. This implicitly encodes a parallel gateway (see Fig. 11) but should be avoided, as described earlier.

Subprocess execution is like activity execution. The upper part of the template generates one rule for each incoming sequence flow. The rule deletes an incoming token and adds a process snapshot representing a subprocess. The created process snapshot is represented with a colored circle on the top left corner of the subprocess with a token at each outgoing sequence flow of its start events (similar to start graph generation). There is a *subprocess* link between the process snapshots to depict the subprocesses relation in Fig. 4. If the subprocess has no start events, a token will be added to every activity and gateway with no incoming sequence flows.

BPMN-FlowNode	Rule Generation Template

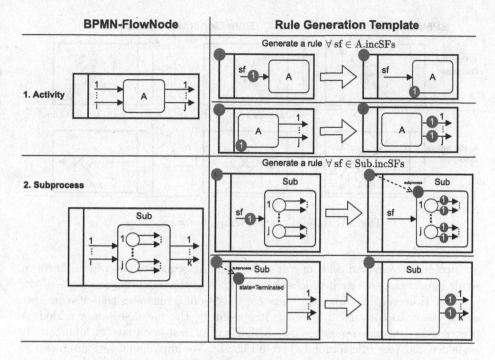

Fig. 10. Rule generation template for activities and subprocesses

The bottom part of the template generates one rule to delete a terminated process snapshot and adds tokens at each outgoing sequence flow. Subprocesses are terminated by the termination rule (see Sect. 3.2).

3.4   Gateways

Figure 11 depicts the rule generation templates for parallel and exclusive gateways (see Fig. 3). A parallel gateway can synchronize and fork the control flow simultaneously. Thus, one rule is generated that deletes one token from each incoming sequence flow and adds one token to each outgoing sequence flow.

Exclusive Gateways are triggered by exactly one incoming sequence flow, and exactly one outgoing sequence flow will be triggered as a result. Thus, one rule must be generated for every combination of incoming and outgoing sequence flows. However, the resulting rule is simple since it only deletes a token from an incoming sequence flow and adds one to an outgoing sequence flow.

3.5   Message Events

Figure 12 depicts the rule generation templates for *message intermediate throw events* and *message intermediate catch events* (MITE and MICE in Fig. 3). The first rule template describes how MITEs interact with MICEs. A MITE deletes

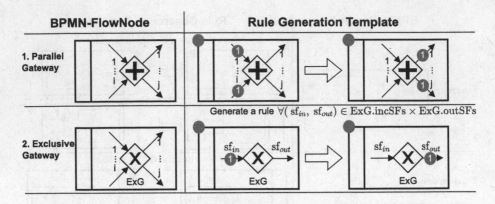

Fig. 11. Rule generation template for gateways

an incoming token and adds one at each outgoing sequence flow. In addition, it sends one message to each process by adding it to the incoming messages of the process. However, sending each message is optional, meaning that if a process is not ready to consume a message immediately, the message is not added. A process can consume a message if its MICE has at least one token at an incoming sequence flow (see rule template two in Fig. 12). We implement optional message sending using nested rules with quantification. Concretely, we use an optional existential quantifier [21] (see blue dotted rectangle marked with optional in Fig. 12) to send a message only if the receiving process is ready to consume it.

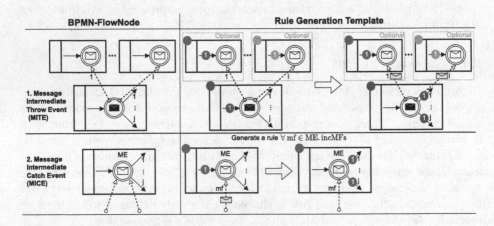

Fig. 12. Rule generation templates for message events

The second rule template in Fig. 12 shows the behavior of MICEs. To trigger a MICE, only one message at an incoming *message flow* is needed. Thus, one rule is generated for each incoming *message flow*. The rule template shows that MICEs delete one message and one token, as well as add a token at each outgoing sequence flow.

4   Model Checking BPMN

Model checking—and verification in general—of BPMN models is necessary to ensure the correctness and reliability of business processes, which ultimately leads to increased efficiency, reduced costs, and user satisfaction. Using our approach, model checking a BPMN model is possible using the generated GT system and behavioral properties based on atomic propositions (see Fig. 1). Behavioral properties are defined using a temporal logic, such as CTL and LTL. In this paper, we will use CTL. An atomic proposition is formalized as a graph and holds in a given state if a match exists from the graph representing the proposition to the graph representing the state [15].

We differentiate between two types of behavioral properties: *general BPMN properties* defined for all BPMN models and *custom properties* tailored towards a particular BPMN model. We do not consider structural properties (like conformance to the syntax of BPMN) since they can be checked using a standard modeling tool without implementing execution semantics. We will now give an example of two predefined general BPMN properties and show how they can be checked using our approach. Then, we describe how custom properties can be defined and checked.

4.1   General BPMN Properties

Safeness and *Soundness* properties are defined for BPMN in [6]. A BPMN model is *safe* if, during its execution, at most one token occurs along the same sequence flow [6]. Soundness is further decomposed into (i) *Option to complete*: any running process instance must eventually complete, (ii) *Proper completion*: after completion, each token of the process instance must be consumed by a different end event, as well as (iii) *No dead activities*: each activity can be executed in at least one process instance [6]. Process completion is synonymous to process termination. In the following, we will describe how to implement the *Safeness* and *Option to complete* properties.

We specify *Safeness* as the CTL property defined in (1). The atomic proposition Unsafe is true if two tokens of one process snapshot point to the same sequence flow. It is shown in Fig. 13 using abstract syntax. We cannot use the concrete syntax to define the Unsafe proposition because the proposition should apply to any BPMN model. Our concrete syntax is always used with a given BPMN model.

Option to complete is specified using the CTL property defined in (2). The atomic proposition AllTerminated is true if there exists no process snapshot in the

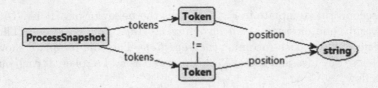

Fig. 13. The atomic proposition *Unsafe* in Groove.

state Running, i.e., all process snapshots are Terminated. AllTerminated is given in [16].

$$AG(\neg\,Unsafe) \qquad (1) \qquad\qquad AF(\text{AllTerminated}) \qquad (2)$$

Checking the properties *Safeness*, *Option to Complete*, and *No Dead Activities* is implemented in our tool [16]. The property *Proper Completion* is not yet implemented, but all the information needed can be found in the GT systems state space.

4.2   Custom Properties

To make model checking user-friendly, we envision modelers defining atomic propositions in the extended BPMN syntax, i.e., the concrete syntax introduced in Fig. 4. Therefore, to define an atomic proposition, a modeler adds process snapshots and tokens to a BPMN model, which we can automatically convert to a graph representing an atomic proposition.

For example, the token distribution shown in Fig. 14 defines two running process snapshots with a token at activity A. Differently colored tokens define different process snapshots. A modeler could use this atomic proposition, for example, to check if, eventually, two processes are executing activity A simultaneously by creating an LTL/CTL property. Thus, a modeler does not need to know the GT semantics used for execution.

However, the modeler must still know the temporal logic, such as LTL and CTL, to express his properties. In the future, a domain-specific property language for BPMN would further lessen the amount of knowledge required from the modeler [17].

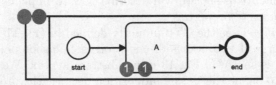

Fig. 14. Token distribution defining an atomic proposition.

5   Implementation

In this section, we will present our tool and then describe experiments regarding its performance.

5.1   BPMN Analyzer Tool

Our approach is implemented in a web-based tool called *BPMN Analyzer*, which is open-source, publicly available, and does not require any installation [16]. Figure 15 depicts a screenshot of the BPMN Analyzer.

The modeler can create or upload a BPMN model, which can then be verified using either BPMN-specific properties or custom CTL properties in the verification section. BPMN Analyzer can generate a GT system for the supplied BPMN model and run model checking in Groove [15]. To evaluate the correctness of our HOT, we have created a comprehensive test suite [16]. It verifies that rules are generated as defined by the rule generation templates in the previous section for each BPMN element. Additionally, we have conducted performance experiments for our approach, as described in the next section.

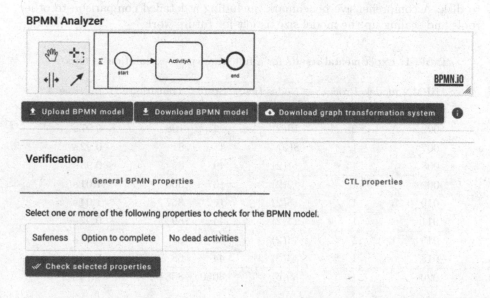

Fig. 15. Screenshot of the BPMN Analyzer tool

5.2   Experiments

Model checking is a useful technique but often falls short in practice due to insufficient performance. Poor performance might have many reasons, most notably large models leading to state space explosion. We experimented with ten different BPMN models from [14] to assess the performance of our implementation. We picked the models at random, besides disregarding some models that were similar. The models include realistic business process models (001, 002, and 020) [14].

To calculate the average runtime, we used the hyperfine benchmarking tool [19] (version 1.15.0), which ran state space exploration for each BPMN model ten times. The experiment was run on Windows 11 (AMD Ryzen 7700X processor, 32 GB RAM) using Groove version 5.8.1 [16].

First, we ran our HOT for the BPMN models. The HOT took less than one second to generate a GT system for each model. Thus, the generation of the GT systems is fast enough.

Second, we ran a full state exploration using the resulting ten GT systems, see Table 1. The exploration takes roughly one second for most of the models. Only model *020* needs nearly two seconds due to its larger state space. Furthermore, up to one second is spent on startup not model checking. For example, Groove reports only 722 ms for state space exploration for model *020*.

We conclude that our approach is sufficiently fast for models of normal size. In addition, there is still room for optimization, such as avoiding costly I/O to disk. A comprehensive benchmark, including a detailed comparison to other tools and scaling up the model size, is left for future work.

Table 1. Experimental results for a full state space exploration in Groove

BPMN model	Processes	Nodes (gw.)	States	Transitions	Total time
001	2	17(2)	68	118	∼ 1.00 s
002	2	16(2)	62	·108	∼ 0.97 s
007	1	8(2)	45	81	∼ 0.92 s
008	1	11(2)	49	85	∼ 0.93 s
009	1	12(2)	137	308	∼ 1.01 s
010	1	15(2)	162	357	∼ 1.04 s
011	1	15(2)	44	69	∼ 0.97 s
015	1	14(2)	53	86	∼ 0.95 s
016	1	14(2)	44	68	∼ 0.94 s
020	1	39(6)	3060	8584	∼ 1.75 s

6   Related Work

The most common formalizations of BPMN use Petri Nets. For example, [7] formalize a subset of BPMN elements by defining a mapping to Petri Nets conceptually close to our HOT-based formalization. Encoding basic BPMN modeling elements into Petri Nets is generally straightforward, but for some advanced elements, it can be complicated to define [13]. For example, representing *termination end events* and *interrupting boundary events*, which interrupt a running process, is usually unsupported because of the complexity of managing the non-local propagation of tokens in Petri Nets [4]. We solve these situations by using nested graph conditions, for example, to remove all tokens when reaching a *termination end event*.

A BPMN formalization based on in-place GT rules is given in [24]. The formalization covers a substantial part of the BPMN specification, including complex concepts such as inclusive gateways and compensation. In addition, the GT rules are visual and thus can be aligned with the informal description of the execution semantics of BPMN. A key difference to our approach is that the rules in [24] are general and can be applied to every BPMN model, while we generate specific rules for each BPMN model using our HOT. Thus, our approach can be seen as a program specialization compared to [24] since we process a concrete BPMN model before its execution. However, they do *not* support property checking since their goal is only formalization.

The tool *BProVe* is based on formal BPMN semantics given in rewriting logic and implemented in the Maude system [4]. Using this formal semantics, they can verify custom LTL properties and general BPMN properties, such as Safeness and Soundness.

The verification framework fbpmn uses first-order logic to formalize and check BPMN models [14]. This formalization is then realized in the TLA^+ formal language and can be model-checked using TLC. Like BProVe, fbpmn allows checking general BPMN properties, such as Safeness and Soundness. Furthermore, they focus on different communication models besides the standard in the BPMN specification and support time-related constructs. We currently disregard time-related constructs [8,14] and data flow [5,10].

Table 2 shows which BPMN elements are supported by our approach and the approaches mentioned above. Compared to other approaches, we cover most BPMN elements. The coverage of BPMN elements greatly impacts how useful each approach is to check properties in practice. In addition, we cover the most important elements found in practice since we come close to the element coverage of popular process engines such as Camunda [1].

The missing elements, when compared to Camunda, are transactions, cancel events, and compensation events. These elements are rather complex, but [24] shows how cancel and compensation events can be formalized. We plan to support these elements by extending our implementation and test suite in the future.

Table 2. BPMN elements supported by different formalizations (based on [24]).

BPMN element/feature	Dijkman et al. [7]	Van Gorp et al. [24]	Corradini et al. [4]	Houhou et al. [14]	This paper
Instantiation and termination					
Start event instantiation	X	X	X	X	X
Exclusive event-based gateway instantiation		X			X
Parallel event-based gateway instantiation					
Receive task instantiation					X
Normal process completion	X	X	X	X	X
Activities					
Activity	X	X	X	X	X
Loop activity	X	X			
Multiple instance activity					
Subprocess	X	X		X	X
Event subprocess					X
Transaction					
Ad-hoc subprocesses					
Gateways					
Parallel gateway	X	X	X	X	X
Exclusive gateway	X	X	X	X	X
Inclusive gateway (split)	X	X	X	X	X
Inclusive gateway (merge)		X		X	X
Event-based gateway			X[1]	X	X
Complex gateway					
Events					
None Events	X	X	X	X	X
Message events	X	X	X	X	X
Timer Events				X	
Escalation Events					X
Error Events	X	X			X
Cancel Events		X			
Compensation Events		X			
Conditional Events					
Link Events		X			X
Signal Events		X			X
Multiple Events					
Terminate Events		X	X	X	X
Boundary Events		X[2]		X[3]	X

[1] Does not support receive tasks after event-based gateways.
[2] Only supports interrupting boundary events on tasks, not subprocesses.
[3] Only supports message and timer events.

7   Conclusion and Future Work

This paper makes two main practical contributions. First, we conceptualize a new approach utilizing a HOT to formalize the semantics of behavioral languages. Our approach moves complexity from the GT rules to the rule templates making up the HOT. Furthermore, the approach can be applied to any behavioral language if one can define its *state structure* and identify its *state-changing elements*.

Second, we apply our approach to BPMN, resulting in a comprehensive formalization regarding element coverage (compared to the literature and industrial process engines) that supports checking behavioral properties. Furthermore, our contribution is implemented in an open-source web-based tool to make our ideas easily accessible to other researchers and practitioners.

Future work targets both of our main contributions. First, we plan a detailed comparison of our HOT approach with approaches that utilize fixed rules. It will be interesting to investigate how the two approaches differ, for example, in runtime during state space generation. Second, we aim to improve our formalization and the resulting tool in multiple ways. We intend to extend our formalization to support the remaining few BPMN elements used in practice and want to turn the modeling environment of our tool into an interactive simulation environment. In addition, we can use this environment to visualize potential counterexamples in cases where behavioral properties are violated.

References

1. Camunda services GmbH: BPMN 2.0 implementation reference. https://docs.camunda.org/manual/7.16/reference/bpmn20/, March 2023
2. Camunda services GmbH: Bpmn-js token simulation. https://github.com/bpmn-io/bpmn-js-token-simulation, March 2023
3. Camunda services GmbH: Bpmnlint. https://github.com/bpmn-io/bpmnlint, March 2023
4. Corradini, F., Fornari, F., Polini, A., Re, B., Tiezzi, F., Vandin, A.: A formal approach for the analysis of BPMN collaboration models. J. Syst. Softw. **180**, 111007 (2021). https://doi.org/10.1016/j.jss.2021.111007
5. Corradini, F., Muzi, C., Re, B., Rossi, L., Tiezzi, F.: Formalising and animating multiple instances in BPMN collaborations. Inf. Syst. **103**, 101459 (2022). https://doi.org/10.1016/j.is.2019.101459
6. Corradini, F., Muzi, C., Re, B., Tiezzi, F.: A classification of BPMN collaborations based on safeness and soundness notions. Electron. Proc. Theor. Comput. Sci. **276**, 37–52 (2018). https://doi.org/10.4204/EPTCS.276.5
7. Dijkman, R.M., Dumas, M., Ouyang, C.: Semantics and analysis of business process models in BPMN. Inf. Softw. Technol. **50**(12), 1281–1294 (2008). https://doi.org/10.1016/j.infsof.2008.02.006
8. Durán, F., Salaün, G.: Verifying timed BPMN processes using Maude. In: Jacquet, J.-M., Massink, M. (eds.) COORDINATION 2017. Lecture Notes in Computer Science, vol. 10319, pp. 219–236. Springer, Cham (2017). https://doi.org/10.1007/978-3-319-59746-1_12
9. Ehrig, H., Heckel, R., Korff, M., Löwe, M., Ribeiro, L., Wagner, A., Corradini, A.: Algebraic approaches to graph transformation – part ii: single pushout approach and comparison with double pushout approach, pp. 247–312. World Scientific, February 1997. https://doi.org/10.1142/9789812384720_0004
10. El-Saber, N.A.S.: CMMI-CM compliance checking of formal BPMN models using Maude. Ph.D. thesis, University of Leicester, January 2015

11. Freund, J., Rücker, B.: Real-Life BPMN: using BPMN and DMN to analyze, Improve, and automate processes in your company. Berlin, 4th edn, Camunda (2019)

12. Heckel, R., Taentzer, G.: Graph transformation for software engineers: with applications to model-based development and domain-specific language engineering. Springer, Cham (2020). https://doi.org/10.1007/978-3-030-43916-3

13. Hofstede, A., Aalst, W.: Workflow patterns: on the expressive power of (petri-net-based) workflow languages. In: Proceedings of Fourth Workshop on the Practical Use of Coloured Petri Nets and CPN Tools (CPN 2002), vol. 560, August 2002

14. Houhou, S., Baarir, S., Poizat, P., Quéinnec, P., Kahloul, L.: A first-order logic verification framework for communication-parametric and time-aware BPMN collaborations. Inf. Syst. **104**, 101765 (2022). https://doi.org/10.1016/j.is.2021.101765

15. Kastenberg, H., Rensink, A.: Model checking dynamic states in GROOVE. In: Valmari, A. (ed.) SPIN 2006. LNCS, vol. 3925, pp. 299–305. Springer, Heidelberg (2006). https://doi.org/10.1007/11691617_19

16. Kräuter, T.: Artifacts - ICGT. https://github.com/timKraeuter/ICGT-2023, October 2023

17. Meyers, B., Deshayes, R., Lucio, L., Syriani, E., Vangheluwe, H., Wimmer, M.: ProMoBox: a framework for generating domain-specific property languages. In: Combemale, B., Pearce, D.J., Barais, O., Vinju, J.J. (eds.) SLE 2014. LNCS, vol. 8706, pp. 1–20. Springer, Cham (2014). https://doi.org/10.1007/978-3-319-11245-9_1

18. Object management group: business process model and notation (BPMN), Version 2.0.2. https://www.omg.org/spec/BPMN/, December 2013

19. Peter, D.: Hyperfine (2022)

20. Rensink, A.: The GROOVE simulator: a tool for state space generation. In: Pfaltz, J.L., Nagl, M., Böhlen, B. (eds.) AGTIVE 2003. LNCS, vol. 3062, pp. 479–485. Springer, Heidelberg (2004). https://doi.org/10.1007/978-3-540-25959-6_40

21. Rensink, A.: Nested quantification in graph transformation rules. In: Corradini, A., Ehrig, H., Montanari, U., Ribeiro, L., Rozenberg, G. (eds.) ICGT 2006. LNCS, vol. 4178, pp. 1–13. Springer, Heidelberg (2006). https://doi.org/10.1007/11841883_1

22. Rensink, A.: How much are your geraniums? Taking graph conditions beyond first order. In: Katoen, J.-P., Langerak, R., Rensink, A. (eds.) ModelEd, TestEd, TrustEd. LNCS, vol. 10500, pp. 191–213. Springer, Cham (2017). https://doi.org/10.1007/978-3-319-68270-9_10

23. Tisi, M., Jouault, F., Fraternali, P., Ceri, S., Bézivin, J.: On the use of higher-order model transformations. In: Paige, R.F., Hartman, A., Rensink, A. (eds.) ECMDA-FA 2009. LNCS, vol. 5562, pp. 18–33. Springer, Heidelberg (2009). https://doi.org/10.1007/978-3-642-02674-4_3

24. Van Gorp, P., Dijkman, R.: A visual token-based formalization of BPMN 2.0 based on in-place transformations. Inf. Softw. Technol. **55**(2), 365–394 (2013). https://doi.org/10.1016/j.infsof.2012.08.014

Computing k-Bisimulations for Large Graphs: A Comparison and Efficiency Analysis

Jannik Rau[1], David Richerby[2], and Ansgar Scherp[1]($\boxtimes$)

[1] University of Ulm, Ulm, Germany
{jannik.rau,ansgar.scherp}@uni-ulm.de
[2] University of Essex, Colchester, UK
david.richerby@essex.ac.uk

Abstract. Summarizing graphs w.r.t. structural features is important
to reduce the graph's size and make tasks like indexing, querying, and
visualization feasible. Our generic parallel BRS algorithm efficiently summarizes large graphs w.r.t. a custom equivalence relation $\sim$ defined on
the graph's vertices V. Moreover, the definition of $\sim$ can be chained $k \geq 1$
times, so the defined equivalence relation becomes a k-bisimulation. We
evaluate the runtime and memory performance of the BRS algorithm for
k-bisimulation with $k = 1, \ldots, 10$ against two algorithms found in the
literature (a sequential algorithm due to Kaushik et al. and a parallel
algorithm of Schätzle et al.), which we implemented in the same software stack as BRS. We use five real-world and synthetic graph datasets
containing 100 million to two billion edges. Our results show that the
generic BRS algorithm outperforms the respective native bisimulation
algorithms on all datasets for all $k \geq 5$ and for smaller k in some cases.
The BRS implementations of the two bisimulation algorithms run almost
as fast as each other. Thus, the BRS algorithm is an effective parallelization of the sequential Kaushik et al. bisimulation algorithm.

Keywords: structural graph summarization · bisimulation · large
labeled graphs

1 Introduction

Storing, indexing, querying, and visualizing large graphs is difficult [7]. One way
to mitigate this challenge is *graph summarization* [9]. Graphs can be summarized
w.r.t. so-called *graph summary models* [6] that define structural features (e. g.,
incoming/outgoing paths), statistical measures (e. g., occurrences of specific vertices), or frequent patterns found in the graph [9]. This gives a *summary graph*,
which is usually smaller than the original but contains an approximation of or
exactly the same information as the original graph w.r.t. the selected features
of the summary model. Tasks that were to be performed on the original graph
can be performed on the summary but much faster. Use cases are optimizing
database queries [22], data visualization [12], and OWL reasoning [28].

© The Author(s), under exclusive license to Springer Nature Switzerland AG 2023
M. Fernández and C. M. Poskitt (Eds.): ICGT 2023, LNCS 13961, pp. 223–242, 2023.
https://doi.org/10.1007/978-3-031-36709-0_12

Blume, Richerby, and Scherp developed a generic structural summarization approach, here referred to as BRS [5,6]. The BRS algorithm summarizes an input graph w.r.t. an arbitrary user-defined equivalence relation specified in its formal language FLUID [6]. The FLUID language supports all features of structural graph summarization found in the literature [6]. There are two groups of these features. The first comprises a vertex's *local* information, e. g., its label set, its direct neighbors, and the labels of its incoming or outgoing edges. The second group considers a vertex's *global* information at distance $k > 1$. This includes, e. g., local information about reachable vertices up to distance k or information about incoming or outgoing paths of length up to k. We use stratified k-bisimulations (formally described in Sect. 3.2) to summarize a graph w.r.t. global information and group together vertices that have equivalent structural neighborhoods up to distance k. Several existing approaches use k-bisimulations to incorporate global information into structural graph summarization [6,9]. The BRS algorithm generalizes these approaches and can chain any definable equivalence relation k times, such that the resulting equivalence classes can be efficiently computed by global information up to distance k [4,24].

However, so far it is not known if a general approach like the BRS algorithm sacrifices performance. We choose two representative algorithms as examples to demonstrate the capabilities of our generic BRS algorithm. First, we have re-implemented the efficient, parallel single-purpose k-bisimulation algorithm of Schätzle, Neu, Lausen, and Przjaciel-Zablocki [25] and investigate whether it outperforms our generic BRS algorithm. Second, we investigate the sequential algorithm for bisimulation by Kaushik, Shenoy, Bohannon, and Gudes [16]. Being sequential, it is naturally disadvantaged against parallel algorithms such as ours. However, we show in this work that the bisimulation of Kaushik et al. [16] can be declaratively specified and executed in the generic BRS algorithm. This effectively parallelizes the algorithm "for free". We evaluate the performance of the BRS-based parallelized computation of the Kaushik et al. graph summary model and compare it with their sequential native algorithm. We also compare both Kaushik et al. variants, native and BRS-based, with the parallel native algorithm of Schätzle et al. [25] and a BRS implementation of Schätzle et al. (see also [4]). Thus, we have four k-bisimulation algorithms. For a fair comparison, we reimplemented the existing native algorithms in the same graph processing framework as the BRS algorithm. We execute the four algorithms on five graph datasets – two synthetic and three real-world – of different sizes, ranging from 100 million edges to billions of edges. We evaluate the algorithms' performance for computing k-bisimulation for $k = 1, \ldots, 10$. We measure running time per iteration and the maximum memory consumption.

The questions we address are: Do the native bisimulation algorithms have an advantage over a generic solution? How well do the native and generic algorithms scale to large real-world and synthetic graphs? Is it possible to effectively scale a sequential algorithm by turning it into a parallel variant by using a general formal language and algorithm for graph summaries?

We discuss related work next. Section 3 defines preliminaries, while the algorithms are introduced in Sect. 4. Section 5 outlines the experimental apparatus. Section 6 describes the results, and these are discussed in Sect. 7.

2   Related Work

Summary graphs can be constructed in several ways. Čebirić et al. [9] classify existing techniques into structural, pattern-mining, statistical, and hybrid approaches. In this paper, we consider only structural approaches based on quotients. Other structural summarization techniques, not based on quotients, are extensively discussed by Čebirić et al. [9]. Structural approaches summarize a graph G w.r.t. an equivalence relation $\sim \subseteq V \times V$ defined on the vertices V of G [7,9]. The resulting summary graph SG consists of vertices VS, each of which corresponds to a equivalence class of the equivalence relation $\sim$.

One can observe that k-bisimulation is a popular feature for structural graph summarization [6]. Bisimulation comes in three forms: backward k-bisimulation classifies vertices based on incoming paths of length up to k, forward bisimulation considers outgoing paths, and backward-forward bisimulation considers both. Bisimulation may be based on edge labels, vertex labels, or both, but this makes no significant difference to the algorithms. A notion of k-bisimulation w.r.t. graph indices is introduced by seminal works such as the k-RO index [21] and the T-index summaries [20]. Milo and Suciu's T-index [20], the $A(k)$-Index by Kaushik et al. [16], and others are examples that summarize graphs using backward k-bisimulation. We chose as representative the sequential algorithm by Kaushik et al. [16], which uses vertex labels, as described in Sect. 4.2. Conversely, the k-RO index, the Extended Property Paths of Consens et al. [11], the SemSets model of Ciglan et al. [10], Buneman et al.'s RDF graph alignment [8], and the work of Schätzle et al. [25] are based on forward k-bisimulation. We note that Schätzle et al. use edge labels, as described in Sect. 4.1. Tran et al. compute a structural index for graphs based on backward-forward k-bisimulation [27]. Moreover, they parameterize their notion of bisimulation to a forward-set L_1 and a backward-set L_2, so that only labels $l \in L_1$ are considered for forward-bisimulation and labels $l \in L_2$ for backward-bisimulation.

Luo et al. examine structural graph summarization by forward k-bisimulation in a distributed, external-memory model [18]. They empirically observe that, for values of $k > 5$, the summary graph's partition blocks change little or not at all. Therefore they state, that for summarizing a graph with respect to k-bisimulation, it is sufficient to summarize up to a value of $k = 5$ [17]. Finally, Martens et al. [19] introduce a parallel bisimulation algorithm for massively parallel devices such as GPU clusters. Their approach is tested on a single GPU with 24 GB RAM, which limits its use on large datasets. Nonetheless, their proposed blocking mechanism could be combined with our vertex-centric approach to further improve performance.

Each of the works proposes a single algorithm for computing a single graph summary based on a bisimulation. Some of the algorithms have bisimulation parameters such as the height and label parameterization in Tran et al. [27].

We suggest a generic algorithm for computing k-bisimulation and show its advantages. Also, our approach allows the bisimulation model to be specified in a declarative way and parallelizes otherwise sequential computations like in Kaushik et al. [16] into a parallel computation.

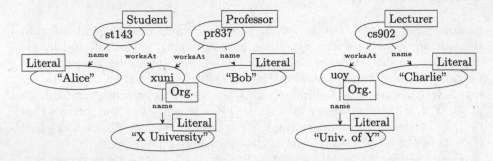

Fig. 1. An example graph G displaying two universities and three employees. Vertices are denoted by ellipses and edges by arrows. Vertex labels are marked with rectangles and edge labels are written on the edge.

3   Preliminaries

3.1   Data Structures

The algorithms operate on multi-relational, labeled graphs $G = (V, E, l_V, l_E)$, where $V = \{v_1, v_2, \ldots, v_n\}$ is a set of vertices and $E \subseteq V \times V$ is a set of directed edges between the vertices in V. Each vertex $v \in V$ has a finite set of labels $l_V(v)$ from a set Σ_V and each edge has a finite set of labels $l_E(u, v)$ from a set Σ_E.

Figure 1 shows an example graph representing two universities and three employees. Vertices are represented by ellipses and edges are labeled arrows. Vertex labels are shown in rectangles. For example, the edges (pr837, xuni) and (st143, xuni) labeled with *worksAt* together with edges (pr837, "Bob"), (st143, "Alice") and (xuni, "X University") labeled with *name* and vertex labels *Professor* (pr837), *Student* (st143) and *Organization* (xuni) state that professor Bob and student Alice both work at the organization X University.

In a graph $G = (V, E, l_V, l_E)$ the *in-neighbors* of a vertex $v \in V$ are the set of vertices $N^-(v) = \{u \mid (u, v) \in E\}$ from which v receives an edge. Similarly, v's *out-neighbors* are the set $N^+(v) = \{w \mid (v, w) \in E\}$ to which it sends edges. For a set $S \subseteq V$, let $N^+(S) = \bigcup_{v \in S} N^+(v)$ be the set of out-neighbors of S.

3.2   Bisimulation

A bisimulation is an equivalence relation on the vertices of a directed graph [16, 27]. Informally, a bisimulation groups vertices with equivalent structural neighborhoods, i. e., the neighborhoods cannot be distinguished based on the vertices'

Table 1. Vertex-labeled backward 2-bisimulation partition of the example graph according to Definition 2.

k	Partition blocks
0	{pr837}, {cs902}, {st143}, {uoy, xuni}, literals
1	{pr837}, {cs902}, {st143}, {uoy}, {xuni}, {"Alice"}, {"Charlie"}, {"Bob"}, {"X University", "Univ. of Y"}
2	{pr837}, {cs902}, {st143}, {uoy}, {xuni}, {"Alice"}, {"Charlie"}, {"Bob"}, {"X University"}, {"Univ. of Y"}

sets of labels and/or edges' labels. *Forward bisimulation (fw)* considers outgoing edges; *backward bisimulation (bw)* uses incoming edges. Vertices u and v are forward-bisimilar if, every out-neighbor u' of u has a corresponding out-neighbor v' of v, and vice versa; furthermore, the two neighbors u' and v' must be bisimilar [16, 27]. Backward-bisimulation is defined similarly but using in-neighbors. This definition corresponds to a *complete* bisimulation. For the neighbors u' and v' to be bisimilar, their in-/out-neighbors must be bisimilar as well. A k-bisimulation is a bisimulation on G that considers features a distance at most k from a vertex when deciding whether it is equivalent to another.

The algorithms of Schätzle et al. [25] and Kaushik et al. [16] compute versions of forward and backward k-bisimulation on labeled graphs $G = (V, E, \Sigma_V, \Sigma_E)$.

Definition 1. *The edge-labeled forward k-bisimulation $\approx_{\mathrm{fw}}^k \subseteq V \times V$ with $k \in \mathbb{N}$ of Schätzle et al. [25] is defined as follows:*

- *$u \approx_{\mathrm{fw}}^0 v$ for all $u, v \in V$,*
- *$u \approx_{\mathrm{fw}}^{k+1} v$ iff $u \approx_{\mathrm{fw}}^k v$ and, for every edge (u, u'), there is an edge (v, v') with $l_E(u, u') = l_E(v, v')$ and $u' \approx_{\mathrm{fw}}^k v'$, and vice-versa.*

For the graph in Fig. 1, $\approx_{\mathrm{fw}}^0$ has the single block V (as for all graphs, all vertices are initially equivalent) and $\approx_{\mathrm{fw}}^1$ has three blocks, i.e., sets of equivalent vertices: {pr837, cs902, st143}, {uoy, xuni}, and the literals. The vertices xuni and uoy are not 1-bisimilar to st143, cs902 and pr837, as they have no outgoing edge labeled *worksAt*. For $k \geq 2$, k-bisimulation in this case makes no more distinctions than 1-bisimulation. Note that Schätzle et al. compute full bisimulations. We have modified their algorithm to stop after k iterations.

Definition 2. *The vertex-labeled backward k-bisimulation $\approx_{\mathrm{bw}}^k \subseteq V \times V$ with $k \in \mathbb{N}$ of Kaushik et al. [16] is defined as follows:*

- *$u \approx_{\mathrm{bw}}^0 v$ iff $l_V(u) = l_V(v)$,*
- *$u \approx_{\mathrm{bw}}^{k+1} v$ iff $v \approx_{\mathrm{bw}}^k u$ and, for every $(u', u) \in E$, there is $(v', v) \in E$ with $u' \approx_{\mathrm{bw}}^k v'$, and vice versa.*

Table 1 shows the vertex-labeled backward 2-bisimulation partitions of the graph in Fig. 1. 0-bisimulation partitions by label. Then, xuni, uoy and the literals are split by their parents' labels. No vertex is 2-bisimilar to any other: every block is a singleton.

3.3   Graph Summaries for Bisimulation

The BRS algorithm summarizes graphs with respect to a *graph summary model* (GSM), a mapping from graphs $G = (V, E)$ to equivalence relations $\sim \subseteq V \times V$. The equivalence classes of $\sim$ partition G. A simple GSM is label equality, i.e., two vertices are equivalent iff they have the same label. Depending on the application, one might want to summarize a graph w.r.t. different GSMs. Therefore, the algorithm works with GSMs defined in our formal language FLUID [6]. To flexibly and quickly define GSMs, the language provides simple and complex schema elements, along with six parameterizations, of which we use two (for details we refer to [6]).

A *complex schema element* $CSE := (\sim_s, \sim_p, \sim_o)$ combines three equivalence relations [6]. Vertices v and v' are equivalent, iff $v \sim_s v'$; and for all $w \in N^+(v)$ there is a $w' \in N^+(v')$ with $l_E(v, w) \sim_p l_E(v', w')$ and $w \sim_o w'$, and vice versa.

The chaining parameterization enables computing k-bisimulations by increasing the neighborhood considered for determining vertex equivalence [6]. It is defined by nesting CSEs. Given a complex schema element $CSE := (\sim_s, \sim_p, \sim_o)$ and $k \in \mathbb{N}_{>0}$, the chaining parameterization CSE^k defines the equivalence relation that corresponds to recursively applying CSE to a distance of k hops. $CSE^1 := (\sim_s, \sim_p, \sim_o)$ and, inductively for $k > 1$, $CSE^k := (\sim_s, \sim_p, CSE^{k-1})$. This results in a summary graph that has one vertex for each equivalence class in each equivalence relation defined within the CSE. Summary vertices v and w are connected via a labeled edge, if all vertices in the input graph represented by v have an edge with this label to a vertex in w. For full details, see [6].

To model backward k-bisimulations, we need to work with incoming edges, whereas the schema elements consider only outgoing edges. In FLUID, this is done with the direction parameterization [6] but, here, we simplify notation. We write SE^{-1} for the schema element defined analogously to SE but using the relation $E^{-1} = \{(y, x) \mid (x, y) \in E\}$ in place of the graph's edge relation E, and the edge labeling $\ell_E^{-1}(y, x) = \ell_E(x, y)$.

Following Definitions 1 and 2 of Schätzle et al. and Kaushik et al., we define CSE_{Sch} and CSE_{Kau} as follows. Here, $id = \{(v, v) \mid v \in V\}$ and $T = V \times V$, and vertices are equivalent in OC_{type} iff they have the same labels [6].

$$CSE_{\text{Sch}} := (T, id, T)^k \tag{1}$$

$$CSE_{\text{Kau}} := \left((OC_{\text{type}}, T, OC_{\text{type}})^{-1}\right)^k. \tag{2}$$

4   Algorithms

We introduce the single-purpose algorithms of Schätzle et al. [25] in Sect. 4.1 and Kaushik et al. [16] in Sect. 4.2. Finally, we introduce the generic BRS algorithm in Sect. 4.3. The first two algorithms compute summaries in a fundamentally different way to the BRS algorithm. At the beginning of the execution, BRS considers every vertex to be in its own equivalence class. During execution, vertices with the same vertex summary are merged. Therefore, BRS can be seen as a *bottom-up* approach. In contrast, Schätzle et al. consider all vertices to be equivalent at the beginning, and Kaushik et al. initially consider all vertices with the same label to be equivalent. The equivalence relation is then successively refined. These two algorithms can be seen as *top-down* approaches.

Here, it is convenient to consider set partitions. A partition of a set V is a set $\{B_1, \ldots, B_\ell\}$ such that: (i) $\emptyset \subsetneq B_i \subseteq V$ for each i, (ii) $\bigcup_i B_i = V$, and (iii) $B_i \cap B_j = \emptyset$ for each $i \neq j$. The sets B_i are known as *blocks*. The equivalence classes of an equivalence relation over V partition the graph's vertices. A key concept is *partition refinement*. A partition $P_i = \{B_{i1}, B_{i2}, \ldots\}$ *refines* $P_j = \{B_{j1}, B_{j2}, \ldots\}$ iff every block B_{ik} of P_i is contained in a block $B_{j\ell}$ of P_j.

4.1 Native Schätzle et al. Algorithm

This algorithm [25] is a distributed MapReduce approach for reducing labeled transition systems. Two fundamental concepts are the *signature* and *ID* of a vertex v with respect to the current iteration's partition P_i.

The *signature* of a vertex v w.r.t. a partition $P_i = \{B_{i1}, B_{i2}, \ldots\}$ of V is given by $\mathrm{sig}_{P_i}(v) = \{(\ell, B_{ij}) \mid (v, w) \in E \text{ with } l_E(v, w) = \ell \text{ and } w \in B_{ij}\}$. That is, v's signature w.r.t. the current iteration's partition P_i is the set of outgoing edge labels to blocks of P_i. By Definition 1, $u \approx_{\mathrm{fw}}^{k+1} v$, iff $\mathrm{sig}_{P_k}(u) = \mathrm{sig}_{P_k}(v)$. Therefore signatures identify the block of a vertex and represent the current bisimulation partition, and the signature of v w.r.t. P_i can be represented as $\mathrm{sig}_{P_{i+1}}(v) = \{(l_E(v, w), \mathrm{sig}_{P_i}(w)) \mid (v, w) \in E\}$. The nested structure of vertex signatures means they can become very large. Thus, we compute a recursively defined hash value proposed by Hellings et al. [13], which is also used by Schätzle et al. [25]. We use this hash function to assign $\mathrm{sig}_{P_i}(v)$ an integer value which we denote by $\mathrm{ID}_{P_i}(v)$. Now the signature of a vertex v w.r.t. the current partition P_i can be represented as $\mathrm{sig}_{P_{i+1}}(v) = \{(l_E(v, w), \mathrm{ID}_{P_i}(w)) \mid (v, w) \in E\}$.

With $\mathrm{sig}_{P_i}(v)$ and $\mathrm{ID}_{P_i}(v)$, the procedure for computing an edge-labeled forward k-bisimulation partition is outlined in Algorithm 1. The initial partition is just V (line 2), as every vertex is 0-bisimilar to every other vertex. Next, the algorithm performs k iterations (lines 3–8). In the ith iteration, the information needed to construct a vertex's signature $\mathrm{sig}_{P_i}(v)$ is sent to every vertex v (line 5). This information is the edge label $l_E(v, w) \in \Sigma_E$ and the block identifier $\mathrm{ID}_{P_{i-1}}(w)$ for every $w \in N^+(v)$. The signature $\mathrm{sig}_{P_i}(v)$ is then constructed using the received information, and the identifiers $\mathrm{ID}_{P_i}(v)$ are updated for all v (line 6). At the end of each iteration, the algorithm checks if any vertex ID was updated, by comparing the number of distinct values in ID_{P_i} and $\mathrm{ID}_{P_{i-1}}$ (line 7). If no vertex ID was updated, we have reached full bisimulation [3,25] and hence can stop execution early. At the end, the resulting k-bisimulation partition P_k is constructed by putting vertices v in one block if they share the same identifier value $\mathrm{ID}_{P_i}(v)$ (line 9).

4.2 Native Kaushik et al. Algorithm

This algorithm [16] sequentially computes vertex-labeled backward k-bisimulations. The following definitions are from [23], modified for backward bisimulation.

Algorithm 1: Bisimulation Algorithm by Schätzle et al. [25]

---

1 **function** BISIMSCHÄTZLE($G = (V, E, l_V, l_E)$, $k \in \mathbb{N}$)
       /* Initially, all $v \in V$ in same block with $\mathrm{ID}_{P_0}(v) = 0$ */
2     $P_0 \leftarrow \{V\}$;
3   **for** $i \leftarrow 1$ **to** k **do**
           /* Map Job */
4   **for** $(v, w) \in E$ **do**
5 | Send $(l_E(v, w), \mathrm{ID}_{P_{i-1}}(w))$
           /* Reduce Job */
6   Construct $\mathrm{sig}_{P_i}(v)$ and update $\mathrm{ID}_{P_i}(v)$;
           /* Check if full bisimulation is reached */
7   **if** $|\mathrm{ID}_{P_i}| = |\mathrm{ID}_{P_{i-1}}|$ **then**
8 | **break**;
9   Construct P_k from ID_{P_i};
10   **return** P_k;

A subset $B \subseteq V$ is *stable* with respect to another subset $S \subseteq V$ if either $B \subseteq N^+(S)$ or $B \cap N^+(S) = \emptyset$. That is, vertices in a stable set B are indistinguishable by their relation to S: either all vertices in B get at least one edge from S, or none do. If B is not stable w.r.t. S, we call S a *splitter* of B.

Building on this, partition P_i of V is *stable with respect to a subset* $S \subseteq V$ if every block $B_{ij} \in P_i$ is stable w.r.t. S. P_i is *stable* if it is stable w.r.t. each of its blocks B_{ij}. Thus, a partition P_i is stable if none of its blocks B_{ij} can be split into a set of vertices that receive edges from some B_{ik} and a set of vertices that do not. A stable partition corresponds to the endpoint of a bisimulation computation: no further distinctions can be made.

This gives an algorithm for bisimulation, due to Paige and Tarjan [23] who refer to it as the "naïve algorithm". The initial partition is repeatedly refined by using its own blocks or unions of them as splitters: if S splits a block B, we replace B in the partition with the two new blocks $B \cap N^+(S)$ and $B - N^+(S)$. When no more splitters exist, the partition is stable [23] and equivalent to the full backward bisimulation of the initial partition P_0 [15].

The algorithm of Kaushik et al., Algorithm 2, modifies this naïve approach. The first difference is that in each iteration $i \in \{1, \ldots, k\}$ the partition is stabilized with respect to each of its own blocks (lines 7–16). This ensures that, after iteration i, the algorithm has computed the i-bisimulation [16], which is not the case in the naïve algorithm. Second, blocks are split as defined above, (lines 9–15). As a result, Algorithm 2 computes the k-backward bisimulation. To check if full bisimulation has been reached, the algorithm uses the Boolean variable wasSplit (lines 6 and 15). If this is false at the end of an iteration i, no block was split, so the algorithm stops early (line 17). Moreover, Algorithm 2 tracks which sets have been used as splitters (line 3), to avoid checking for stability against sets w.r.t. which the partition is already known to be stable. The algorithm provided by Kaushik et al. does not include this. If a partition P is

Algorithm 2: Bisimulation Algorithm by Kaushik et al. [16]

```
1  function BISIMKAUSHIK(G = (V, E, l_V, l_E), k ∈ ℕ)
      /* P := {B_1, B_2, ..., B_t} */
2     P ← partition V by label;
3     usedSplitters ← ∅;
4     for i ← 1, ..., k do
5        P^copy ← P;
6        wasSplit ← false;
7        for B^copy ∈ P^copy − usedSplitters do
            /* Use blocks of copy partition to stabilize blocks of
               original partition */
8           for B ∈ P do
9              succ ← B ∩ N^+(B^copy);
10             nonSucc ← B − N^+(B^copy);
               /* Split non-stable blocks */
11             if succ ≠ ∅ and nonSucc ≠ ∅ then
12                P.add(succ);
13                P.add(nonSucc);
14                P.delete(B);
15                wasSplit ← true;
16          usedSplitters.add(B^copy);
17       if ¬wasSplit then
18          break;
19    return P;
```

stable w.r.t. a block B, each refinement of P is also stable w.r.t. B [23]. So after the partition P is stabilized w.r.t. a block copy B^{copy} (lines 7–16), we can add B^{copy} to the usedSplitters set and not consider it in subsequent iterations.

4.3 Generic BRS Algorithm

The parallel BRS algorithm is not specifically an implementation of k-bisimulation. Rather, one can define a graph summary model in a formal language FLUID (see Sect. 3.3). This model is denoted by $\sim$ and input to the BRS algorithm, which then summarizes a graph w.r.t. $\sim$. In particular, the k-bisimulation models of Schätzle et al. and Kaushik et al. can be expressed in FLUID, as shown in Sect. 3.3. Thus, the BRS algorithm can compute k-bisimulation partitions. In other words, k-bisimulation can be incorporated into any graph summary model defined in FLUID. The BRS algorithm summarizes the graph w.r.t. $\sim$ in parallel and uses the Signal/Collect paradigm. In Signal/Collect [26], vertices collect information from their neighbors, sent over the edges as signals. Details of the use of the Signal/Collect paradigm in our algorithm can be found in Blume et al. [5]. Briefly, the algorithm builds equivalence classes by starting with every vertex in its own singleton set and forming unions of equivalent vertices. Before outlining the algorithm, we give a necessary definition.

Definition 3. *Suppose we have a graph summary GS of $G = (V, E, l_V, l_E)$ w.r.t. some GSM $\sim$. For each $v \in V$, the* vertex summary *vs is the subgraph of GS that defines v's equivalence class w.r.t. $\sim$.*

We give the pseudocode of our version of the BRS algorithm in Algorithm 3 and briefly describe it below. [4] gives a step-by-step example run.

Initialization (lines 3–5). For each vertex $v \in V$, VERTEXSCHEMA computes the local schema information w.r.t. $\sim_s$ and $\sim_o$ of the graph summary model $(\sim_s, \sim_p, \sim_o)^k$. The method also takes into account $\sim_p$, i. e., the equivalence relation defined over the edge labels. Order-invariant hashes of this schema information are stored as the identifiers $id_{\sim_s}$ and $id_{\sim_o}$ (lines 4–5).

At the end of the initialization, every vertex has identifiers $id_{\sim_s}$ and $id_{\sim_o}$. Two vertices v and v' are equivalent w.r.t. $(\sim_s, \sim_p, \sim_o)$, iff $v.id_{\sim_s} = v'.id_{\sim_s}$. Thus, this initialization step can be seen as iteration $k = 0$ of bisimulation.

Case of $k = 1$ bisimulation (lines 6–11). Every vertex v sends, to each in-neighbor w, its $id_{\sim_o}$ value and the label set $\ell(w, v)$ of the edge (w, v) (line 9).

Every vertex v sends, to each in-neighbor w, its $id_{\sim_o}$ value and the label set $L = \ell(w, v)$ of the edge (w, v) (line 9). Thus, each vertex receives a set of schema $\langle L, id_{\sim_o} \rangle$ pairs from its out-neighbors, which are collated into the set M_o (line 10) and merged with an order-invariant hash to give v's new $id_{\sim_s}$ (line 11). Here, the MERGEANDHASH(M_o) function first merges the elements of the schema message M_o received from vertex o and hashes it with an order-independent hash function. This hash is then combined with the existing hash value $v.id_{\sim_s}$ using the xor ($\oplus$) operator.

Case of $k > 1$ bisimulation (lines 13–32). In the first iteration (lines 13–19), every vertex v sends a message to each of its out-neighbors w. The message contains v's $id_{\sim_s}$ and $id_{\sim_o}$ values, and the edge label set $\ell(w, v)$ (line 15). Subsequently, the incoming messages of the vertex are merged into a set of tuples with the received information $\langle \ell(w, v), id_{\sim_s} \rangle$ and $\langle \ell(w, v), id_{\sim_o} \rangle$ (lines 16 and 17). Finally, the identifiers $id_{\sim_s}$ and $id_{\sim_o}$ of v are updated by hashing the corresponding set (lines 18 and 19). Note that, whenever an update of an identifier value $v.id_{\sim_s}$ of vertex v is performed, the algorithm combines the old $v.id_{\sim_s}$ with the new hash value, indicated by $\oplus$.

In the remaining iterations, the algorithm performs the same steps (lines 20–27), but excludes the edge label set $\ell(w, v)$ when merging messages for $id_{\sim_o}$ (line 25). When merging the messages in $id_{\sim_o}$, it is not necessary to consider $\ell(w, v)$, as in the iterations 2 to $k - 1$ it is only needed to update the $id_{\sim_o}$ values using the hash function as described above. This is possible as $id_{\sim_o}$ by definition already contains the edge label set $\ell(w, v)$, computed in the first iteration.

In the final iteration, the identifiers are updated w.r.t. the final messages (lines 28–32). The final messages received are the values stored in the out-neighbors' $id_{\sim_o}$ values. Each vertex signals its $id_{\sim_o}$ value to its in-neighbors (line 30). The messages a vertex receives are merged (line 31) and hashed to update the final $id_{\sim_s}$ value (line 32). Equivalence between any two vertices v and v' can now be defined. Vertices with the same $id_{\sim_s}$ value are merged (line 33), ending the computation.

Algorithm 3: Parallel BRS algorithm

---

1. **function** PARALLELSUMMARIZE$(G, (\sim_s, \sim_p, \sim_o)^k)$
2.   **returns** *graph summary SG*

    /* Initialization */
3.   **for all** $v \in V$ **do in parallel**
4. $v.id_{\sim_s} \leftarrow$ hash(VERTEXSCHEMA$(v, G, \sim_s, \sim_p)$);
5. $v.id_{\sim_o} \leftarrow$ hash(VERTEXSCHEMA$(v, G, \sim_o, \sim_p)$);

    /* If $k = 1$, only signal edge labels and $v.id_{\sim_o}$ */
6.   **if** $k = 1$ **then**
7.   **for all** $v \in V$ **do in parallel**
8.   **for all** $w \in N^-(v)$ **do**
9.   SENDMSGS$(w, \langle \ell(w,v), 0, v.id_{\sim_o}\rangle)$;
10. $M_o \leftarrow \{\langle L, id_{\sim_o}\rangle \mid \langle L, id_{\sim_s}, id_{\sim_o}\rangle$ was received$\}$;
11. $v.id_{\sim_s} \leftarrow v.id_{\sim_s} \oplus$ MERGEANDHASH(M_o);
12.   **else**

        /* Signal initial messages. Update $v.id_{\sim_s}$ and $v.id_{\sim_o}$ */
13.   **for all** $v \in V$ **do in parallel**
            /* Message each in-neighbor */
14.   **for all** $w \in N^-(v)$ **do**
15.   SENDMSG$(w, \langle \ell(w,v), v.id_{\sim_s}, v.id_{\sim_o}\rangle)$;
            /* Collect all incoming messages of v */
16. $M_s \leftarrow \{\langle L, id_{\sim_s}\rangle \mid \langle L, id_{\sim_s}, id_{\sim_o}\rangle$ was received$\}$;
17. $M_o \leftarrow \{\langle L, id_{\sim_o}\rangle \mid \langle L, id_{\sim_s}, id_{\sim_o}\rangle$ was received$\}$;
            /* Update identifiers by hashing the messages */
18. $v.id_{\sim_s} \leftarrow v.id_{\sim_s} \oplus$ MERGEANDHASH(M_s);
19. $v.id_{\sim_o} \leftarrow v.id_{\sim_o} \oplus$ MERGEANDHASH(M_o);

        /* Signal messages $k - 2$ times. As above, but we do not
            include L when updating $v.id_{\sim_o}$. (See text.) */
20.   **for** $i \leftarrow 2$ **to** $k - 1$ **do**
21.   **for all** $v \in V$ **do in parallel**
22.   **for all** $w \in N^-(v)$ **do**
23.   SENDMSG$(w, \langle \ell(w,v), v.id_{\sim_s}, v.id_{\sim_o}\rangle)$;
24. $M_s \leftarrow \{\langle L, id_{\sim_s}\rangle \mid \langle L, id_{\sim_s}, id_{\sim_o}\rangle$ received$\}$;
25. $M_o \leftarrow \{\langle id_{\sim_o}\rangle \mid \langle L, id_{\sim_s}, id_{\sim_o}\rangle$ received$\}$;
26. $v.id_{\sim_s} \leftarrow v.id_{\sim_s} \oplus$ MERGEANDHASH(M_s);
27. $v.id_{\sim_o} \leftarrow v.id_{\sim_o} \oplus$ MERGEANDHASH(M_o);

        /* Signal final messages. Update $v.id_{\sim_s}$ */
28.   **for all** $v \in V$ **do in parallel**
29.   **for all** $w \in N^-(v)$ **do**
30.   SENDMSG$(w, \langle \emptyset, 0, v.id_{\sim_o}\rangle)$;
31. $M_o \leftarrow \{\langle id_{\sim_o}\rangle \mid \langle L, id_{\sim_s}, id_{\sim_o}\rangle$ was received$\}$;
32. $v.id_{\sim_s} \leftarrow v.id_{\sim_s} \oplus$ MERGEANDHASH(M_o);

33. $SG \leftarrow$ FINDANDMERGE(SG, V);
34.   **return** SG;

Table 2. Statistics of the datasets.

Graph	$|V|$	$|E|$	$|\Sigma_V|$	$r(l_V)$	$\mu(|l_V(v)|)$	$|\Sigma_E|$
Laundromat100M	30 M	88 M	33, 431	7, 373	0.93 ± 44	5, 630
BTC150M	5 M	145 M	69	137	1.04 ± 0.26	10, 750
BTC2B	80 M	1.92 B	113, 365	576, 265	0.95 ± 1.82	38, 136
BSBM100M	18 M	90 M	1, 289	2, 274	1.02 ± 0.13	39
BSBM1B	172 M	941 M	6, 153	27, 306	1.03 ± 0.18	39

We show in [4] that Algorithm 3 computes k-bisimulation of a graph with m edges in time $O(km)$. As a modification of the algorithm in [6], the algorithm is correct, as long as hash collisions are avoided.

5   Experimental Apparatus

5.1   Datasets

We experiment with smaller and larger as well as real-world and synthetic graphs. Table 2 lists statistics of these datasets, where $r(l_V) = |\{l_V(v) \mid v \in V\}|$ is the number of different label sets (range) and $\mu(|l_V(v)|)$ is the average number of labels of a vertex $v \in V$.

Three real-world datasets were chosen. The *Laundromat100M* dataset contains 100 M edges of the LOD Laundromat service [1]. This service automatically cleaned existing linked datasets and provided the cleaned version on a publicly accessible website. The *BTC150M* and *BTC2B* datasets contain, respectively, around 150 million and 1.9 billion edges of the Billion Triple Challenge 2019 (BTC2019) dataset [14]. 93% of the total edges originate from Wikidata [29]. BTC150M is the first chunk of the 1.9 billion edges.   For synthetic datasets, two versions of the Berlin SPARQL Benchmark (BSBM) [2] were used. The BSBM data generator produces RDF datasets that simulate an e-commerce use case. *BSBM100M* was generated with $284, 826$ products and has about 17.77 million vertices and 89.54 million edges. *BSBM1B* was generated with $2, 850, 000$ products and has about 172 million vertices and 941 million edges.

5.2   Procedure

An experiment consists of the algorithm to run, the dataset to summarize, and the bisimulation degree k. In case of the BRS algorithm, it additionally consists of the graph summary model to use. We have two different graph summary models defined by Schätzle et al. [25] and Kaushik et al. [16]. Each model comes in two implementations, one native implementation as defined by the original authors and a generic implementation through our hash-based version of the BRS algorithm. We chose the bisimulation algorithms of Kaushik et al. and Schätzle et al. as they represent two typical variants of backward and forward k-bisimulation models as found in the literature (cf. Sect. 2). This choice of algorithms allows us to demonstrate that our algorithm can be applied to different settings. We use the terms *BRS-Schätzle* and *BRS-Kaushik* to refer to our implementations of the two GSMs BRS algorithm; we refer to our single-purpose implementations of these two GSMs as *native Schätzle* and *native Kaushik*.

The four algorithms are applied on the five datasets, giving 20 experiments. Each experiment is executed with a bisimulation degree of $k = 1, \ldots, 10$, using the following procedure. We run the algorithms six times with the specific configuration. We use the first run as a warm-up and do not account it for our measurements. The next five runs are used to measure the variables.

5.3 Implementation

All algorithms, i.e., the native algorithms of Schätzle et al. and Kaushik et al. and their generic BRS-variants are implemented using the same underlying framework and paralellization approach, i.e., are implemented in Scala upon the Apache Spark Framework. This API offers flexible support for parallel computation and message passing, which enables implementation of Map-Reduce and Signal-Collect routines. We use an Ubuntu 20 system with 32 cores and 2 TB RAM. The Apache Spark contexts were given the full resources. Time and memory measurements were taken using the Apache Spark Monitoring API.

5.4 Measures

We evaluate the algorithms' running time and memory consumption. For every run of an experiment, we report the total run time, the run time of each of the k iterations, and the maximum JVM on-heap memory consumption.

6 Results

We present full results for each algorithm, iteratively calculating k-bisimulation for every value of $k = 1, \ldots, 10$ and every dataset, in Fig. 2. Table 3 summarizes the average total run time (minutes) for each experiment, for the computation of 10-bisimulation. The BRS algorithm takes an additional initialization step (see Algorithm 3, lines 13–19). Table 4 reports the maximum JVM on-heap memory in GB for each experiment. The BRS-Schätzle algorithm computes the 10-bisimulation the fastest on all datasets, except for BSBM100M, where BRS-Kaushik is fastest. Native Schätzle consumes the least memory on all smaller datasets. Native Schätzle consumes slightly more memory on BSBM1B than BRS, whereas on BTC2B, the memory consumption is about the same.

Smaller Datasets (100M+ Edges). Figure 2 shows the average run time (minutes) for each of the ten iterations on the smaller datasets. The native Kaushik experiments take much longer than the others (Figs. 2a, 2c, and 2e), so we provide plots without native Kaushik (Figs. 2b, 2d, and 2f), to allow easier comparison between the other algorithms.

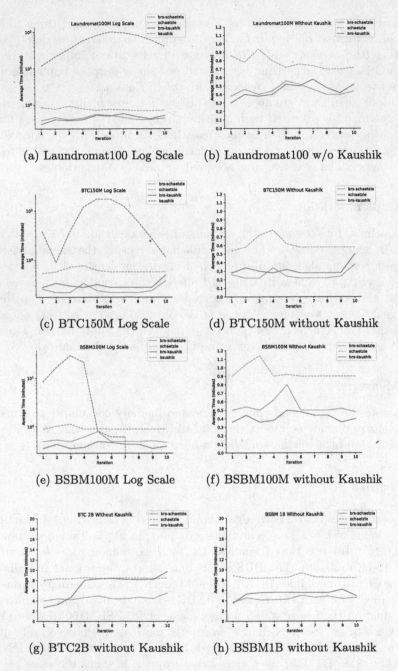

Fig. 2. Average iteration times (minutes) on the smaller datasets in (a) to (f) and larger datasets in (g) and (h).

Table 3. Total run time (in minutes) needed for computing the $k = 1, \ldots, 10$ bisimulation (average and standard deviation over 5 runs).

	Schätzle et al.		Kaushik et al.	
	BRS	Native	BRS	Native
Laundromat100M	$5.56_{(0.50)}$	$7.72_{(0.07)}$	$5.60_{(0.15)}$	$586.66_{(21.09)}$
BTC150M	$4.08_{(1.55)}$	$6.14_{(0.38)}$	$4.54_{(1.34)}$	$78.02_{(3.71)}$
BTC2B	$61.96_{(11.38)}$	$83.74_{(1.96)}$	$85.92_{(13.6)}$	out of time
BSBM100M	$6.46_{(0.08)}$	$9.40_{(0.06)}$	$5.20_{(0.50)}$	$77.84_{(2.41)}$
BSBM1B	$54.44_{(4.35)}$	$85.98_{(3.83)}$	$64.00_{(3.22)}$	out of memory

BRS-Schätzle and native Schätzle have relatively constant iteration time on all smaller datasets. For example, on Laundromat100M (Fig. 2b), native Schätzle computes an iteration in about 0.7 to 0.95 minutes. BRS is slightly faster on all smaller datasets, but uses more memory (Table 4). BRS-Kaushik shows similar results to BRS-Schätzle. Again iteration time is relatively constant on all datasets. As before, it is fastest on BTC150M.

Table 4. Maximal JVM on-heap memory (in GB) used for $k = 1, \ldots, 10$ bisimulation. We provide the maximal memory usage determined over five runs, instead of the average and standard deviation, in order to demonstrate what memory is needed to perform the bisimulations without running out of memory.

	Schätzle et al.		Kaushik et al.	
	BRS	Native	BRS	Native
Laundromat100M	211.5	147.9	210.5	335.1
BTC150M	140.6	107.7	130.1	181.7
BTC2B	1,249.1	1,249.7	1,249.3	out of time
BSBM100M	248.6	113.1	172.0	327.0
BSBM1B	1,248.2	1,335.4	1,249.2	out of memory

Native Kaushik shows a different behavior across the datasets. Iteration time varies on all datasets: the iteration time increases in early iterations until it reaches a maximum value, and then decreases. The only exception to this behavior occurs on BTC150M, where the iteration time decreases from iteration one to two before following the described pattern for the remaining iterations. For example, on Laundromat100M (Fig. 2a) an iteration takes about 10 to 100 minutes. The maximum iteration time is reached in iteration six. Native Kaushik runs fastest on BSBM100M, taking about 77.84 minutes on average. BRS-Kaushik is much faster on all smaller datasets (Table 3) and uses slightly less memory (Table 4).

Larger Datasets (1B+ Edges). BRS-Schätzle computes the $k = 1, \ldots, 10$ bisimulation on BTC2B and BSBM1B with a relatively constant iteration time. The total run time (Table 3), is lowest on BSBM1B: 54.44 minutes on average. Native Schätzle also has nearly constant running time across iterations. The average running time for all ten iterations is lowest on BTC2B: 83.74 minutes on average, compared to 85.98 minutes on BSBM1B. Comparing the run times of the two implementations, BRS is 37% on the BSBM1B dataset and 26% faster on BTC2B. Both algorithms require about the same memory during execution on BTC2B. On BSBM1B, native Schätzle uses slightly more memory than BRS. Native Kaushik did not complete one iteration on BTC2B in 24 hours, so execution was canceled. The algorithm ran out of memory on BSBM1B (Table 4).

Finally, BRS-Kaushik also has similar iteration times. On BSBM1B (Fig. 2h), computing a bisimulation iteration ranges from about 4 to 6 minutes. On BTC2B (Fig. 2g), the execution times for iterations one to three range from 2 to 4 minutes. Iterations four to nine take about 8 minutes each. For the final iteration ten, the running time slightly increases to about 9.5 minutes.

7   Discussion

Main Results. Our results show that, on all datasets, the generic BRS algorithm outperforms the native bisimulation algorithms of Schätzle et al. and Kaushik et al. for $k = 10$ (see Table 3). Since the BRS algorithm has an initialization phase, which is not present in the two native algorithms, we examine the data more closely to see at which value of k BRS begins to outperform the native implementations. We provide the exact numbers per iteration together with the standard deviation and averaged over five executions in [24]. Our generic BRS-Kaushik outperforms the native Kaushik et al. for all k: this is unsurprising, as BRS parallelizes the original sequential algorithm. The time taken to compute k-bisimulation is the total time for iterations up to, and including, k. The comparison of BRS-Schätzle with the native Schätzle et al. algorithm shows that our generic algorithm begins to outperform the native one at $k = 3$ for the Laundromat100M, BSBM100M, and BSBM1B; at $k = 4$ for BTC150M; and at $k = 5$ for BTC2B. Again, we emphasize that BRS is a generic algorithm supporting all graph summary models definable in FLUID (see introduction), whereas native algorithms compute only one version of bisimulation each.

Note that, for each dataset and each algorithm, we iteratively compute in total ten bisimulation-based graph summaries using every value of k from 1 to 10. As can be seen in Figure 2, the execution times per bisimulation iteration are fairly constant over all iterations for all datasets for BRS and Schätzle et al. The native Kaushik et al. algorithm sequentially computes the refinement of the partitions, so execution time directly relates to the number of splits per k-bisimulation iteration. This can be seen by the curve in the plots of the smaller datasets (see Figs. 2a, 2c, and 2e). In particular, Kaushik et al. detect that full bisimulation has been reached and do not perform further computations: this happens after computing the 7th iteration (7-bisimulation, see Fig. 2e) on the

BSBM100M dataset but not on the other datasets on which the execution of the Kaushik et al. algorithm completed. This kind of termination check could also be added to the BRS algorithm but this is nontrivial, as we discuss below.

This consistent per-k iteration runtime is explained by each iteration of the BRS algorithm processing every vertex, considering all its neighbors. Thus, the running time of the iteration depends primarily on the size of the graph, and not on how much the k-bisimulation that is being computed differs from the $(k-1)$-bisimulation that was computed at the previous iteration. The native Schätzle et al. algorithm behaves in the same way. There is some variation in per-iteration execution time but this may be because, as each vertex is processed, duplicate incoming messages must be removed. The time taken for this will depend on the distribution of the incoming messages, which will vary between iterations.

We consider experimenting with values between $k = 1$ and 10 to be appropriate. The native Kaushik et al. algorithm terminates when full bisimulation is reached, which shows that full bisimulation is reached at $k = 7$ on the BSBM100M dataset, but has not been reached up to $k = 10$ for Laundromat100 and BTC150M. The rapidly decreasing per-iteration running times for native Kaushik et al. for BTC150M (Fig. 2c) suggests that full bisimulation will be reached in a few more iterations. A similar curve can be observed for Laundromat100 (Fig. 2a), but the running time for the $k = 10$ iteration is higher, suggesting that full bisimulation will not be reached for several iterations. Thus, on these datasets, computing 10-bisimulations is a reasonable thing to do, as full bisimulation has not yet been reached. Note that, for the other algorithms that we consider (native Schätzle, and the two instances of BRS), per-iteration running time is largely independent of k and of whether full bisimulation as been reached (see the per-iteration measurements on BSBM100M in [24]).

Conversely, our primary motivation is graph summarization. When two vertices in a graph are 10-bisimilar, this means they have equivalent neighborhoods out to distance 10. This means that they are already "largely similar" for that value of k. Thus, we feel that, having computed k-bisimulation for a relatively large value of k, there is little advantage in going to $k+1$. In other words, adding another iteration of $k + 1$ still leaves us with vertices that are "largely similar".

Scalability to Graph Size. From the execution time of computing k-bisimulation for $k = 1, \ldots, 10$, we observe that BRS-Schätzle, BRS-Kaushik, and native Schätzle scale linearly. Here, we consider the scalability of the algorithms with respect to graph size. To this end, we fix on iterations $k = 1, \ldots, 10$ and compare the total time taken to compute these bisimulations for input graphs of different (large) sizes. BTC150M contains approximately 5M vertices and 145M edges. BTC2B has around 80M vertices and 2B edges, which is a factor of about 18 and 14 larger, respectively. BRS-Schätzle takes 4.08 minutes, BRS-Kaushik 4.54 minutes, and native Schätzle 6.14 minutes on BTC150M. On BTC2B, they take 15 times, 19 times, and 14 times longer. BSBM1B contains about 10 times as many vertices and edges as BSBM100M. The experiments on BSBM1B took about 10 times longer for BRS-Schätzle, 12 times for BRS-Kaushik, and 9 times for native Schätzle. Thus, the scaling factor of the execution times is approximately equal to that of the graph's size. Finally, the total runtimes of native

Kaushik on the different graphs indicate that the algorithm does not scale linearly with the input graph's size. The initial partition P_0 has one block per label set and Laundromat100M has many different label sets (Definition 2). The algorithm checks stability of every block against every other, leading to a run time that is quadratic in the number of blocks.

Generalization and Threats to Validity. We use synthetic and real-world graphs, which is important to analyze the practical application of an algorithm [5,17]. Two GSMs were used for evaluation of our hash-based BRS algorithm. The GSM of Schätzle et al. computes a forward k-bisimulation, based on edge labels (Definition 1). The GSM of Kaushik et al. computes a backward k-bisimulation based on vertex labels (Definition 2). Hence, the two GSMs consider different structural features for determining vertex equivalence. Regardless, for both GSMs, the BRS algorithm scales linearly with the number of bisimulation iterations and the input graph's size and computes the aggregated $k = 1, \ldots, 10$-bisimulations the fastest on every dataset.

All algorithms are implemented in Scala, in the same framework. Each algorithm was executed using the same procedure on the same machine with exclusive access during the experiments. Each experimental configuration was run six times. The first run is discarded in the evaluations to address side effects.

8   Conclusion and Future Work

We focus on the performance (runtime and memory use) of our generic, parallel BRS algorithm for computing different bisimulation variants, and how this performance compares to specific algorithms for those bisimulations, due to Schätzle et al. and Kaushik et al. Our experiments comparing $k = 1, \ldots, 10$ bisimulations on large synthetic and real-world graphs show that our generic, hash-based BRS algorithm outperforms the respective native bisimulation algorithms on all datasets for all $k \geq 5$ and for smaller k in some cases. The experimental results indicate that the parallel BRS algorithm and native Schätzle et al. scale linearly with the number of bisimulation iterations and the input graph's size. Our experiments also show that our generic BRS-Kaushik algorithm effectively parallelizes the original sequential algorithm of Kaushik et al., showing runtime performance similar to BRS-Schätzle. Overall, we recommend using our generic BRS algorithm over implementations of specific algorithms of k-bisimulations. Due to the support of a formal language for defining graph summaries [6], the BRS approach is flexible and easily adaptable, e. g., to changes in the desired features and user requirements, without sacrificing performance.

Future work includes incorporating a check (similar to Kaushik et al.) for whether full bisimulation has been reached. This is nontrivial as the algorithm of Kaushik et al. is based on splitting partitions, whereas BRS is based on describing the equivalence class in which each vertex lives. It is easy to check that no classes have been split, but harder to check that every pair of vertices that had the same description at the previous iteration have the same description at the current iteration, since the description of every vertex changes at each iteration.

Acknowledgments. This paper is the result of Jannik's Bachelor's thesis supervised by David and Ansgar. We thank Till Blume for discussions and his support in implementing the k-bisimulation algorithms in FLUID [5].

References

1. Beek, W., Rietveld, L., Bazoobandi, H.R., Wielemaker, J., Schlobach, S.: LOD laundromat: a uniform way of publishing other people's dirty data. In: Mika, P., Tudorache, T., Bernstein, A., Welty, C., Knoblock, C., Vrandečić, D., Groth, P., Noy, N., Janowicz, K., Goble, C. (eds.) ISWC 2014. LNCS, vol. 8796, pp. 213–228. Springer, Cham (2014). https://doi.org/10.1007/978-3-319-11964-9_14
2. Bizer, C., Schultz, A.: The Berlin SPARQL benchmark. Int. J. Semant. Web Inf. Syst. **5**(2), 1–24 (2009). https://doi.org/10.4018/jswis.2009040101
3. Blom, S., Orzan, S.: A distributed algorithm for strong bisimulation reduction of state spaces. Electron. Notes Theor. Comput. Sci. **68**(4), 523–538 (2002). https://doi.org/10.1016/S1571-0661(05)80390-1
4. Blume, T., Rau, J., Richerby, D., Scherp, A.: Time and memory efficient parallel algorithm for structural graph summaries and two extensions to incremental summarization and k-bisimulation for long k-chaining. CoRR abs/2111.12493 (2021). arxiv:2111.12493
5. Blume, T., Richerby, D., Scherp, A.: Incremental and parallel computation of structural graph summaries for evolving graphs. In: CIKM, pp. 75–84. ACM (2020). https://doi.org/10.1145/3340531.3411878
6. Blume, T., Richerby, D., Scherp, A.: FLUID: a common model for semantic structural graph summaries based on equivalence relations. Theor. Comput. Sci. **854**, 136–158 (2021). https://doi.org/10.1016/j.tcs.2020.12.019
7. Bonifati, A., Dumbrava, S., Kondylakis, H.: Graph summarization. CoRR abs/2004.14794 (2020). https://arxiv.org/abs/2004.14794
8. Buneman, P., Staworko, S.: RDF graph alignment with bisimulation. Proc. VLDB Endow. **9**(12), 1149–1160 (2016). https://doi.org/10.14778/2994509.2994531
9. Čebirić, Š, Goasdoué, F., Kondylakis, H., Kotzinos, D., Manolescu, I., Troullinou, G., Zneika, M.: Summarizing semantic graphs: a survey. VLDB J. **28**(3), 295–327 (2018). https://doi.org/10.1007/s00778-018-0528-3
10. Ciglan, M., Nørvåg, K., Hluchý, L.: The semsets model for ad-hoc semantic list search. In: WWW, pp. 131–140. ACM (2012). https://doi.org/10.1145/2187836.2187855
11. Consens, M.P., Fionda, V., Khatchadourian, S., Pirrò, G.: S+EPPs: construct and explore bisimulation summaries, plus optimize navigational queries; all on existing SPARQL systems. Proc. VLDB Endow. **8**(12), 2028–2031 (2015). https://doi.org/10.14778/2824032.2824128
12. Goasdoué, F., Guzewicz, P., Manolescu, I.: RDF graph summarization for first-sight structure discovery. VLDB J. **29**(5), 1191–1218 (2020). https://doi.org/10.1007/s00778-020-00611-y
13. Hellings, J., Fletcher, G.H.L., Haverkort, H.J.: Efficient external-memory bisimulation on dags. In: SIGMOD, pp. 553–564. ACM (2012). https://doi.org/10.1145/2213836.2213899
14. Herrera, J.-M., Hogan, A., Käfer, T.: BTC-2019: the 2019 billion triple challenge dataset. In: Ghidini, C., et al. (eds.) ISWC 2019. LNCS, vol. 11779, pp. 163–180. Springer, Cham (2019). https://doi.org/10.1007/978-3-030-30796-7_11

15. Kanellakis, P.C., Smolka, S.A.: CCS expressions, finite state processes, and three problems of equivalence. Inf. Comput. **86**(1), 43–68 (1990). https://doi.org/10.1016/0890-5401(90)90025-D

16. Kaushik, R., Shenoy, P., Bohannon, P., Gudes, E.: Exploiting local similarity for indexing paths in graph-structured data. In: ICDE, pp. 129–140. IEEE (2002). https://doi.org/10.1109/ICDE.2002.994703

17. Luo, Y., Fletcher, G.H.L., Hidders, J., Bra, P.D., Wu, Y.: Regularities and dynamics in bisimulation reductions of big graphs. In: Workshop on Graph Data Management Experiences and Systems, p. 13. CWI/ACM (2013). https://doi.org/10.1145/2484425.2484438

18. Luo, Y., Fletcher, G.H.L., Hidders, J., Wu, Y., Bra, P.D.: External memory k-bisimulation reduction of big graphs. In: CIKM, pp. 919–928. ACM (2013). https://doi.org/10.1145/2505515.2505752

19. Martens, J., Groote, J.F., van den Haak, L., Hijma, P., Wijs, A.: A linear parallel algorithm to compute bisimulation and relational coarsest partitions. In: Salaün, G., Wijs, A. (eds.) FACS 2021. LNCS, vol. 13077, pp. 115–133. Springer, Cham (2021). https://doi.org/10.1007/978-3-030-90636-8_7

20. Milo, T., Suciu, D.: Index structures for path expressions. In: Beeri, C., Buneman, P. (eds.) ICDT 1999. LNCS, vol. 1540, pp. 277–295. Springer, Heidelberg (1999). https://doi.org/10.1007/3-540-49257-7_18

21. Nestorov, S., Ullman, J.D., Wiener, J.L., Chawathe, S.S.: Representative objects: concise representations of semistructured, hierarchical data. In: ICDE, pp. 79–90. IEEE Computer Society (1997). https://doi.org/10.1109/ICDE.1997.581741

22. Neumann, T., Moerkotte, G.: Characteristic sets: Accurate cardinality estimation for RDF queries with multiple joins. In: ICDE, pp. 984–994. IEEE (2011). https://doi.org/10.1109/ICDE.2011.5767868

23. Paige, R., Tarjan, R.E.: Three partition refinement algorithms. SIAM J. Comput. **16**(6), 973–989 (1987). https://doi.org/10.1137/0216062

24. Rau, J., Richerby, D., Scherp, A.: Single-purpose algorithms vs. a generic graph summarizer for computing k-bisimulations on large graphs. CoRR abs/2204.05821 (2022). https://doi.org/10.48550/arXiv.2204.05821

25. Schätzle, A., Neu, A., Lausen, G., Przyjaciel-Zablocki, M.: Large-scale bisimulation of RDF graphs. In: Workshop on Semantic Web Information Management, pp. 1:1–1:8. ACM (2013). https://doi.org/10.1145/2484712.2484713

26. Stutz, P., Strebel, D., Bernstein, A.: Signal/Collect12. Semant. Web **7**(2), 139–166 (2016). https://doi.org/10.3233/SW-150176

27. Tran, T., Ladwig, G., Rudolph, S.: Managing structured and semistructured RDF data using structure indexes. IEEE Trans. Knowl. Data Eng. **25**(9), 2076–2089 (2013). https://doi.org/10.1109/TKDE.2012.134

28. Vaigh, C.B.E., Goasdoué, F.: A well-founded graph-based summarization framework for description logics. In: Description Logics, vol. 2954. CEUR-WS.org (2021). http://ceur-ws.org/Vol-2954/paper-8.pdf

29. Wikimedia Foundation: Wikidata (2022). https://www.wikidata.org/

Dominant Eigenvalue-Eigenvector Pair Estimation via Graph Infection

Kaiyuan Yang[1], Li Xia[2], and Y. C. Tay[3(✉)]

[1] Department of Quantitative Biomedicine, University of Zurich, 8057 Zurich, Switzerland
[2] Department of Statistics and Data Science, National University of Singapore, Singapore 117546, Singapore
[3] Department of Computer Science, National University of Singapore, Singapore 117417, Singapore
dcstayyc@nus.edu.sg

Abstract. We present a novel method to estimate the dominant eigenvalue and eigenvector pair of any non-negative real matrix via graph infection. The key idea in our technique lies in approximating the solution to the first-order matrix ordinary differential equation (ODE) with the Euler method. Graphs, which can be weighted, directed, and with loops, are first converted to its adjacency matrix A. Then by a naive infection model for graphs, we establish the corresponding first-order matrix ODE, through which A's dominant eigenvalue is revealed by the fastest growing term. When there are multiple dominant eigenvalues of the same magnitude, the classical power iteration method can fail. In contrast, our method can converge to the dominant eigenvalue even when same-magnitude counterparts exist, be it complex or opposite in sign. We conduct several experiments comparing the convergence between our method and power iteration. Our results show clear advantages over power iteration for tree graphs, bipartite graphs, directed graphs with periods, and Markov chains with spider-traps. To our knowledge, this is the first work that estimates dominant eigenvalue and eigenvector pair from the perspective of a dynamical system and matrix ODE. We believe our method can be adopted as an alternative to power iteration, especially for graphs.

Keywords: Dominant Eigenvalue · Eigenvector Centrality · First-order Matrix ODE · Euler Method · Graph Infection

1 Introduction

Graph epidemic models seek to describe the dynamics of contagious disease transmission over networks [5,11]. The infectious disease transmits from a node to its neighbors via connecting edges over the network. Spread of the epidemic is affected by multiple factors such as the infection rate, the recovery duration,

---

K. Yang and L. Xia—Equal Contribution.

and infection severity, and particularly to graphs, the network topology and the mobility of the network structure. Graph epidemic models can encode richer and more sophisticated architectures than traditional compartmental epidemic models [5,11].

For disease spread on networks, the principal eigenvalue of the network's adjacency matrix has long been shown to be an important factor on the dynamics of disease [8,14]. In fact, the well-known basic reproduction number 'R0' is itself the dominant eigenvalue of the next generation matrix [2,3]. Interestingly, despite the clear connection between the dominant eigenvalue and the network epidemics, few have attempted to approach the principal eigenvalue in the reverse manner. Here we seek to answer the question: Can we elucidate the principal eigenvalue from the progression of the infection spread over the associated network?

Another motivation of ours comes from the extant issues on eigenvalue computation from the classical power iteration method. Power iteration method [9], or power method, has been widely used for computation of the principal eigenvalue. However, when multiple dominant eigenvalues of equal modulus exist, power iteration method cannot converge [12]. Note that such failed convergence of power method is not uncommon, especially for graphs. One prominent failure is for any bipartite graph: all non-zero eigenvalues of a bipartite graph's adjacency matrix come in real number pairs that are of opposite sign. Thus if there exists a dominant eigenvalue λ for the adjacency matrix of the bipartite graph, then so is $-\lambda$. Power iteration method can also fail in Markov chains with periods. For example, there are as many eigenvalues equally spaced around the unit circle as the period of the periodic unichain [7]. Unless the Markov chain is an ergodic unichain which has only one dominant eigenvalue of 1, problematic convergence to the dominant eigenvector from power iteration should be expected.

In this paper, we aim to develop an epidemic-based method to estimate the principal eigenvalue and eigenvector of a network, and compare its applications with the classical power iteration method. Our perspective of using the fastest growing term in general solution of graph infection ordinary differential equation (ODE) to 'reverse engineer' the principal eigenvalue is new. Our proposed alternative method can also overcome several limitations of power iteration method. In particular, our method works better when there are multiple dominant eigenvalues of opposite sign or are complex conjugates.

2   Our Method: Inspired by Euler Method for Graph Infection ODE

In this section we present our proposed method. See Table 1 for the notations and symbols. The key idea in our technique lies in solving the matrix ODE that describes the graph infection process. The fastest growing term of the general solution to the ODE will reveal the adjacency matrix's dominant eigenvalue. This idea comes from a naive infection model for a graph, **but note that our estimation technique is numerical and does not simulate infection.**

Table 1. Table of Notations

Notation	Description
$\mathcal{G}$	Graph, $\mathcal{G} = \{\mathcal{V}, \mathcal{E}\}$. Graphs can be **weighted, directed, and with loops**
$\mathcal{V}$	Set of vertices of the graph, $\mathcal{V} = \{v_1, ..., v_N\}$, $|\mathcal{V}| = N$
$\mathcal{E}$	Set of edges of the graph
$\mathcal{N}(i)$	Set of in-degree neighbour nodes of node v_i
A	Adjacency matrix of the graph, $A \in \mathbb{R}_{\geq 0}^{N \times N}$
β	Ratio of infection severity transmitted per unit time, $\beta \in \mathbb{R}_{>0}$
$x_i(t)$	Infection severity of node v_i at time t, $x_i(t) \in \mathbb{R}_{\geq 0}$
$\mathbf{x}(t)$	Vector form of the infection severity of each node at time t
$I(t)$	Total infection of the graph at time t, $I(t) = \sum_{i=1}^{N} x_i(t)$
λ_k	k-th eigenvalue of A, $\lambda_k \in \mathbb{C}$. $|\lambda_1| \geq |\lambda_2| \geq ... \geq |\lambda_N|$
$\boldsymbol{\mu}_k$	k-th eigenvector of A, $A\boldsymbol{\mu}_k = \lambda_k \boldsymbol{\mu}_k$ and $\boldsymbol{\mu}_k \neq \mathbf{0}$
r_k	Algebraic multiplicity of the k-th eigenvalue of A
p_k	Geometric multiplicity of the k-th eigenvalue of A
d_k	Deficit of the k-th eigenvalue of A, $d_k = r_k - p_k$
C_k	Coefficient of $\boldsymbol{\mu}_k$ in the linear combination of eigenvectors for initial condition $\mathbf{x}(0)$
m	Slope of the **secant line** with step-size Δt for log-scale plot $I(t)$ vs time

2.1 Graph Infection and Adjacency Matrix

For our setup, **our graph is static**, i.e., the network structure does not vary as the infection process unfolds. For a given graph $\mathcal{G}$, the edges $\mathcal{E}$ connecting nodes $\mathcal{V}$ over the network do not rewire or change weights once the infection starts. The graph can be **weighted, directed, and with loops**. For such a graph network, the adjacency matrix A describes the connections between its nodes, in which each entry a_{ij} represents the edge from node j to node i. Matrix A can be symmetric for undirected networks, and potentially asymmetric for directed networks. Elements of A can be zero and one for unweighted connections. Or in the case of weighted connections, A may be an arbitrary non-negative matrix. Thus our $A \in \mathbb{R}_{\geq 0}^{N \times N}$, where N is the number of nodes.

A basic example demonstrating the mechanism of graph infection for one node is shown in Fig. 1. The rate of change of the infection severity $x_i(t)$ at node v_i can be intuitively expressed as a difference equation shown in the figure. In order to describe the infection change for the graph, we turn to matrix ODE.

2.2 General Solution to Graph Infection ODE

The setup for our numerical method is based on the change of severity of infection for each node in the network over time. Conceptually, we can view this model as a description of infection severity of nodes, where nodes with severely infected neighbors would receive more severe transmission. See Fig. 2 for a sketch of the infection severity ODE idea. Alternatively, one may

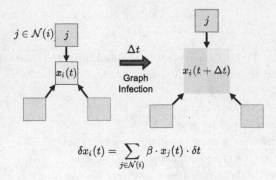

$$\delta x_i(t) = \sum_{j \in \mathcal{N}(i)} \beta \cdot x_j(t) \cdot \delta t$$

Fig. 1. Graph infection with infection rate of β and the difference equation of the infection severity $x_i(t)$ at node v_i. Infection spreads from infected neighbouring nodes $\mathcal{N}(i)$ via in-degree edges. Infection severity accumulates, think of virus count in a host.

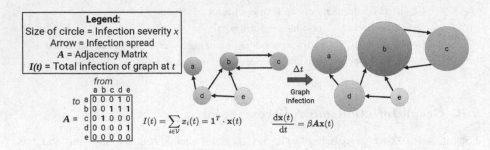

Fig. 2. An example network and its evolution of infection severity.

perceive this model as a pest which continuously multiplies and invades neighboring nodes over a network. **Note that we make the evolution of severity deterministic, and thus remove stochastic elements in the model.**

The severity of node i at time t, $x_i(t)$, is related to its neighboring nodes' severity. $x_i(t)$ changes with time according to this differential equation:

$$\frac{dx_i(t)}{dt} = \beta \sum_{j=1}^{N} A_{ij} x_j(t) \tag{1}$$

Using $\mathbf{x}(t)$ to represent the column vector of each node's infection severity at time t, Eq. 1 becomes:

$$\frac{d\mathbf{x}(t)}{dt} = \beta A \mathbf{x}(t) \tag{2}$$

Note the general solution to the above matrix ODE of Eq. 2 can be divided into *four* cases based on the eigenvalues of A. Since the characteristic equation of A is an Nth-order polynomial with real coefficients, A has exactly N eigenvalues (including repetitions if any) that are either a real number or a pair of complex conjugate numbers. Here we distinguish the four cases based on whether the eigenvalue is real or complex and on whether it is repeated.

Case One: Real Distinct Eigenvalues. The first case is when the adjacency matrix A has N real and distinct eigenvalues. Then we can decompose the initial condition into N linearly independent eigenvectors,

$$\mathbf{x}(0) = \sum_{k=1}^{N} C_k \boldsymbol{\mu}_k$$

In this case, the matrix ODE Eq. 2 has general solution:

$$\mathbf{x}(t) = \sum_{k=1}^{N} C_k e^{\beta \lambda_k t} \boldsymbol{\mu}_k \tag{3}$$

where $\lambda_1, \lambda_2, ..., \lambda_N$ are the eigenvalues of the adjacency matrix A, and $\boldsymbol{\mu}_1, \boldsymbol{\mu}_2, ..., \boldsymbol{\mu}_N$ are the corresponding eigenvectors.

Case Two: Complex Distinct Eigenvalues. In the second case, A still has N distinct eigenvalues, however, some of the eigenvalues are complex. Since A is real, if a complex number $\lambda_k = a + ib$ is an eigenvalue of our A, then its complex conjugate $\overline{\lambda_k} = a - ib$ is also an eigenvalue, with $a, b \in \mathbb{R}$.

Let $\lambda_k = a + ib$ be a complex eigenvalue of A, and $\boldsymbol{\mu}_k = \mathbf{a} + i\mathbf{b}$ be the corresponding eigenvector where vectors $\mathbf{a}$ and $\mathbf{b}$ have only real entries. Since A still has N linearly independent eigenvectors, the general solution to the matrix ODE Eq. 2 remains in the form of Eq. 3. However, the general solution will contain the following two terms in place of the conjugate complex eigenvalues λ_k and $\overline{\lambda_k}$ and their eigenvectors:

$$C_p e^{\beta a t} \left(\mathbf{a} \cos(\beta b t) - \mathbf{b} \sin(\beta b t)\right) + C_q e^{\beta a t} \left(\mathbf{a} \sin(\beta b t) + \mathbf{b} \cos(\beta b t)\right) \tag{4}$$

where C_p and C_q are some scalars determined by the initial condition $\mathbf{x}(0)$.

Case Three: Repeated Complete Eigenvalues. When the characteristic equation of A has a multiple root, the associated eigenvalue is called a repeated eigenvalue. Suppose λ_k is a real multiple-root of the characteristic equation of A, then the algebraic multiplicity of λ_k, defined as r_k, is the number of times λ_k appears as a root. On the other hand, geometric multiplicity of λ_k, defined as p_k, is the number of linearly independent eigenvectors associated with λ_k. Here we distinguish whether the algebraic multiplicity and geometric multiplicity are equal or not.

In Case Three, the algebraic multiplicity and geometric multiplicity for the repeated eigenvalue are equal, or $r_k = p_k$. And we say the repeated eigenvalue is

complete, as in without deficit. Let λ_k be the repeated eigenvalue with geometric multiplicity p_k, where $r_k = p_k$, and $\boldsymbol{\mu}_k, \mathbf{w}_1, \mathbf{w}_2, ..., \mathbf{w}_{p_k-1}$ are the p_k linearly independent ordinary eigenvectors associated with λ_k. Thus despite repeated eigenvalues in $\boldsymbol{A}$, there are still N linearly independent eigenvectors. The general solution to the matrix ODE Eq. 2 remains in the form of Eq. 3, but contains the following term for the repeated complete eigenvalue λ_k:

$$C_1 e^{\beta \lambda_k t} \boldsymbol{\mu}_k + C_2 e^{\beta \lambda_k t} \mathbf{w}_1 + ... + C_{p_k} e^{\beta \lambda_k t} \mathbf{w}_{p_k-1} \qquad (5)$$

where C_1 to C_{p_k} are some scalars determined by the initial condition $\mathbf{x}(0)$.

Case Four: Repeated Defective Eigenvalues. In Case Four, there exists a repeated eigenvalue λ_k for which the geometric multiplicity is smaller than the algebraic multiplicity, or $r_k > p_k$. Hence, there are fewer than N linearly independent eigenvectors for $\boldsymbol{A}$. We call such a repeated eigenvalue λ_k as **defective**, and $d_k = r_k - p_k$ is the **deficit** of λ_k.

In order to compensate for the defect gap in the multiplicity, we need to construct d_k number of *generalized eigenvectors* for the ODE general solution. These generalized eigenvectors are associated with the defective repeated eigenvalue λ_k and based on the p_k linearly independent ordinary eigenvectors.

Let λ_k be the repeated eigenvalue with deficit of d_k, and $\boldsymbol{\mu}_k, \mathbf{w}_1, \mathbf{w}_2, ..., \mathbf{w}_{p_k-1}$ are the p_k linearly independent ordinary eigenvectors associated with λ_k. Note that the eigenspace must have dimension of at least one. Therefore $\boldsymbol{\mu}_k$ is guaranteed to be available for us to construct the generalized eigenvectors. In the simplest two scenarios, we can have either one generalized eigenvector to find or we have only one ordinary eigenvector available.

In the first scenario of only needing to find one generalized eigenvector, or when we have $d_k = 1$ and $r_k = p_k + 1$, the general solution to the matrix ODE Eq. 2 will have the following term for the defective repeated eigenvalue λ_k:

$$C_1 e^{\beta \lambda_k t} \boldsymbol{\mu}_k + C_2 e^{\beta \lambda_k t} \mathbf{w}_1 + ... + C_{p_k} e^{\beta \lambda_k t} \mathbf{w}_{p_k-1} + C_{p_k+1} e^{\beta \lambda_k t} (t\boldsymbol{\mu}_k + \mathbf{g}) \qquad (6)$$

where C_1 to C_{p_k+1} are some scalars determined by the initial condition $\mathbf{x}(0)$, and vector $\mathbf{g}$ is a generalized eigenvector such that $(\boldsymbol{A} - \lambda_k \boldsymbol{I})\mathbf{g} = \boldsymbol{\mu}_k$.

In the second scenario of having only one ordinary eigenvector $\boldsymbol{\mu}_k$ associated with λ_k, or when $p_k = 1$ and $r_k = d_k + 1$, the general solution to the matrix ODE Eq. 2 will have the following term for the defective repeated eigenvalue λ_k:

$$C_1 e^{\beta \lambda_k t} \boldsymbol{\mu}_k + C_2 e^{\beta \lambda_k t} (t\boldsymbol{\mu}_k + \tilde{\mathbf{g}}_1) + C_3 e^{\beta \lambda_k t} \left(\frac{t^2}{2!} \boldsymbol{\mu}_k + t\tilde{\mathbf{g}}_1 + \tilde{\mathbf{g}}_2 \right) +$$
$$... + C_{d_k+1} e^{\beta \lambda_k t} \left(\frac{t^{d_k}}{d_k!} \boldsymbol{\mu}_k + \frac{t^{d_k-1}}{(d_k-1)!} \tilde{\mathbf{g}}_1 + ... + \frac{t^2}{2!} \tilde{\mathbf{g}}_{d_k-2} + t\tilde{\mathbf{g}}_{d_k-1} + \tilde{\mathbf{g}}_{d_k} \right)$$
$$(7)$$

where C_1 to C_{d_k+1} are some scalars determined by the initial condition $\mathbf{x}(0)$, and vectors $\tilde{\mathbf{g}}_1$ to $\tilde{\mathbf{g}}_{d_k}$ are generalized eigenvectors such that $(\boldsymbol{A} - \lambda_k \boldsymbol{I})\tilde{\mathbf{g}}_1 = \boldsymbol{\mu}_k$ and $(\boldsymbol{A} - \lambda_k \boldsymbol{I})\tilde{\mathbf{g}}_{j+1} = \tilde{\mathbf{g}}_j$ for $j \in \{1, ..., d_k\}$.

In between the two extreme scenarios, when the deficit $1 < d_k < r_k - 1$, the chains of generalized eigenvectors can be made up of some combinations of the p_k ordinary eigenvectors associated with the defective repeated eigenvalue λ_k. Although the structure of the chains of generalized eigenvectors can be complicated, we will still end up with r_k number of linearly independent solution vectors involving chains of generalized eigenvectors. Therefore the general solution to the matrix ODE Eq. 2 can contain r_k number of terms of the ordinary and generalized eigenvectors arranged in chains, resembling the terms of Eq. 6 and Eq. 7.

Note that it is possible for A to have repeated complex conjugate pair of eigenvalues. The method involving generalized eigenvectors as discussed above works for defective complex eigenvalue as well. For more details on the multiple eigenvalue solutions to matrix ODE, please refer to Chapter 5 in [6]. However, the repeated complex eigenvalues are not a major concern for us because of the Perron-Frobenius theorem as we will discuss in the following section.

2.3   Fastest Growing Term and Perron-Frobenius Theorem

We first examine the graph infection ODE's general solution from the perspective of the long-term behavior of the network. According to the **Perron-Frobenius theorem** extended to **non-negative real matrices**, there exists a real non-negative eigenvalue that is greater than or equal to the absolute values of all other eigenvalues of that matrix. So

$$\exists \lambda_1. \ \lambda_1 \in \mathbb{R}_{\geq 0}$$
$$\forall \lambda_k. \ k > 1 \Rightarrow \lambda_1 \geq |\lambda_k|$$

In the Case One of real and distinct eigenvalues, the dominance of the principal real eigenvalue will mean the following:

$$\lambda_1 > \lambda_k \Rightarrow e^{\lambda_1 t} > e^{\lambda_k t}$$

Thus $\beta \lambda_1 t$ becomes the dominant exponent term for Eq. 3. **Since it is exponential,** $\exp(\beta \lambda_1 t)$ **very quickly overwhelms all the other terms in the summation in the Case One**, as the eigenvalues are real and distinct. The dynamics along the principal eigenvalue will dominate the long-term behavior in Eq. 2. Therefore the fastest growing term in the general solution for Eq. 2 corresponds to:

$$\mathbf{x}(t) \approx C_1 e^{\beta \lambda_1 t} \boldsymbol{\mu}_1 \tag{8}$$

i.e., λ_1 and $\boldsymbol{\mu}_1$ are revealed through the change in $\mathbf{x}(t)$.

Note that this dominant exponent also works for complex eigenvalues in the Case Two of having complex eigenvalues. Since for $k > 1$, $\lambda_k \in \mathbb{C}$, let $\lambda_k = a + ib$, for some real and non-zero numbers a and b. Then

$$\lambda_1 \geq |\lambda_k| = |a + ib| \Rightarrow \lambda_1 > a \Rightarrow e^{\lambda_1 t} > e^{at}$$

Hence, the exponent with principal real eigenvalue will still dominate the long-term behavior of the general solution to the matrix ODE, even when there are complex eigenvalues with the same norm as the principal eigenvalue in Eq. 4. Thus we can derive λ_1 and $\boldsymbol{\mu}_1$ using Eq. 8 similar to Case One.

When there are repeated eigenvalues as in Case Three and Case Four, unless the principal real eigenvalue has algebraic multiplicity greater than 1, the asymptotic behavior of the ODE's general solution is still Eq. 8. Now suppose the repeated eigenvalue is the principal real eigenvalue λ_1 and $r_1 > 1$. In Case Three, since the algebraic multiplicity and geometric multiplicity are the same, $r_1 = p_1$, the fastest growing term in the general solution to ODE is Eq. 5, which is a scalar multiple of $\exp(\beta\lambda_1 t)$, and converges to a vector spanned by the p_k linearly independent ordinary eigenvectors associated with λ_1. Similarly for Case Four, the fastest growing term in the general solution to ODE is a scalar multiple of $\exp(\beta\lambda_1 t)$. However, the associated vector in Case Four is not convergent.

Table 2 gives a summary of the fastest growing term to the ODE for Case One to Four. Suppose there are multiple eigenvalues of the same magnitude or modulus, say λ_k and λ_1. As shown in Table 2, in all four Cases, the fastest growing term converges to the Perron-Frobenius real dominant eigenvalue $\lambda_1 \in \mathbb{R}_{\geq 0}$. In Case One and Two, despite the presence of a λ_k of opposite sign or is a complex conjugate, the fastest growing term is convergent to both the λ_1 and $\boldsymbol{\mu}_1$. As we shall see later, this marks the major advantage over the classical power iteration method.

Table 2. Fastest Growing Term in the general solution to ODE Eq. 2

Case	Dominant Eigenvalues $|\lambda_k| = \lambda_1$	Fastest Growing Term
1: Real Distinct Eigenvalues	$-\lambda_k = \lambda_1.\ \lambda_k \in \mathbb{R}_{\leq 0}$	$C_1 e^{\beta\lambda_1 t}\boldsymbol{\mu}_1$
2: Complex Distinct Eigenvalues	$|\lambda_k| = \lambda_1.\ \lambda_k \in \mathbb{C}$	$C_1 e^{\beta\lambda_1 t}\boldsymbol{\mu}_1$
3: Repeated Complete Eigenvalues	$\lambda_k = \lambda_1.\ \lambda_k \in \mathbb{R}_{\geq 0}$	$e^{\beta\lambda_1 t}(C_1\boldsymbol{\mu}_1 + C_k\boldsymbol{\mu}_k)$
4: Repeated Defective Eigenvalues	$\lambda_k = \lambda_1.\ \lambda_k \in \mathbb{R}_{\geq 0}$	$e^{\beta\lambda_1 t}(C_1\boldsymbol{\mu}_1 + C_k(t\boldsymbol{\mu}_1 + \mathbf{g}))^*$

*For Case Four, here we only show the simplest scenario of $d_1 = 1, r_1 = 2$, and $\mathbf{g}$ is a generalized eigenvector such that $A\mathbf{g} = \lambda_1\mathbf{g} + \boldsymbol{\mu}_1$.

2.4 Euler Method to Approximate Matrix ODE

Let $I(t)$ be the total severity of infection for the graph at time t:

$$I(t) = \sum_{i=1}^{N} x_i(t) = \mathbf{1}^T \cdot \mathbf{x}(t)$$

We turn to **Euler method**. From matrix differential Eq. 2, we can estimate change of $\mathbf{x}(t)$ by the **finite difference approximation**:

$$\mathbf{x}(t + \Delta t) \approx \mathbf{x}(t) + \frac{d\mathbf{x}(t)}{dt}\Delta t$$
$$= \mathbf{x}(t) + \beta A\mathbf{x}(t)\Delta t \quad \text{by Eq. 2} \tag{9}$$

Then we rewrite the finite difference for severity of graph as:

$$
\begin{aligned}
I(t + \Delta t) &= \mathbf{1}^T \cdot \mathbf{x}(t + \Delta t) \\
&\approx \mathbf{1}^T \cdot (\mathbf{x}(t) + \beta \mathbf{A} \mathbf{x}(t) \Delta t) \\
&\approx \mathbf{1}^T \cdot C_1 e^{\beta \lambda_1 t} \boldsymbol{\mu}_1 \cdot (1 + \beta \lambda_1 \Delta t) \\
&= \mathbf{1}^T \cdot \mathbf{x}(t) \cdot (1 + \beta \lambda_1 \Delta t) \\
&= I(t) \cdot (1 + \beta \lambda_1 \Delta t)
\end{aligned}
\tag{10}
$$

Note that here we substitute $\mathbf{x}(t)$ with the fastest growing term for Case One and Case Two from Table 2. But it should be obvious that when there are repeated and complete dominant eigenvalues, the fastest growing term of Case Three from Table 2 substituting $\mathbf{x}(t)$ will also give the same derivation here at Eq. 10.

2.5   Secant Line of Euler Method

Next we make use of the finite difference of the graph infection severity with secant line. Figure 3 shows a graphical explanation of the role of secant line and Euler method in our approximation. We plot the **log of total severity** of the graph at time t, $\ln(I(t))$, versus time in number of discrete time-steps, as shown in Fig. 3. Take two points along the $\ln(I(t))$ curve separated by Δt, and **draw a secant line**, that secant line will have slope m defined by:

$$
\begin{aligned}
m &= \frac{\ln(I(t + \Delta t)) - \ln(I(t))}{\Delta t} \\
&= \frac{\ln(1 + \beta \lambda_1 \Delta t)}{\Delta t} \quad \text{by Eq. 10}
\end{aligned}
$$

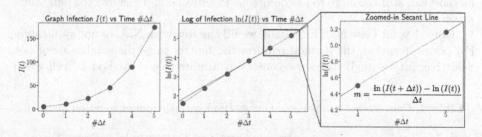

Fig. 3. Secant line of graph infection plot using the adjacency matrix $\mathbf{A}$ from the network in Fig. 2, with $\mathbf{x}(0) = \mathbf{1}$, $\beta = 1$, $\Delta t = 1$. **(Left)** Plot of graph infection $I(t)$ vs time in number of discrete time-steps $\#\Delta t$. Blue dotted lines connecting the data points are the secant lines. **(Middle)** Plot the log of graph infection $\ln(I(t))$ vs discrete time-steps. Yellow segment is a highlighted secant line connecting two graph infection measurements. Red line is the slope m of the yellow secant line. **(Right)** Zoomed-in view of the secant line. The slope m is used to estimate the dominant eigenvalue and eigenvector. (Color figure online)

Note that the slope m of the secant line can be measured for a given time-step Δt as shown in Fig. 3. Therefore the slope m of the secant line and the true dominant eigenvalue λ_1 is related by:

$$e^{m\Delta t} = 1 + \beta\lambda_1\Delta t$$

$$\therefore \lambda_1 = \frac{e^{m\Delta t} - 1}{\beta\Delta t} \tag{11}$$

The associated dominant eigenvector $\boldsymbol{\mu}_1$ is revealed by Eq. 9 along with the iterative process. Since our method depends on the long-term behavior of the dynamical system, we iteratively calculate the secant line slope m after each time step is taken. After sufficient number of time steps, we start to zoom into large t when the matrix ODE solution is dominated by the fastest growing term. This secant-line-based method thus allows us to obtain a fair estimation of the dominant eigenvalue λ_1 via measuring m for a chosen β and Δt as elucidated in Eq. 11. *(Note that infection rate β is introduced to help set up the context and perspective of epidemic and infection adopted by our method. However, β can be assumed to be always 1 for simplicity, i.e. 100% of a node's infection severity will be transmitted through a connecting outward edge per unit time.)*

3   Theoretical Comparison with Power Iteration

Table 3 shows the comparison between our method with the classical power iteration method. In Case One where there are N real and distinct eigenvalues, power iteration is not guaranteed to converge in particular when there is a λ_k such that $\lambda_1 = -\lambda_k$. Our method on the other hand is not affected by the presence of an opposite sign dominant eigenvalue, which is the case for all bipartite graphs such as tree graphs. In Case Two with complex distinct eigenvalues, our method can still converge to the dominant eigenvalue and eigenvector pair when power iteration will fail.

Note that for Case Four in Table 3, we fill the row with N.A. or not applicable. For power iteration, the method requires the matrix to be diagonalizable or not defective. Interestingly, the same constraint applies to our method as well when

Table 3. Convergence comparison: Our method with the power iteration method

Case	Dominant Eigenvalues	Converge to λ_1?		Converge to $\boldsymbol{\mu}_1$?	
		Ours	Power Iter.	**Ours**	Power Iter.
1: Real Distinct Eigenvalues	$-\lambda_k = \lambda_1.\ \lambda_k \in \mathbb{R}_{\leq 0}$	✓	✗	✓	✗
2: Complex Distinct Eigenvalues	$|\lambda_k| = \lambda_1.\ \lambda_k \in \mathbb{C}$	✓	✗	✓	✗
3: Repeated Complete Eigenvalues	$\lambda_k = \lambda_1.\ \lambda_k \in \mathbb{R}_{\geq 0}$	✓	✓	✗	✗
4: Repeated Defective Eigenvalues	$\lambda_k = \lambda_1.\ \lambda_k \in \mathbb{R}_{\geq 0}$	N.A.	N.A.	N.A.	N.A.

the dominant eigenvalue is defective. This can be observed from the derivation in Eq. 10 where we factor $I(t + \Delta t)$ with $I(t)$. When we try to plug in the fastest growing term from Table 2, the above derivation only works for Case One to Three, i.e. when the matrix is diagonalizable.

4   Experimental Results

In this section, we compare our method with power iteration method experimentally. We demonstrate our method's advantages for tree graphs, bipartite graphs, directed graphs with periods, and Markov chains with spider-traps. For simplicity, we use infection rate of $\beta = 100\%$, time step-size of $\Delta t = 1$, and initial graph infection condition of $\mathbf{x}(0) = \mathbf{1}$. Thus the Eq. 11 to estimate the dominant eigenvalue for each secant line segment will simply be $\lambda_1 = e^m - 1$. Same configurations are used for both our method and power iteration. The number of iterations for our method is defined as the number of time steps used. We measure the convergence of eigenvectors by calculating the angle between them based on their dot product.

4.1   Tree Graph

For unweighted tree graphs, the eigenvalues of the adjacency matrix are all real since the matrix is symmetric. Figure 4 contains a tree graph with nine nodes. The true dominant eigenvalue and eigenvector pair for this tree graph can be fairly estimated by our method after ten iteration steps. However, since there always exists another eigenvalue of the same norm as λ_1 but of opposite sign, power iteration does not converge to the true dominant eigenpair, as shown in the figure's green plots.

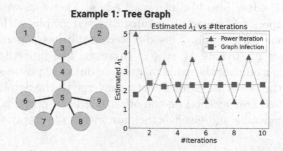

Fig. 4. The dominant eigenvalue for this **tree graph** $\lambda_1 = \sqrt{4 + \sqrt{2}} \approx 2.327$. The estimated dominant eigenvalue by our method after ten iteration steps converges to 2.321 approximately. Our estimated eigenvector also converges to ground truth eigenvector to near $0°$ angle. Power iteration does not converge for either eigenvalue or eigenvector because of the presence of another eigenvalue of opposite sign $\lambda_2 = -\lambda_1$. (Color figure online)

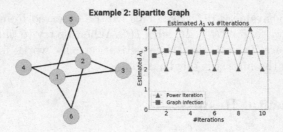

Fig. 5. The true dominant eigenvalue for this **bipartite graph** $\lambda_1 = 2\sqrt{2} \approx 2.828$. The estimated dominant eigenvalue by our method after ten iteration steps converges to approximately 2.829. Our estimated eigenvector also converges to ground truth within near 0° angle. Power iteration does not converge for either eigenvalue or eigenvector because there is an eigenvalue of the same magnitude but of opposite sign $\lambda_2 = -\lambda_1$. (Color figure online)

4.2   Bipartite Graph

In fact, not just for tree graphs, our method works better than power iteration for the superset of tree graphs: bipartite graphs. Figure 5 gives an example of a bipartite graph. The bipartite graph with six nodes have a true dominant eigenvalue of around 2.83, which can be estimated by our method within ten iterations. The associated eigenvector is also well estimated to within 0° angle by dot product. Compare that to the failed convergence pattern from power iteration method plotted in green.

4.3   Directed Graph with Period of Two

Bipartite graphs can be abstracted as directed graphs with period of two when we convert all the edges into bi-directional edges. Therefore we further experiment with the superset of bipartite graphs: directed graphs with period of two. Figure 6 shows one such graph. This directed graph has a period of two, and the true dominant eigenvalue $\lambda_1 = \sqrt{2}$. However, like bipartite graphs, there exists another eigenvalue of the same magnitude but of opposite sign $\lambda_2 = -\sqrt{2}$. Because of this, the power iteration cannot converge to the dominant eigenvalue nor the eigenvector. Our method on the other hand can give a good estimate of the dominant eigenvalue and eigenvector pair within ten iterations.

4.4   Markov Chain with Spider-Trap

Another prominent class of examples where our method outperforms the power iteration method is for transition probability graphs. In particular, we look into Markov chains with 'spider-traps', a name coined to describe parts of the network from which a crawler cannot escape [13]. When a Markov chain contains a spider-trap, the equilibrium distribution will be determined by the spider-trap in the long term. Since there are as many eigenvalues equally spaced around the unit

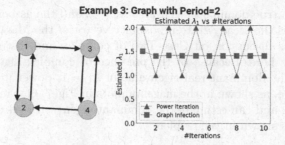

Fig. 6. The true dominant eigenvalue for this **directed graph with period of two** $\lambda_1 = \sqrt{2} \approx 1.414$. The estimated dominant eigenvalue by our method after ten iteration steps converges to around 1.414. Our estimated eigenvector also converges to the true dominant eigenvector to within near $0°$ angle. Power iteration does not converge for either eigenvalue or eigenvector because there are two eigenvalues of the same magnitude $\lambda_2 = -\lambda_1$.

circle as the period of the Markov unichain [7], the period of the spider-trap will affect whether the power iteration can converge. If the period of the spider-trap is more than one, then there are multiple eigenvalues of the same magnitude as the dominant eigenvalue $\lambda_1 = 1$. Thus the power iteration method will not converge in Markov chains with periodic spider-traps.

Note that the dominant eigenvector μ_1 associated with the dominant eigenvalue $\lambda_1 = 1$ is an important attribute of the network that describes the equilibrium distribution of the Markov chain.

Markov Chain with Period of Three. For a Markov chain with period of three, see Fig. 7. There are five nodes in the Markov chain example. Nodes 1–3 constitute a spider-trap with period of three. This Markov chain has three eigenvalues of the same magnitude of 1, two of which are complex conjugates. The

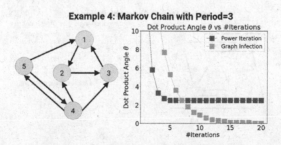

Fig. 7. Performance on Markov chain with period of three. With our method, the dot product angle θ between the estimated dominant eigenvector and the true dominant eigenvector μ_1 converges to almost $\theta \approx 0°$. Power iteration estimated dominant eigenvector fails to converge to true dominant eigenvector, with difference of $\theta \approx 2°$, because there are two other complex conjugate eigenvalues of the same magnitude as λ_1. (Color figure online)

remaining is the true dominant eigenvalue $\lambda_1 = 1$, and the associated dominant eigenvector μ_1 is the steady-state probability vector of the Markov chain.

Our previous discussion on Case Two of complex and distinct eigenvalues comes in handy here. Because of the presence of complex eigenvalues of the same norm as the dominant real eigenvalue of $\lambda_1 = 1$, power iteration method will not converge, as shown in the magenta plots in Fig. 7. Our method does not have such issues and can estimate the dominant eigenvector μ_1 associated with λ_1 accurately as shown in yellow plots.

Markov Chain with Period of Four. There are seven nodes in the Markov chain depicted in Fig. 8, out of which Nodes 1–4 constitute a spider-trap with period of four. For this Markov chain with a spider-trap of period four, there are four eigenvalues of the same magnitude. One of them is the $\lambda_1 = 1$ with the associated μ_1 steady-state vector. The other three eigenvalues are spaced equally around the unit circle with values of $-1, i, -i$ respectively.

As discussed in the Case Two scenario, the power iteration will not converge to the dominant eigenvector μ_1. However, our method can estimate the eigenvector μ_1 as shown in the yellow vs magenta plots in Fig. 8.

5   Discussion

There are a few key ideas that play important roles in our method. One being the perspective of 'reverse-engineering' the principal eigenvalue by solving a dynamical system based on graph infection epidemic model (i.e. using infection dynamics to estimate a graph's eigenvalue). Euler method is used as a viable way to iteratively approximate the solution to the matrix ODE that describes the dynamical system. In particular, we apply the Perron-Frobenius theorem extended to non-negative matrices to derive the relationship between the non-negative real Perron-Frobenius eigenvalue and the fastest growing term of the ODE general

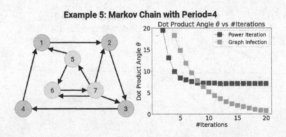

Fig. 8. Performance on Markov chain with period of four. The dot product angle θ between the estimated dominant eigenvector and the true dominant eigenvector μ_1 converges to almost $\theta \approx 0°$ with our method. Power iteration estimated dominant eigenvector fails to converge to true dominant eigenvector, with difference of $\theta \approx 7.1°$, because there are four eigenvalues of the same magnitude of 1, $\lambda_1 = 1, \lambda_2 = -1, \lambda_3 = i, \lambda_4 = -i$. (Color figure online)

solution. Finally, it is interesting to see that when we plot the trends on total infection severity, the slopes of secant line segments that arise from the Euler method help us reveal the matrix's dominant eigenvalue and eigenvector pair.

5.1 Relationship with NetworkX Eigenvector-Centrality Implementation (Workaround on Power Iteration)

NetworkX [10] is a hugely influential Python package for network analysis. In NetworkX, there is a built-in eigenvector centrality function that returns the dominant eigenvector of a graph. This eigenvector-centrality feature has been available in NetworkX from as early as 2010 based on the NetworkX official GitHub repository git log. However, until the release of NetworkX 2.0 beta 1 in August 2017, the implementation of NetworkX's eigenvector-centrality algorithm has been the classical power iteration method which cannot converge when Case One or Case Two occurs, as we have discussed.

Interestingly, based on the GitHub discussions dating back to August 2015[1] from their maintenance team of the NetworkX, it was observed that addition of A by a positive multiple of the adjacency matrix, such as $(A+I)$, could alleviate the convergence problems for power iteration for graphs. Since the NetworkX 2.0 beta 1 release in August 2017, the eigenvector-centrality has thus been modified to be using power iteration on $(A + I)$.

Here we want to highlight that NetworkX's workaround power iteration implementation can be thought of as a special case in our method. Recall our Euler method and the finite difference approximation, where we estimate the change of $\mathbf{x}(t)$ by

$$\mathbf{x}(t + \Delta t) \approx \mathbf{x}(t) + \frac{d\mathbf{x}(t)}{dt}\Delta t$$
$$= \mathbf{x}(t) + \beta A\mathbf{x}(t)\Delta t \quad \text{by Eq. 2}$$

When we take time step size of 1 unit, $\Delta t = 1$, infection rate to be $\beta = 100\%$, the above finite difference equation becomes

$$\mathbf{x}(t + 1) \approx \mathbf{x}(t) + \frac{d\mathbf{x}(t)}{dt}$$
$$= \mathbf{x}(t) + A\mathbf{x}(t)$$
$$= (A + I)\mathbf{x}(t)$$

In other words, NetworkX's workaround power iteration on $(A + I)$ is in fact operating similar to a special context of our graph infection method. Note that the consideration in NetworkX's shifted power iteration implementation is to shift the graph spectrum along the positive real axis direction. Whereas our motivation stems from trying to solve an epidemic dynamical system iteratively. Therefore, it is interesting that two conceptually different starting points eventually reached solutions that take a similar mathematical form!

---

[1] NetworkX GitHub issue #1704: https://github.com/networkx/networkx/issues/1704

5.2   Matrix-Free Implementation

We remark on the matrix-free nature of our method. Implementation-wise, our method just sums all the node's infection severity after each time-step, $I(t) = \sum_{i=1}^{N} x_i(t)$, and computes the slope of the secant line m by dividing the difference in log of total infection severity, $\ln(I(t))$, over Δt. Therefore our method is matrix-free, and does not require the whole network matrix to be stored inside the memory, which is important for large graphs and can be parallelized. Please refer to section Source Code 6 for our example implementation, only requiring a few lines of code, in Python or R language.

5.3   Extension to Dynamic Graphs

It is possible to expand our scope to non-static graphs. For example, there is a theoretical bound on the changes of the dominant eigenvector for strongly connected graphs under perturbation [4]. Furthermore, Chen and Tong [1] proposed an algorithm that can effectively monitor how the dominant eigenvalue (computed by our method, say) can be incrementally updated if the graph's A is perturbed, which can be applicable for fast-changing large graphs. More generally on the stability of eigenvalues for graphs, Zhu and Wilson [15, 16] provide some experiments on the stability of the spectrum to perturbations in the graph, which can be useful heuristics in dynamic graphs.

5.4   Limitations

As can be seen from the row on Case Three in Table 3, when the dominant eigenvalue λ_1 is repeated and complete, our method cannot guarantee the convergence to the associated dominant eigenvector μ_1. But this is the same limitation faced by power iteration method for repeated complete dominant eigenvalue, as the converged vector is spanned by the p_k number of eigenvectors corresponding to the geometric multiplicity.

When the dominant eigenvalue is not only repeated but also defective, as shown from the row on Case Four in Table 3, due to the complicated arrangements with generalized eigenvectors, our derivation involving the fastest growing term and slope of secant line no longer holds when the dominant eigenvalue is defective. Two things to note on the Case Four though. First is that the same limitation imposed by defective dominant eigenvalue also applies to power iteration. The Jordan form of the defective dominant eigenvalue also renders power iteration ineffective. Secondly, note that our method does not require the underlying matrix to be diagonalizable. As long as the defective eigenvalue does not happen to be the dominant eigenvalue λ_1, our method can converge to the dominant eigenvalue (Case Three), and to its dominant eigenvector (Case One and Case Two).

6   Conclusion

We have proposed a novel method to estimate the dominant eigenvalue and eigenvector pair of any non-negative real matrix via graph infection. To our knowledge, this is the first work that estimates dominant eigen-pair from the perspective of a dynamical system and matrix ODE. Our method overcomes several limitations of the classical power iteration when the matrix has multiple dominant eigenvalues that are complex conjugates or are of opposite sign. We believe our method can be adopted as an alternative to power iteration, especially for graphs. It is our hope that this fresh perspective of 'reverse-engineering' the dominant eigenvalue and eigenvector from matrix ODE of epidemic dynamical system can not only be of some practical use, but also to leave some food for thought in the eigenvalue algorithm literature.

Source Code

We provide example implementations of our method in Python or R language. Source code is available at GitHub: https://github.com/FeynmanDNA/Dominant_EigenPair_Est_Graph_Infection.

Acknowledgement. We thank Olafs Vandans, Chee Wei Tan, Johannes C. Paetzold, Houjing Huang, and Bjoern Menze for helpful comments and discussions.

References

1. Chen, C., Tong, H.: On the eigen-functions of dynamic graphs: fast tracking and attribution algorithms. Stat. Anal. Data Min. ASA Data Sci. J. **10**(2), 121–135 (2017)
2. Diekmann, O., Heesterbeek, J., Roberts, M.G.: The construction of next-generation matrices for compartmental epidemic models. J. Roy. Soc. Interface **7**(47), 873–885 (2010)
3. Diekmann, O., Heesterbeek, J.A.P., Metz, J.A.: On the definition and the computation of the basic reproduction ratio R_0 in models for infectious diseases in heterogeneous populations. J. Math. Biol. **28**(4), 365–382 (1990). https://doi.org/10.1007/BF00178324
4. Dietzenbacher, E.: Perturbations of matrices: a theorem on the Perron vector and its applications to input-output models. J. Econ. **48**(4), 389–412 (1988)
5. Dobson, S.: Epidemic Modelling - Some Notes, Maths, and Code. Independent Publishing Network (2020)
6. Edwards, C., Penney, D., Calvis, D.: Differential Equations and Boundary Value Problems: Computing and Modeling. Pearson Prentice Hall (2008). https://books.google.ch/books?id=qi6ePwAACAAJ
7. Gallager, R.G.: Stochastic Processes: Theory for Applications. Cambridge University Press, Cambridge (2013)
8. Ganesh, A., Massoulié, L., Towsley, D.: The effect of network topology on the spread of epidemics. In: Proceedings IEEE 24th Annual Joint Conference of the IEEE Computer and Communications Societies, vol. 2, pp. 1455–1466. IEEE (2005)

9. Golub, G.H., Van der Vorst, H.A.: Eigenvalue computation in the 20th century. J. Comput. Appl. Math. **123**(1–2), 35–65 (2000)

10. Hagberg, A.A., Schult, D.A., Swart, P.J.: Exploring network structure, dynamics, and function using networkx. In: Varoquaux, G., Vaught, T., Millman, J. (eds.) Proceedings of the 7th Python in Science Conference, Pasadena, CA, USA, pp. 11–15 (2008)

11. Kiss, I., Miller, J., Simon, P.: Mathematics of Epidemics on Networks: From Exact to Approximate Models. Springer, Cham (2017). https://doi.org/10.1007/978-3-319-50806-1

12. Quarteroni, A., Sacco, R., Saleri, F.: Numerical Mathematics, vol. 37. Springer, Heidelberg (2010). https://doi.org/10.1007/b98885

13. Rajaraman, A., Ullman, J.D.: Link analysis, pp. 139–175. Cambridge University Press (2011). https://doi.org/10.1017/CBO9781139058452.006

14. Wang, Y., Chakrabarti, D., Wang, C., Faloutsos, C.: Epidemic spreading in real networks: an eigenvalue viewpoint. In: 22nd International Symposium on Reliable Distributed Systems 2003. Proceedings, pp. 25–34. IEEE (2003)

15. Wilson, R.C., Zhu, P.: A study of graph spectra for comparing graphs and trees. Pattern Recogn. **41**(9), 2833–2841 (2008)

16. Zhu, P., Wilson, R.C.: Stability of the eigenvalues of graphs. In: Gagalowicz, A., Philips, W. (eds.) CAIP 2005. LNCS, vol. 3691, pp. 371–378. Springer, Heidelberg (2005). https://doi.org/10.1007/11556121_46

Tool Presentation

Implementing the λ_{GT} Language: A Functional Language with Graphs as First-Class Data

Jin Sano$^{(\boxtimes)}$ and Kazunori Ueda

Waseda University, Tokyo 169–8555, Japan
{sano,ueda}@ueda.info.waseda.ac.jp

Abstract. Several important data structures in programming are beyond trees; for example, difference lists, doubly-linked lists, skip lists, threaded trees, and leaf-linked trees. They can be abstracted into graphs (or more precisely, port hypergraphs). In existing imperative programming languages, these structures are handled with destructive assignments to heaps as opposed to a purely functional programming style. These low-level operations are prone to errors and not straightforward to verify. On the other hand, purely functional languages allow handling data structures declaratively with type safety. However, existing purely functional languages are incapable of dealing with data structures other than trees succinctly. In earlier work, we proposed a new purely functional language, λ_{GT}, that handles graphs as immutable, first-class data structures with pattern matching and designed a new type system for the language. This paper presents a prototype implementation of the language constructed in about 500 lines of OCaml code. We believe this serves as a reference interpreter for further investigation, including the design of full-fledged languages based on λ_{GT}.

Keywords: Functional language · reference interpreter · graph transformation

1 Introduction

λ_{GT} is a purely functional language that handles graphs as immutable, first-class data structures with pattern matching [23]. Graph transformation has been established as an expressive computational paradigm for manipulating graphs, but how to incorporate its ideas into standard programming languages is far from obvious. λ_{GT} was designed as an attempt to integrate ideas developed for graph rewriting into a purely functional language. λ_{GT} does not rely on destructive rewriting but employs immutable composition and decomposition (with pattern matching) of graphs.

J. Sano—Currently with NTT Laboratories.

© The Author(s), under exclusive license to Springer Nature Switzerland AG 2023
M. Fernández and C. M. Poskitt (Eds.): ICGT 2023, LNCS 13961, pp. 263–277, 2023.
https://doi.org/10.1007/978-3-031-36709-0_14

Algebraic data types in purely functional languages are concerned with tree structures. When matching these data structures in standard functional languages, we use patterns that allow the matching of a bounded region near the root of the structure. On the other hand, λ_{GT} enables more powerful matching as will be described in Sect. 2.

To establish the foundation of λ_{GT}, we incorporated the graphs and the axioms for their equivalence (structural congruence) of HyperLMNtal [21, 22, 27] into the call-by-value λ-calculus. HyperLMNtal is a graph transformation formalism that can handle hypergraphs and rewriting. Unlike conventional graph transformation formalisms [7, 18], HyperLMNtal is a term-based language, whose syntax and semantics follow the style of process algebra [20]. HyperLMNtal handles *port hypergraphs*. Port graphs are graphs in which links are attached to nodes at specific points called *ports* [10]. We have extended the links to hyperlinks that can connect any number of vertices. It is important to note that graphs in HyperLMNtal may have *free (hyper)links*, through which two or more graphs may be interconnected. This makes it easier to incorporate graphs into the λ-calculus and to construct its type system.

We have also proposed a type system for λ_{GT} [23]. The type system restricts the shapes of the graphs using a graph grammar. However, (untyped) λ_{GT} is capable of handling any port hypergraphs allowed by the syntax.

Since hypergraphs can be more complex than trees, λ_{GT} requires non-trivial formalism and implementation. While standard graph transformation systems rewrite one global graph with rewrite rules, graphs in λ_{GT} are immutable local values that can be bound to variables, decomposed by pattern matching with possibly multiple wildcards, in which the matched subgraphs may be used separately, passed as inputs to functions, and composed to construct larger graphs.

To study the feasibility of the language, we built a concise implementation of the untyped λ_{GT}. We believe that this serves as a *reference interpreter* for further investigation including in the design of full-fledged languages based on λ_{GT}. The interpreter is written in about 500 lines of OCaml [14] code.

The rest of this paper is organized as follows. Section 2 outlines the formal syntax and the operational semantics of λ_{GT}. Section 3 describes examples. Section 4 explains the implementation. Section 5 discusses related work and indicates our expected future work.

2   The Syntax and Semantics of λ_{GT}

We briefly review the syntax and the semantics of the λ_{GT} language. For further details, the readers are referred to [23].

Throughout the paper, for some syntactic entity E, $\overrightarrow{E}$ stands for a sequence $E_1, \ldots, E_n$ for some n (≥ 0). The length of the sequence $\overrightarrow{E}$ is denoted as $|\overrightarrow{E}|$. A set whose elements are $E_1, \ldots, E_n$ is denoted as $\{\overrightarrow{E}\}$.

$$\text{Value } G ::= \mathbf{0} \mid p(\overrightarrow{X}) \mid X \bowtie Y \mid (G, G) \mid \nu X.G$$
$$\text{Atom Name } p ::= C \mid \lambda x[\overrightarrow{X}].e$$
$$\text{Expression } e ::= T \mid (\textbf{case } e \textbf{ of } T \to e \mid \textbf{otherwise} \to e) \mid (e\ e)$$
$$\text{Graph Template } T ::= \mathbf{0} \mid p(\overrightarrow{X}) \mid X \bowtie Y \mid (T, T) \mid \nu X.T \mid x[\overrightarrow{X}]$$

Fig. 1. Syntax of λ_{GT}.

2.1 Syntax of λ_{GT}

The λ_{GT} language is composed of three lexical categories: (i) *Link Names* $(X, Y, Z, \ldots)$, (ii) *Constructor Names* (C), and (iii) *Graph Context Names* $(x, y, z, \ldots)$. We use roman for the constructor names in the concrete syntax (e.g., Cons, Leaf, $1, 2, \ldots$).

λ_{GT} is designed to be a small language focusing on handling graphs. The syntax of λ_{GT} is given in Fig. 1.

Graphs handled by λ_{GT} are called *values* and correspond to graphs in Hyper-LMNtal. They are written in a term-based syntax to ensure high affinity with programming languages [22]. The term $\mathbf{0}$ denotes the empty graph. An *atom* $p(\overrightarrow{X})$ stands for a *node* of a data structure with the label p and totally ordered links $\overrightarrow{X}$. An atom name is either a constructor (as in $\mathrm{Nil}(X)$ and $\mathrm{Cons}(Y, Z, X)$) or a λ-abstraction. That is, functions can be embedded in graphs as the names of atoms. A λ-abstraction atom has the form $(\lambda x[\overrightarrow{X}].e)(\overrightarrow{Y})$. Here, $x[\overrightarrow{X}]$, where $\overrightarrow{X}$ is a sequence of different links, denotes a *graph context* (detailed below), which corresponds to and extends a variable in functional languages. We need this extension because, unlike standard tree structures with single roots, a graph may have an arbitrary number of roots that are "access points" to the graph from outside.

Embedding a λ-abstraction into a graph node and allow it to take a graph with free links are two new important design choices of λ_{GT}.

A *fusion*, $X \bowtie Y$, fuses the link X and the link Y into a single link. A *molecule*, (G, G), stands for the composition or gluing of graphs. A *hyperlink creation*, $\nu X.G$, hides the link name X in G. In λ_{GT}, all links are free links as long as they are not hidden. The set of the free links in T is denoted as $fn(T)$. Links that do not occur free are called *local links*.

A λ_{GT} program is called an *expression*. T is a graph with zero or more *graph contexts*. A graph context, $x[\overrightarrow{X}]$, serves as a wildcard in pattern matching. It matches any graph with free links $\overrightarrow{X}$. (**case** e_1 **of** $T \to e_2 \mid$ **otherwise** $\to e_3$) evaluates e_1, checks whether it matches the graph template T, and reduces to e_2 or e_3. $(e_1\ e_2)$ is an application.

2.2 Structural Congruence

We define the relation $\equiv$ on graphs as the minimal equivalence relation satisfying the rules shown in Fig. 2. Here, $G\langle \overrightarrow{Y}/\overrightarrow{X}\rangle$ is a *hyperlink substitution* that replaces all free occurrences of $\overrightarrow{X}$ with $\overrightarrow{Y}$ [23].

(E1) $(\mathbf{0}, G) \equiv G$ (E2) $(G_1, G_2) \equiv (G_2, G_1)$ (E3) $(G_1, (G_2, G_3)) \equiv ((G_1, G_2), G_3)$

(E4) $\dfrac{G_1 \equiv G_2}{(G_1, G_3) \equiv (G_2, G_3)}$ (E5) $\dfrac{G_1 \equiv G_2}{\nu X.G_1 \equiv \nu X.G_2}$

(E6) $\nu X.(X \bowtie Y, G) \equiv \nu X.G\langle Y/X \rangle$ where $X \in \mathit{fn}(G) \vee Y \in \mathit{fn}(G)$

(E7) $\nu X.\nu Y.X \bowtie Y \equiv \mathbf{0}$ (E8) $\nu X.\mathbf{0} \equiv \mathbf{0}$ (E9) $\nu X.\nu Y.G \equiv \nu Y.\nu X.G$

(E10) $\nu X.(G_1, G_2) \equiv (\nu X.G_1, G_2)$ where $X \notin \mathit{fn}(G_2)$

Fig. 2. Structural congruence on graphs.

Two graphs related by $\equiv$ are essentially the same and are convertible to each other in zero steps. (E1), (E2) and (E3) are the characterization of molecules as multisets. (E4) and (E5) are structural rules that make $\equiv$ a congruence. (E6) and (E7) are concerned with fusions. (E7) says that a closed fusion is equivalent to $\mathbf{0}$. (E6) is an absorption law of $\bowtie$, which says that a fusion can be absorbed by connecting hyperlinks. Because of the symmetry of $\bowtie$, (E6) says that a graph can emit a fusion as well. (E8), (E9) and (E10) are concerned with hyperlink creation. We have shown in [22] that the symmetry of $\bowtie$ and α-conversion of local link names can be derived from the rules of Fig. 2.

Using structural congruence (syntax-directed approach) instead of graph isomorphism or bisimulation is not common in graph transformation formalisms. However, we believe that a syntax-directed approach has higher affinity with modern general-purpose programming languages and makes the semantics of λ_{GT} easier to define than adopting other approaches.

2.3   Operational Semantics of λ_{GT}

A capture-avoiding substitution θ of a graph context $x[\vec{X}]$ with a template T in an expression e is called a *graph substitution* and is written as $e[T/x[\vec{X}]]$. The definition of θ must handle the case where the graph context x in e has links other than $\vec{X}$; the readers are referred to [23]. We say that T matches a graph G if there exists a graph substitution θ such that $G \equiv T\theta$.

We give the reduction relation in Fig. 3. In order to define the small-step reduction relation, we extend the syntax with evaluation contexts defined as

$$E ::= [\,] \mid (\mathbf{case}\ E\ \mathbf{of}\ T \to e \mid \mathbf{otherwise} \to e) \mid (E\ e) \mid (G\ E).$$

As usual, $E[e]$ stands for E whose hole is filled with e.

Rd-Case1 and Rd-Case2 are for matching graphs. If the matching has succeeded (Rd-Case1), we apply to e_2 the graph substitution obtained in the matching. This roughly corresponds to double pushout in ordinary graph transformation, but λ_{GT} allows flexible handling of graph contexts, e.g., the creation of multiple copies of them. The pattern matching can be non-deterministic due to its expressive power. It is left to future work to put constraints over the graph templates in case expressions to ensure deterministic matching.

Rd-β applies a function to a value. The definition is standard except that (i) we need to check the correspondence of free links and (ii) we use graph substitution instead of the standard substitution in the λ-calculus.[1] The Rd-Ctx is the same as in the call-by-value λ-calculus.

$$\frac{G \equiv T\theta}{(\textbf{case } G \textbf{ of } T \rightarrow e_2 \mid \textbf{otherwise} \rightarrow e_3) \longrightarrow_{\texttt{val}} e_2\theta} \text{ Rd-Case1}$$

$$\frac{\neg \exists \theta.(G \equiv T\theta)}{(\textbf{case } G \textbf{ of } T \rightarrow e_2 \mid \textbf{otherwise} \rightarrow e_3) \longrightarrow_{\texttt{val}} e_3} \text{ Rd-Case2}$$

$$\frac{fn(G) = \{\overrightarrow{X}\}}{((\lambda x[\overrightarrow{X}].e)(\overrightarrow{Y}) \, G) \longrightarrow_{\texttt{val}} e[G/x[\overrightarrow{X}]]} \text{ Rd-}\beta \qquad \frac{e \longrightarrow_{\texttt{val}} e'}{E[e] \longrightarrow_{\texttt{val}} E[e']} \text{ Rd-Ctx}$$

Fig. 3. Reduction relation of λ_{GT}.

3 Examples

We first introduce the following abbreviation schemes: (i) a nullary atom $p()$ can be simply written as p; (ii) *Term Notation*: $\nu Y.(p_1(\overrightarrow{X}, Y, \overrightarrow{Z}), p_2(\overrightarrow{W}, Y))$, where $Y \notin \{\overrightarrow{X}, \overrightarrow{Z}, \overrightarrow{W}\}$, can be written as $p_1(\overrightarrow{X}, p_2(\overrightarrow{W}), \overrightarrow{Z})$; (iii) we abbreviate $\nu X_1. \ldots .\nu X_n.G$ to $\nu X_1 \ldots X_n.G$; (iv) application is left-associative; (v) parentheses can be omitted if there is no ambiguity. Here, (i) and (ii) apply also to graph contexts.

We first show that the untyped λ_{GT} can easily encode single-step graph rewriting using a HyperLMNtal rewrite rule. The rule $G_1 \rightarrow G_2$ matches a graph G, which has a subgraph G_1 and rewrites the subgraph to G_2 [23]. One-step rewriting using this rule can be encoded into a function $(\lambda g[\overrightarrow{Y}].\ \textbf{case } g[\overrightarrow{Y}] \textbf{ of } (G_1, x[\overrightarrow{X}]) \rightarrow (G_2, x[\overrightarrow{X}]) \mid \textbf{otherwise} \rightarrow g[\overrightarrow{Y}])(Z)$, where $\{\overrightarrow{Y}\} \stackrel{\text{def}}{=} fn(G)$ and $\{\overrightarrow{X}\} \stackrel{\text{def}}{=} fn(G_1)$. This function takes G and returns the graph obtained by applying the rule (or the original graph if the rule is not applicable). Repeated rule application can be encoded in a standard manner using a fixed-point combinator (or $\texttt{let rec}$ provided by our implementation).

λ_{GT} naturally extends existing functional languages to support graphs as first-class data. For example, Fig. 4 shows a program for concatenating two singleton difference lists (a.k.a. list segments). The λ-abstraction atom is a function that takes two difference lists, both having X (head) and Y (tail) as free links, and returns their concatenation also having X and Y as its free links.

---

[1] It may appear that the $\overrightarrow{Y}$ of $(\lambda x[\overrightarrow{X}].e)(\overrightarrow{Y})$ does not play any role here. However, such a link becomes necessary when the λ-abstraction is made to appear in a data structure (e.g., as in $\nu Y.(Cons(Y, W, Z), (\lambda x[\ldots].\ldots)(Y)))$. This is why λ-abstraction atoms are allowed to have argument links. Once such a function is accessed and β-reduction starts, the role of $\overrightarrow{Y}$ ends, while the free links *inside* the abstraction atom start to play key roles.

$$(\lambda\,x[Y,X].\ (\lambda\,y[Y,X].\ \nu Z.(x[Z,X],y[Y,Z]))(Z))(Z)\quad \mathrm{Cons}(1,Y,X)\quad \mathrm{Cons}(2,Y,X)$$
$$\longrightarrow_{\mathbf{val}}\quad (\lambda\,y[Y,X].\ \nu Z.(\mathrm{Cons}(1,Z,X),y[Y,Z]))(Z)\quad \mathrm{Cons}(2,Y,X)$$
$$\longrightarrow_{\mathbf{val}}\quad \nu Z.(\mathrm{Cons}(1,Z,X),\mathrm{Cons}(2,Y,Z))$$

Fig. 4. An example of a reduction: concatenation of difference lists.

$$(\lambda\,x[Y,X].\ \mathbf{case}\ x[Y,X]\ \mathbf{of}$$
$$y[\mathrm{Cons}(z,Y),X]\to \mathrm{Cons}(z,y[Y],X)\quad |\quad \mathbf{otherwise}\to x[Y,X])(Z)$$

Fig. 5. Rotate function on a difference list.

Figure 5 shows the function that rotates a difference list by taking the last node of the difference list and reconnecting it to the head. If $\mathrm{Cons}(1,\mathrm{Cons}(2,Y),X)$ is applied to the function, it will return $\mathrm{Cons}(2,\mathrm{Cons}(1,Y),X)$.

Without the term notation, the graph template used in the matching (on the second line of Fig. 5) can be written as $\nu W.(y[W,X],\mathrm{Cons}(Z,Y,W),z[Z])$. The matching against this template proceeds as follows:

$$\nu WZ.(\mathrm{Cons}(1,W,X),\mathrm{Cons}(Z,Y,W),2(Z))$$
$$\equiv \nu WZ.(y[W,X],\mathrm{Cons}(Z,Y,W),z[Z])\,[\mathrm{Cons}(1,W,X)/y[W,X],\ 2(Z)/z[Z]]$$

In order to implement the language correctly, we need to handle corner cases, most notably matching that needs to exploit fusion. Consider the case where a singleton list $\nu Z.(\mathrm{Cons}(Z,Y,X),1(Z))$ is given to the function. We need a subgraph with free links W and X, the free links of the graph context $y[W,X]$ in the graph template, which do not exist in the list. Thus the matching would not proceed without supplying subgraphs.

In this case, we need to first supply a fusion atom using (E6) of Fig. 2 and match $y[W,X]$ against the supplied fusion atom as follows:

$$\mathrm{Cons}(1,Y,X)$$
$$\equiv \nu WZ.(W\bowtie X,\mathrm{Cons}(Z,Y,W),1(Z))$$
$$= \nu WZ.(y[W,X],\mathrm{Cons}(Z,Y,W),z[Z])\,[W\bowtie X/y[W,X],\ 1(Z)/z[Z]]$$

Therefore, the program will result in $\nu W.(\mathrm{Cons}(1,W,X),Y\bowtie W)$, which is congruent to $\mathrm{Cons}(1,Y,X)$. Since pattern matching involving fusions does not appear in ordinary ADT matching in functional languages, and formalization and implementation of fusions is not common in ordinary graph transformation, it has been a major challenge to deal with fusions properly.

4   Implementation

We built a concise reference interpreter, which is around 500 lines of purely functional OCaml code. In addition to the syntax of Fig. 1, we implemented standard

```
1 let check_link σ X Y =
2    match (X, Y) with
3    (F_X, F_Y) → if X = Y then Some σ else None
4    (F_X, L_i) → None
5    (L_i, Y) →
6       if L_i ↦ Z ∈ σ then if Y = Z then Some σ else None
7       else Some (σ ∪ {L_i ↦ Y})
```

Fig. 6. Link name matching.

constructs including `let` and `let rec` expressions and arithmetic operations, as shown in Appendix. We also implemented a visualiser using the interpreter. We allow `Log` atoms, which are identity functions serving as breakpoints. The tool visualises graphs applied to the atom at each breakpoint. The demo in Appendix shows visualising each step of function application to the leaves of a leaf-linked tree. To run a code, users press the `Run` button and then press `Proceed` to go to the next breakpoint.

In our implementation, values, i.e., graphs, are represented with lists of atoms without link creations inside; i.e., in prenex normal form. We call them *host graphs*. Links are classified into free links with `string` names and local links α-converted to fresh `integer` ids. From now on, L_i denotes a local link with the unique id i and F_X denotes a free link with the name X. Fusions with local link(s) are absorbed beforehand. The interpreter transforms a graph template to a pair of a list of the atoms and a list of the graph contexts. For example, the graph template $y[\mathrm{Cons}(z, Y), X]$, the pattern in the Fig. 5, and the host graph $\mathrm{Cons}(1, Y, X)$ are transformed into the following, respectively:

$$\langle [\mathrm{Cons}(L_{1001}, F_Y, L_{1000})], [y\,[L_{1000}, F_X], z\,[L_{1001}]] \rangle \tag{1}$$

$$[\mathrm{Cons}(L_0, F_Y, F_X), 1(L_0)] \tag{2}$$

The interpreter first tries to match all the atoms in the template to the host graph and then match the graph contexts in the remaining host graph. The interpreter backtracks if the consequent matching fails. As a reference interpreter, optimization of matching strategies is left to future implementation.

To match an atom, we need to check that they have the same name and the correspondence of links. Since local link names are α-convertible, it is not sufficient to check that they have the same link names. We maintain a *link environment* $σ$, a mapping from the local link names in a template to the link names in a host graph. Figure 6 shows the matching of a link name X in a template with a link name Y in a host graph using a link environment $σ$. Free links in the template match links with the same names in the host graph (line 3). On the other hand, the matching of local links in the template (line 5–7) is more flexible since we can α-convert them.

For example, the atom $\text{Cons}(L_{1001}, F_Y, L_{1000})$ in the template (1) matches the atom $\text{Cons}(L_0, F_Y, F_X)$ in the host graph (1) with the environment $\sigma_1 \overset{\text{def}}{=} \{L_{1000} \mapsto F_X, L_{1001} \mapsto L_0\}$, and the remaining host graph will be $G_1 \overset{\text{def}}{=} [1(L_0)]$.

After all the atoms in the template have matched, we substitute link names in the host graph with inverse of the obtained link environment. For example, the remaining host graph G_1 becomes $G_2 \overset{\text{def}}{=} [1(L_{1001})]$.

If the mapping is not injective, we supply fusions to the host graph. For example, if we obtain the link environment $\{L_{1000} \mapsto L_0, L_{1001} \mapsto L_0\}$, we should supply $L_{1000} \bowtie L_{1001}$. If a local link in the template is mapped to a free link in the host graph, we also need to supply a fusion since a local link does not match a free link without such a treatment. For example, in the link environment σ_1, we have $L_{1000} \mapsto F_X$. Thus, we should supply $L_{1000} \bowtie F_X$ and obtain $G_3 \overset{\text{def}}{=} [1(L_{1001}), L_{1000} \bowtie F_X]$ as the remaining subgraph.

To match graph contexts, the interpreter collects connected graphs (atoms) that have free links of the contexts. For example, $y[L_{1000}, F_X]$ and $z[L_{1001}]$, the graph contexts in (1), can match host graphs $L_{1000} \bowtie F_X$ and $1(L_{1001})$, the subgraphs in G_3, respectively. After the matching, we obtain the graph substitution

$$[L_{1000} \bowtie F_X / y [L_{1000}, F_X], 1(L_{1001}) / z [L_{1001}]] . \tag{3}$$

Graph substitution can be done by adding the matched atom(s) G to the host graph, updating link names. If we have matched G with $x[\overrightarrow{X}]$ and the template has $x[\overrightarrow{Y}]$, we update G with the substitution $\langle \overrightarrow{Y} / \overrightarrow{X} \rangle$. We also reassign ids for the local links because composing graphs may result in a conflict of ids.

With the obtained graph substitution (3), we can instantiate the template on the right-hand side of $\rightarrow$ in Fig. 5, which will result in $[L_{1000} \bowtie F_X, \text{Cons}(L_{1001}, F_Y, L_{1000}), 1(L_{1001})]$. Our implementation absorbs fusions as much as possible after graph substitutions. Thus, it will be normalized, and the ids are reassigned, to obtain $[\text{Cons}(L_0, F_Y, F_X), 1(L_0)]$.

The rest of the evaluator is implemented just like those for functional languages. It takes an environment, a mapping from graph contexts to the matched subgraphs, and evaluates an expression with the environment.

To use it as the backend of the visualiser with breakpoints, we have slightly changed the evaluator and implemented it in a continuation-passing style. This allows us to stop the evaluation at each breakpoint and proceed from the point after getting a signal from a user.

5   Related and Future Work

There have been languages based on graph transformation [1,9,11–13,15,19, 24,25,27], but integrating them into functional languages seems still an open problem. There have been several attempts to support graphs in functional languages. Functional programming with structured graphs [17] can express cyclic graphs using recursive definitions, i.e., `let rec`. Some libraries support graphs and their operations [8]. However, these are not programming languages but external supports. We have laid out fundamental theories for λ_{GT}, including the equivalence of graphs, pattern matching, and reduction. We also designed a new type system, though this paper focused on the untyped λ_{GT}.

Rho-calculus is a rewriting calculus that aims to give uniform, generalised semantics of matching [3,5]. Rho-calculus enables us to define equational theories for custom matching. This makes it possible for rho-calculus to encode the λ-calculus. This is in contrast with the goal of λ_{GT}: Rather than pursuing a general framework, we focused on giving a simple implementation of our calculus that we designed to be the simplest to handle graphs in a functional language.

Clean is a language for functional graph rewriting [4]. Clean can handle term graphs (data graphs), which may contain sharing and cycles, using user-specified rewrite rules. The rules may contain variables that match nodes. However, the data graphs must be closed, i.e., they cannot contain variables. In λ_{GT}, not only templates but graphs may contain free links, which correspond to variables in Clean, and they can be used in composing graphs as we have done in Fig. 4 (difference list concatenation).

FUnCAL [16] is a functional language for graph-based databases whose equivalence is defined with bisimulation, which is different from the congruence of λ_{GT} graphs. FUnCAL is implemented in Haskell as an embedded DSL. The form of the query is limited compared to the patterns in λ_{GT}.

GrapeVine is a tool to support functional graph rewriting with persistent data structures [28]. Although this is a tool rather than a functional language, the implementation technique is insightful and useful for the future implementation of λ_{GT}.

For future work, we are planning to develop and implement a full-fledged language based on λ_{GT}. Graphs are treated as immutable data structures because they are incorporated into a pure functional language, but generating code that allows destructive updating of graphs when possible is an issue for future research. To do this without sacrificing safety, it is necessary to develop static analysis. We are planning to extend the type system to analyse the direction (or polarity) of links [26] and then perform ownership checking [6]. Then, the next step is to develop a method for transpiling into lower-level code using reference types in functional languages or imperative code with pointers.

Acknowledgements. We would like to thank the anonymous reviewers for their valuable comments. This work was supported by JSPS Grant-in-Aid for Scientific Research (23K11057).

Appendix

Our implementation consists of about 500 lines of OCaml code as shown in Table 1.[2] This is about half the size of the reference interpreter of the graph transformation-based language GP 2 [1], which is about 1,000 lines of Haskell code [2].

Table 1. LOC of the interpreter.

parser/parser.mly	84	eval/preprocess.ml	34	parser/parse.ml	10
parser/lexer.mll	57	eval/pushout.ml	34	eval/match.ml	9
eval/eval.ml	55	util/list_extra.ml	29	util/option_extra.ml	9
eval/syntax.ml	48	eval/postprocess.ml	24	util/combinator.ml	7
eval/match_atoms.ml	38	parser/syntax.ml	20	bin/main.ml	3
eval/match_ctxs.ml	35	util/util.ml	19	Total	515

Figure 7 gives the concrete syntax of the language we have implemented. Comments start at %. As usual, let defines a variable that has a value on the right-hand side of =, if $n = 0$. If $n \geq 1$, this defines a curried n-ary function. Recursive n-ary functions can be defined with let recs. We designed a syntax that is easy to parse and did not yet implement the term-notation abbreviation. We do not intend to provide a full-fledged language but to study the feasibility of the language with a concise implementation.

$$
\begin{aligned}
e ::= & \ \{T\} \mid \mathbf{case}\ e\ \mathbf{of}\ \{T\} \mathbin{\texttt{->}} e \mid \mathbf{otherwise} \mathbin{\texttt{->}} e \mid e\,e \mid e\ op\ e \\
& \mid\ \mathtt{let}\ x[\overrightarrow{X}]\ x_1[\overrightarrow{X_1}] \ldots\ x_n[\overrightarrow{X_n}] = e\ \mathtt{in}\ e \qquad\qquad (n \geq 0) \\
& \mid\ \mathtt{let\ rec}\ x[\overrightarrow{X}]\ x_1[\overrightarrow{X_1}] \ldots\ x_n[\overrightarrow{X_n}] = e\ \mathtt{in}\ e \qquad (n \geq 1) \\
op ::= & \ \texttt{+} \mid \texttt{*} \mid \texttt{-} \mid \texttt{<} \mid \texttt{=} \\
T ::= & \ C(\overrightarrow{X}) \mid (\backslash x[\overrightarrow{X}].e)(\overrightarrow{Y}) \mid X \mathbin{\texttt{><}} Y \mid T,T \mid \mathbf{nu}\ X.T \mid x[\overrightarrow{X}] \\
X ::= & \ \texttt{\_[A-Z][a-zA-Z0-9\_']*} \\
C ::= & \ \texttt{[A-Z][a-zA-Z0-9\_']*} \mid \texttt{[1-9][0-9]*} \mid \texttt{0} \\
x ::= & \ \texttt{[a-z][a-zA-Z0-9\_']*}
\end{aligned}
$$

Fig. 7. Concrete syntax of λ_{GT}.

The examples in Sect. 3 are concerned with difference lists, which are powerful and useful data structures, and we have shown the composition and the decomposition of these structures. However, the graphs and their operations in λ_{GT} are not limited to them.

---

[2] The interpreter is available at https://github.com/sano-jin/lambda-gt-alpha/tree/icgt2023. The visualiser is available at https://sano-jin.github.io/lambda-gt-playground.

For operations that cannot be performed by simply decomposing and composing graphs, we can use some atoms as markers for matching. For example, a map function that applies a function to the elements of the leaves of a leaf-linked tree can be expressed as Fig. 8. If is a successor function and we have bound the graph of Fig. 9 (a) to , evaluating the program will return the graph of Fig. 9 (b). Figure 11 shows the demo of the program. Figure 11 (a) is the view obtained by pressing the Run button after filling in the code. Figure 11 (b)–(g)

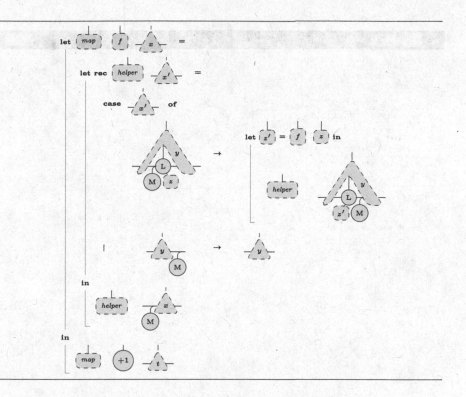

Fig. 8. Map a function to the leaves of a leaf-linked tree.

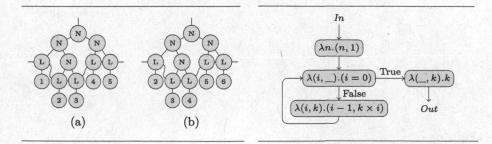

Fig. 9. Leaf-linked trees.   **Fig. 10.** Dataflow graph (factorial of n).

shows each graph at the breakpoints, from which computation can be resumed with the **Proceed** button.

λ_{GT} is capable of handling graphs that include functions. Figure 10 illustrates a dataflow graph that calculates the factorial of an input number. The graph has input and output links from and to the outside and functions at the nodes. There are two kinds of nodes; nodes with an input and an output, and a node with an input and two outputs (a branch). This can be written as Fig. 12, lines 2–28, in λ_{GT}.

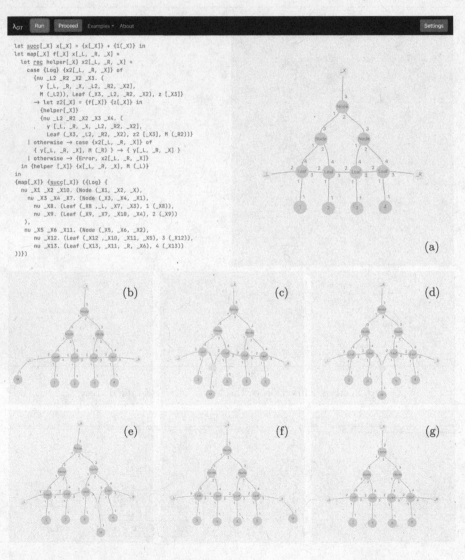

Fig. 11. Demo of mapping a function to the leaves of a leaf-linked tree.

```
1    % let f1 n = (n,1)
2    let f1[_Z] n[_Z] = {nu _N1 _N2.(T2(_N1,_N2,_Z),n[_N1],1(_N2))} in
3
4    % let pred1 (i,_) = i = 0
5    let pred1[_Z] v[_Z] = case {v[_Z]} of
6      {nu _I _K.(T2(_I,_K,_Z),i[_I],k[_K])} -> {i[_Z]} = {0(_Z)}
7      | otherwise -> {Error1} in
8
9    % let f2 (i,k) = (i - 1, k * i)
10   let f2[_Z] v[_Z] = case {v[_Z]} of
11     {nu _I _K.(T2(_I,_K,_Z),i[_I],k[_K])} ->
12       let i'[_Z] = {i[_Z]} - {1(_Z)} in
13       let k'[_Z] = {k[_Z]} * {i[_Z]} in
14       {nu _I _K.(T2(_I,_K,_Z),i'[_I],k'[_K])}
15     | otherwise -> {Error2} in
16
17   % let f3 (_,k) = k
18   let f3[_Z] v[_Z] = case {v[_Z]} of
19     {nu _I _K.(T2(_I,_K,_Z),i[_I],k[_K])} -> {k[_Z]}
20     | otherwise -> {Error3} in
21
22   % The dataflow to calculate the factorial of an input number.
23   let dataflow[_In,_Out] = {nu _X1 _X2 _X3 _F1 _F2 _P.(
24     N2(_F1,_In,_X1),f1[_F1],
25     N3(_P,_X1,_X3,_X2),pred1[_P],
26     N2(_F2,_X2,_X1),f2[_F2],
27     N2(_F3,_X3,_Out),f3[_F3]
28   )} in
29
30   % The evaluator
31   let rec proceed[_Z] g[_In,_Out] =
32     case {Log} {g[_In,_Out]} of
33       {nu _X _Y _F _V.(N2(_F,_X,_Y),f[_F],
34         M(_V,_X),v[_V],rest[_X,_Y,_In,_Out])} ->
35       let v'[_Z] = {f[_Z]} {v[_Z]} in
36       {proceed[_Z]} {nu _X _Y _F _V.(N2(_F,_X,_Y),f[_F],
37                     M(_V,_Y),v'[_V],rest[_X,_Y,_In,_Out])}
38     | otherwise -> case {g[_In,_Out]} of
39       {nu _X _Y1 _Y2 _P _V.(N3(_P,_X,_Y1,_Y2),pred[_P],
40         M(_V,_X),v[_V],rest[_X,_Y1,_Y2,_In,_Out])} ->
41       {proceed[_Z]} (case {pred[_Z]} {v[_Z]} of
42         {True(_Z)} -> {nu _X _Y1 _Y2 _P _V.(N3(_P,_X,_Y1,_Y2),pred[_P],
43                       M(_V,_Y1),v[_V],rest[_X,_Y1,_Y2,_In,_Out])}
44         | otherwise -> {nu _X _Y1 _Y2 _P _V.(N3(_P,_X,_Y1,_Y2),pred[_P],
45                       M(_V,_Y2),v[_V],rest[_X,_Y1,_Y2,_In,_Out])})
46     | otherwise -> case {g[_In,_Out]} of
47       {nu _V.(M(_V,_Out),v[_V],rest[_In,_Out])} -> {v[_Z]}
48     | otherwise -> {Error4} in
49
50   % Initialise with a marker
51   let run[_Z] v[_Z] g[_In,_Out] =
52     {proceed[_Z]} {nu _V.(g[_In,_Out],M(_V,_In),v[_V])} in
53
54   % The main code
55   {run[_Z]} {5(_Z)} {dataflow[_In,_Out]} % 5!
```

Fig. 12. Embedding a dataflow language in λ_{GT}.

The evaluator of such a dataflow graph is shown in Fig. 12, lines 31–52. We use a marker M to trace the value on the flow (lines 53–54). Lines 33–37 deal with a node with an input and an output (N2). They apply the embedded function to the value. Lines 39–45 deal with a node with an input and two outputs (N3). They decide which way to go by applying the embedded function to the value. Line 47 terminates the evaluator when the marker is pointing at the output link _Out. We have run the code on our interpreter and checked that the interpreter returns {120(_Z)} ($= 5!$) after printing 14 intermediate graphs.

We can even build a larger dataflow from components. For example, we can compose the graph connecting the output and the input of them: {nu _X.

276 J. Sano and K. Ueda

(dataflow[_In,_X],dataflow[_X,_Out])}. Applying {run[_Z]} to {3(_Z)} and the graph returns {720(_Z)} (= (3!)!).

References

1. Bak, C.: GP 2: efficient implementation of a graph programming language. Ph.D. thesis, Department of Computer Science, The University of York (2015), https://etheses.whiterose.ac.uk/12586/
2. Bak, C., Faulkner, G., Plump, D., Runciman, C.: A reference interpreter for the graph programming language GP 2. In: Rensink, A., Zambon, E. (eds.) Proceedings Graphs as Models, GaM@ETAPS 2015. EPTCS, vol. 181, pp. 48–64 (2015). https://doi.org/10.4204/EPTCS.181.4
3. Baldan, P., Bertolissi, C., Cirstea, H., Kirchner, C.: A rewriting calculus for cyclic higher-order term graphs. Math. Struct. Comput. Sci. **17**(3), 363–406 (2007). https://doi.org/10.1017/S0960129507006093
4. Brus, T.. H.., van Eekelen, M.. C.. J.. D.., van Leer, M.. O.., Plasmeijer, M.. J..: Clean — a language for functional graph rewriting. In: Kahn, Gilles (ed.) FPCA 1987. LNCS, vol. 274, pp. 364–384. Springer, Heidelberg (1987). https://doi.org/10.1007/3-540-18317-5_20
5. Cirstea, Horatiu, Kirchner, Claude, Liquori, Luigi: Matching power. In: Middeldorp, Aart (ed.) RTA 2001. LNCS, vol. 2051, pp. 77–92. Springer, Heidelberg (2001). https://doi.org/10.1007/3-540-45127-7_8
6. Dietl, W., Müller, P.: Universes: lightweight ownership for JML. J. Object Technol. **4**(8), 5–32 (2005). https://doi.org/10.5381/jot.2005.4.8.a1
7. Ehrig, H., Ehrig, K., Prange, U., Taentzer, G.: Fundamentals of Algebraic Graph Transformation. Monographs in Theoretical Computer Science. An EATCS Series, Springer, Berlin (2006). https://doi.org/10.1007/3-540-31188-2
8. Erwig, M.: Inductive graphs and functional graph algorithms. J. Funct. Program. **11**(5), 467–492 (sep 2001). https://doi.org/10.1017/S0956796801004075
9. Fernández, Maribel, Kirchner, Hélène., Mackie, Ian, Pinaud, Bruno: Visual modelling of complex systems: towards an abstract machine for PORGY. In: Beckmann, Arnold, Csuhaj-Varjú, Erzsébet, Meer, Klaus (eds.) CiE 2014. LNCS, vol. 8493, pp. 183–193. Springer, Cham (2014). https://doi.org/10.1007/978-3-319-08019-2_19
10. Fernández, M., Pinaud, B.: Labelled port graph - a formal structure for models and computations. Electr. Notes Theor. Comput. Sci. **338**, 3–21 (2018). https://doi.org/10.1016/j.entcs.2018.10.002
11. Fradet, P., Métayer, D.L.: Structured gamma. Sci. Comput. Program. **31**(2), 263–289 (1998). https://doi.org/10.1016/S0167-6423(97)00023-3
12. Ghamarian, A., de Mol, M., Rensink, A., Zambon, E., Zimakova, M.: Modelling and analysis using GROOVE. Int. J. Softw. Tools Technol. Transfer **14**(1), 15–40 (2012). https://doi.org/10.1007/s10009-011-0186-x
13. Jakumeit, E., Buchwald, S., Kroll, M.: GrGen.NET: the expressive, convenient and fast graph rewrite system. Int. J. Softw. Tools Technol. Transfer **12**(3), 263–271 (2010). https://doi.org/10.1007/s10009-010-0148-8
14. Leroy, X., Doligez, D., Frisch, A., Garrigue, J., Rémy, D., Vouillon, J.: The OCaml system release 4.14. INRIA (2022)
15. Manning, G., Plump, D.: The GP programming system. In: Proceedings of Graph Transformation and Visual Modelling Techniques (GT-VMT 2008). Electronic Communications of the EASST, vol. 10 (2008). https://doi.org/10.14279/tuj.eceasst.10.150

16. Matsuda, K., Asada, K.: A functional reformulation of UnCAL graph-transformations: or, graph transformation as graph reduction. In: Proceedings of POPL 1997. pp. 71–82. ACM (2017). https://doi.org/10.1145/3018882.3018883

17. Oliveira, B.C., Cook, W.R.: Functional programming with structured graphs. SIGPLAN Not. **47**(9), 77–88 (2012). https://doi.org/10.1145/2398856.2364541

18. Rozenberg, G.: Handbook of Graph Grammars and Computing by Graph Transformation, Volume 1: Foundations. World Scientific (1997). https://doi.org/10.1142/3303

19. Runge, O., Ermel, C., Taentzer, G.: AGG 2.0 - new features for specifying and analyzing algebraic graph transformations. In: Schürr, A., Varró, D., Varró, G. (eds.) Proc. AGTIVE 2011. LNCS, vol. 7233, pp. 81–88. Springer, Berlin, Heidelberg (2012). https://doi.org/10.1007/978-3-642-34176-2_8

20. Sangiorgi, D., Walker, D.: The Pi-Calculus: A Theory of Mobile Processes. Cambridge University Press (2001)

21. Sano, J.: Implementing G-Machine in HyperLMNtal. Bachelor's thesis, Waseda University (2021). https://doi.org/10.48550/arXiv.2103.14698

22. Sano, J., Ueda, K.: Syntax-driven and compositional syntax and semantics of Hypergraph Transformation System (in Japanese). In: Proceedings of 38nd JSSST Annual Conference (JSSST 2021) (2021). https://jssst.or.jp/files/user/taikai/2021/papers/45-L.pdf

23. Sano, J., Yamamoto, N., Ueda, K.: Type checking data structures more complex than trees. J. Inf. Process. **31**, 112–130 (2023). https://doi.org/10.2197/ipsjjip.31.112

24. Schürr, A., Winter, A.J., Zündorf, A.: The PROGRES approach: language and environment. In: Handbook of Graph Grammars and Computing by Graph Transformation: Volume 2: Applications, Languages and Tools, chap. 13, pp. 487–550. World Scientific (1997). https://doi.org/10.1142/9789812384720_0002

25. Ueda, K.: LMNtal as a hierarchical logic programming language. Theoret. Comput. Sci. **410**(46), 4784–4800 (2009). https://doi.org/10.1016/j.tcs.2009.07.043

26. Ueda, K.: Towards a substrate framework of computation. In: Concurrent Objects and Beyond, LNCS, vol. 8665, pp. 341–366. Springer (2014). https://doi.org/10.1007/978-3-662-44471-9_15

27. Ueda, K., Ogawa, S.: HyperLMNtal: an extension of a hierarchical graph rewriting model. KI - Künstliche Intelligenz **26**(1), 27–36 (2012). https://doi.org/10.1007/s13218-011-0162-3

28. Weber, J.H.: Tool support for functional graph rewriting with persistent data structures - GrapeVine. In: Behr, N., Strüber, D. (eds.) Proc. ICGT 2022. LNCS, vol. 13349, pp. 195–206. Springer, Cham (2022). https://doi.org/10.1007/978-3-031-09843-7_11

Blue Skies

A Living Monograph for Graph Transformation

Nicolas Behr[1] and Russ Harmer[2](✉)

[1] Université Paris Cité, CNRS, IRIF, 8 Place Aurélie Nemours,
75205 Paris Cedex 13, France
`nicolas.behr@irif.fr`
[2] Université de Lyon, ENS de Lyon, UCBL, CNRS, LIP, 46 allée d'Italie,
69364 Lyon Cedex 07, France
`russell.harmer@ens-lyon.fr`

Abstract. A preliminary account of the notion of a *living monograph* for the field of graph transformation, and the reasons that led us to it, is given. The advantages of such a system are discussed along with the technical problems that will need to be overcome in order to build it.

Keywords: graph transformation · collaborative mathematics · formalised mathematics

1 Towards a Living Monograph

In this paper, we outline our notion of *living monograph* that we plan to develop in the context of a new French national research agency project[1]. A key aspect of this project is to provide theoretical and technological tools for the representation and expression of diagrammatic reasoning in the proof assistant Coq [9] with the goal of enabling the user to interact with the proof assistant by manipulating graphical depictions of diagrams.

The notion of living monograph for graph transformation presented in this paper is not a specific deliverable of this project but we anticipate that it will follow as a natural consequence to showcase certain aspects of the project: an online, collaborative system for reading and writing the style of mathematics required for graph transformation that is fully integrated with the possibility of an underlying formalisation in Coq. Such a system could even be built in the absence of full success of the project in providing purely graphical interaction with Coq, although the result would clearly be less ergonomic, and the success of the living monograph itself would depend principally on the degree of support and engagement of the graph transformation community.

In the remainder of this introduction, we explain the reasons that led us to this notion. In the subsequent sections, we explain the conceptual advantages

---

[1] CoREACT: COq-based Rewriting: towards Executable Applied Category Theory (ANR-22-CE48-0015).

© The Author(s), under exclusive license to Springer Nature Switzerland AG 2023
M. Fernández and C. M. Poskitt (Eds.): ICGT 2023, LNCS 13961, pp. 281–291, 2023.
https://doi.org/10.1007/978-3-031-36709-0_15

of such a system as well as the specific technical hurdles that we will need to overcome in the course of the project. We conclude with some remarks about the more general significance of these ideas beyond the graph transformation community.

1.1   Surveys in the Graph Transformation Literature

The field of graph transformation has a rich and extensive research literature which is generally considered to begin with the work of Pfaltz & Rosenfeld [21] and Schneider [23] in the late 1960s. The early community and literature was notably structured by a series of six quadrennial international workshops, the first of which was held in 1978; the first and third proceedings thereof included significant survey material providing up-to-date, comprehensive entry points to the field in 1978 [10] and 1986 [14] respectively.

The publication in the 1990s of a three-volume handbook [12,13,22] marked a significant maturation of the field; its approximately 1800 pages provided a comprehensive account of the theory and applications of graph transformation after some twenty-five years of active development. In hindsight, the appearance of this monumental text seems to identify the period in time where the field had grown sufficiently to support the emergence of more specialised subfields.

Indeed, in 2002, the workshop series mutated into the biennial conference ICGT (which became annual from 2014 onwards) accompanied by various, more specialised, satellite events. It is noteworthy that, other than an introduction to the field from the perspective of software engineering [5] and a short introduction to its application to DNA computing [17], both included in the 2002 proceedings, there have been no further survey articles in subsequent proceedings.

This is perhaps unsurprising, because this modern style of 'conference plus satellite events' precisely emphasises the rapid dissemination of new research over the production of up-to-date surveys, but we argue that such surveys remain important. Clearly, they serve as entry points to a field for young researchers, although tutorial satellite events at least partially fulfil this need; but they can also serve as a means for a community to retain a certain degree of global cohesion in the face of an ever growing, ever more specialised primary literature.

The final surveys, to date, in the field of graph transformation were published in 2006 and 2020 in the form of traditional research monographs. The first [11] provided a full account of the theory of graph transformation, particularly those developments subsequent to the first volume of the handbook [22], as well as an account of the application of this theory to model transformation. The second [18] gave a comprehensive account of the application of graph transformation to a variety of problems in software engineering, including an updated treatment of model transformation.

1.2   Shortcomings of Surveys in an Active Literature

The above discussion already makes it clear that, beyond a certain point of development, the writing of a comprehensive survey of an entire research field requires too much time and effort to be worthwhile: the two final monographs discussed above both have restricted scope—the first provides full theoretical coverage, for its time, but only a limited treatment of applications—and it seems reasonable to expect that any future such manuscripts will similarly target subfields and/or focus exclusively on developments since a previous survey.

A second aspect, which remained implicit in the above discussion, is that any such survey—no matter how long or comprehensive—will inevitably become out-of-date sooner or later: the developments on the theoretical side, in the decade that followed the production of the first volume of the handbook [22], were thoroughly documented in the later monograph [11]; but, at the time of writing, the wide-ranging theoretical developments in the field since the latter's publication have yet to be treated in the form of a monograph. Similarly, much of the more application-oriented material in the second and third volumes of the handbook [12,13] has never been revisited in such a form. Perhaps the very nature of a traditional survey (be it an article, a handbook chapter or a monograph) is actually rather ill-suited to the purpose it purports to address: even for a well-coordinated team of experts, there is an irreconcilable tension between the sheer amount of time it takes to provide comprehensive coverage and the need for the text to remain up-to-date.

A further problem arises in the form of the errors that inevitably creep into a long manuscript. These generally take the form of incorrect, or more often simply incomplete, proofs that escape the proof-reading process. For example, in graph transformation, the complex diagrammatic reasoning increasingly employed can be difficult to convey in an accurate, yet concise, fashion in a traditional text.

This analysis leads us to envisage the possibility of a system that enables the collaborative writing of mathematics—and, in particular, the diagrammatic reasoning prevalent in graph transformation—in a way that intrinsically supports the formalisation of said mathematics in a proof assistant. By these means, we seek to address the shortcomings identified above: by opening the writing of such a *living monograph* to an entire community, the different members of that community can contribute, according to their specialised knowledge, and as such provide better coverage, and far greater reactivity in the event that updates or corrections become necessary, than any small group of experts; and, by enabling access to the benefits of proof assistants without requiring expertise on the part of the user, the bottleneck of technical proof-reading should be reduced with a concomitant improvement in the quality of the resulting monograph.

In effect, we wish to combine the well-known advantages of a wiki-like system, based on distributed collaboration, for writing down mathematics with those of a proof assistant, that aids and informs the writing process, in order to provide an online resource that can be readily maintained up-to-date and rapidly corrected upon discovery of errors or omissions.

2   Technical Challenges of a Living Monograph

The input format of current proof assistants is one-dimensional and text-based. In order to be able to express higher-dimensional notations, such as categorical diagrams, some kind of encoding is therefore required. For our purposes, this raises two initial questions: (i) how to represent an individual diagram and prove desired properties of that diagram; and (ii) how to combine diagrams so as to perform diagrammatic reasoning that is valid in some desired logic?

However, these questions still ultimately presuppose a classical interaction with the proof assistant whereby the diagrammatic reasoning, in the head of the user, is only ever 'seen' in encoded form by the system. This allows the proof assistant to verify the correctness and completeness of a proof proposed by the user but does not enable the interactive development of a proof between the user and the machine. Ideally, we would like an augmented notion of interaction with the proof assistant where the user, and potentially the system, can perform diagrammatic reasoning by manipulating graphical depictions of diagrams.

We now examine these questions in greater detail to clarify the technical—and conceptual—obstacles to the construction of a system capable of supporting the diagrammatic reasoning found in graph transformation, e.g. the formalisation of axiomatic settings, such as adhesive categories, the proofs therein of results such as confluence, normalisation or the concurrency theorem and the proofs that concrete settings satisfy those axioms and thus enjoy those results.

2.1   Categorical Reasoning in a Proof Assistant

Diagrams. The most natural approach to representing commutative diagrams in a proof assistant consists in encapsulating the objects and arrows of the diagram, together with the required equalities of paths, in a type. Such a representation can subsequently be instantiated into any specific concrete category that we have defined. We can therefore delegate the composition and decomposition, i.e. basic diagram chasing, of such commutative diagrams to the proof assistant.

In order to identify commutative diagrams that satisfy a universal property, we can encapsulate the statement of that universal property together with the commutative diagram into a new type. For example, the property

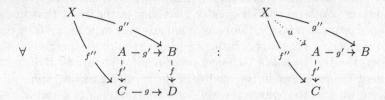

of *being a pullback square* could be represented in Coq in the following manner:

```
Record Pullback {C: Category} [A B C D: C]
       (g': A->B) (f: B->D) (f': A->C) (g: C->D) := {
    square_commutes: fog' ≈ gof';
    pullback_up: ∀ X (g'': X->B) (f'': X->C),
        fog'' ≈ gof'' -> ∃! u: X->A, f'' ≈ f'ou ∧ g'' ≈ g'ou }.
```

We define the *property* of being a pullback square rather than the more usual notion of the *construction* of a pullback span given a starting co-span, since many possible such pullback spans can be constructed. As a result, that latter style of definition is highly inefficient for use with a proof assistant whereas, given the former definition, it would be straightforward to define an operation that constructs a pullback span, given a co-span, using existential quantifiers [4].

This definition is sufficient and convenient for proving typical properties of pullbacks that depend on the manipulation of their universal property, e.g. their uniqueness up to unique isomorphism or the pasting lemma. However, this style of definition forces us to prove uniqueness up to unique isomorphism for every limit (and co-limit) construction individually [4]. An alternative approach is possible, based on defining constructions such as pullbacks directly as (co-)limits, which establishes uniqueness up to unique isomorphism, once and for all, for all such constructions. This has the disadvantage of forcing us to prove equivalence of those definitions with the original ones, but does provide some useful flexibility as different formalisations typically lend themselves to different uses.

Diagrammatic Reasoning. Much of the basic diagrammatic reasoning used in category theory involves composing and decomposing commutative diagrams in order to propagate properties known to hold in some part of a diagram to another part. For example, the *composition* lemma for pullbacks

$$
\begin{array}{ccccc}
A & -h'\!\rightarrow & B & -f'\!\rightarrow & E \\
\downarrow g'' & PB & \downarrow g' & PB & \downarrow g \\
C & -h\!\rightarrow & D & -f\!\rightarrow & F
\end{array}
\quad\Longrightarrow\quad
\begin{array}{ccc}
A & -\!f'\circ h'\!\longrightarrow & E \\
\downarrow g'' & PB & \downarrow g \\
C & -\!f\circ h\!\longrightarrow & F
\end{array}
$$

provides the means to propagate the property of *being a pullback* from the two inner squares to the outer rectangle. Clearly, many other such lemmata exist, e.g. the preservation of monomorphisms by pullback or the stability of final pullback complements under pullbacks, whose proofs—essentially just repeated application of universal properties plus diagram chasing—can be performed in category theory with no additional axioms required.

In a formalisation in a proof assistant, we would clearly not want to inline such proofs in order to justify each successive step in a chain of diagrammatic reasoning. Instead, it would be more natural to formulate these various means of propagating properties as the *reasoning rules* of a basic diagrammatic logic whose soundness would be established, once and for all, by formalising the proof justifying each reasoning rule of the logic. A key early goal of the CoREACT project is precisely to develop such a logical formalism within Coq.

This basic diagrammatic logic would serve as a universal core that could be extended, in many different ways, with additional reasoning rules whose proofs of soundness would rely on additional axioms. For example, in the field of graph transformation, various properties[2] observed to hold in concrete settings—and

---

[2] e.g. the preservation of monomorphisms under pushout or the stability of pushouts under pullbacks.

frequently used in the proofs of important results—have been subsumed into a number of closely-related axiom systems, e.g. adhesive and quasi-adhesive categories [20], $\mathcal{M}$-adhesive categories [11] and rm-quasi-adhesive categories [16] among many others. A key second goal of the CoREACT project is to provide the means to define such extensions of the core logical formalism in Coq so as to enable fully-fledged diagrammatic reasoning for graph transformation as found in [6,11,20] and many others.

An interesting aspect of this methodology is that one could formalise such a logic in a proof assistant without necessarily formally proving the soundness of its rules. From this perspective, a collection of reasoning rules could be viewed as an autonomous layer that could, but need not, be connected to an underlying formalisation of category theory (plus some optional axiom system). Of course, such an approach might not make any sense—in the case that the collection of reasoning rules is badly chosen—but, if done sensibly, would provide a *light-weight* approach to formalisation that might be highly appropriate for certain purposes: a user wishing to prove some complicated theorems using a well-chosen collection of reasoning rules would probably be happy to accept them as given to focus exclusively on formalising the higher-level logic of those theorems.

More generally, such a separation of concerns seems to be a pragmatic course of action. On the one hand, the higher-level logic enables the succinct proof of theorems *about* graph transformation; on the other hand, the lower-level logic that determines reasonable collections of reasoning rules, on the basis of the various axiom systems proposed in the literature, has more of a *meta-theoretic* character with the purpose of analysing the relative expressive power of different axiom systems.

Moreover, our recent work on compositional rewriting theories [7] suggests that this separation of concerns could be taken further still. The fibrational framework introduced in that work provides higher-level reasoning *macros* that intrinsically induce and structure entire collections of the lower-level reasoning rules. In the same way that those rules enable the user to focus on higher-level logic, rather than repeated application of low-level universal properties, these reasoning macros seem to enable a further level of succinctness that remains to be investigated.

2.2 Categorical Reasoning with a Proof Assistant

The discussion so far has focussed on questions concerning the representation of diagrams, and of diagrammatic reasoning, in a proof assistant such as Coq. Assuming an adequate resolution of the various technical difficulties identified, this would still require all formalised content to be expressed in the language of the proof assistant. However, in order for a living monograph system to provide a comfortable and ergonomic interface to the proof assistant for the non-expert, it should ideally allow for diagrammatic input and the expression of diagrammatic reasoning in that interface. It would then be the responsibility of the system to translate the input to, and other events in, the interface to the one-dimensional input format of the proof assistant itself.

These aspects of our proposed living monograph are currently dependent on certain tasks of the CoREACT project, but we do not exclude alternative future approaches. The idea is to extend the Coq document model with a *diagram* object of which the jsCoq framework [3] could handle the visualisation and, in concert with the YADE commutative diagram editor, provide a web-based interface to create and manipulate commutative diagrams and generate their underlying Coq representations.

This is a complicated but—we believe—realistic integration task that would provide the first prototype of a system capable of creating a wiki-like document, whose text, diagrams and other mathematical content are provided by users, connected to an underlying formalisation in Coq.

The CoREACT project also seeks to incorporate diagrammatic reasoning into the interface with the aim of providing assistance to the user in *developing* proofs, not merely formalising proofs already worked out on paper. This has two principal aspects: a database of combinations of reasoning steps found in already-formalised proofs for the system to query in order to *suggest* pertinent next steps in a proof; and the use of gestures, in the general sense of proof-by-pointing [8], to enable the user to invoke diagrammatic reasoning steps.

The first of these aspects aims to provide means to ease the cognitive burden of reasoning with the large and complicated diagrams that occur in proofs in graph transformation, e.g. simply identifying that certain squares within such a diagram are pullbacks could suggest plausible next steps to the user—and, in the longer term, enable the system to make pertinent suggestions. The second aspect is already under active investigation in the context of the Actema system, that currently provides a graphical interface for the manipulation of one-dimensional Coq syntax, which will also play a key role in the CoREACT project.

3   Advantages of a Living Monograph

Our initial motivation to develop this project grew out of the observations made in Sect. 1 concerning the shortcomings of a traditional research monograph in a large, but still dynamic, area of research: the intrinsic tension between providing broad coverage, of maintaining the text up to date and of providing some assurance of the completeness, and correctness, of the results presented.

We hope to have convinced the reader that our notion of living monograph, as outlined in this paper, does address these issues: the use of wiki-like systems for providing broad coverage and fast reactivity to required updates is well known. Coupled with the benefits of a proof assistant, once enabled with ergonomic means to create and manipulate commutative diagrams and perform associated reasoning, the resulting system would indeed constitute a highly novel kind of on-line mathematical editor. We anticipate that this would not only aid the writing process but also assist the writer in the very development of the mathematics.

Although not part of our original motivation, we might also envisage certain advantages of a living monograph for the reader. Beyond the obvious convenience of hyperlinking, to navigate easily within the text, let us note, in particular, the possible application of the Coq knowledge graph. Given a Coq document, this details the dependencies between its different definitions and proofs. This could be a powerful tool for understanding as it would allow the reader, at any moment, to inspect the provenance of a particular result in order to understand better the overall structure of the mathematical theory described in the text. In effect, it would provide a finer-grained and on-demand version of the chapter and section dependencies often given at the start of research monographs and textbooks.

Indeed, let us emphasise that this use of the knowledge graph is representative of how we imagine the living monograph would make use of its underlying proof assistant. The aim would be to exploit and display the insights that the assistant can bring, and which would be practically inaccessible without it, in order to augment the typical content of a traditional monograph. In particular, we would not—at least by default—wish to display the formal proofs of results as part of the standard reading experience; instead, the reader should be able to click in order to access the formalised proof should they wish to do so—but we imagine that the simple existence of the underlying formalisation would suffice for the majority of readers.

Let us conclude by noting some more general potential consequences of the notion of living monograph. If a research community succeeds in organising itself around a comprehensive living monograph, this would obviously provide clear entry points, for the newcomer, into the primary literature. However, it could also give rise to a new mode of attribution, finer-grained than the traditional scientific paper and previously typically restricted to lecture notes or theses, whereby small, but significant, improvements to definitions or results could be acknowledged by their incorporation into the living monograph—as a kind of *micro-publication*. In a similar vein, the existence of the living monograph would hopefully encourage the formalisation of existing results, major or minor, as a legitimate and attributed activity in and of itself.

At a time when the traditional publication model of academia is increasingly under question, the ability to acknowledge such work—generally ignored by the traditional model—might help to advance further the debate on how a scientist's contribution to their field might best be identified. It is to be hoped that it would at least provide a more nuanced picture than does an H-index. Finally, let us note that the notion of living monograph is clearly not restricted to the field of graph transformation or to categorical reasoning per se. The basic concept could apply to many fields of mathematics and perhaps foster better communication between neighbouring fields by enabling hyperlinking *between* living monographs.

Related work

We are aware of two existing lines of work on the formalisation of certain aspects of graph transformation. The first, initiated by Strecker and summarised in [25], uses the Isabelle proof assistant to formalise certain concrete notions of graphs, homomorphisms and rewriting steps in order to provide formal proofs of the preservation of certain properties under transformation. More recently, Söldner and Plump have formalised, again in Isabelle, the equivalence of the operational definition of direct derivations (by deletion and glueing operations) with that of a double-pushout diagram [24] for a certain class of concrete graphs and their homomorphisms.

The notion of micro-publication outlined above bears some similarity to the idea of the Archive of Formal Proofs although the content in the AFP is typically more substantial in size and scope in that it typically presents entire libraries or non-trivial results rather than individual lemmata.

The idea of a *living review* [26], i.e. a survey article that is *required* to be maintained up-to-date as a condition of publication, was pioneered in the late 1990s by the Max Planck Institute for Gravitational Physics in the form of its *Living Reviews in Relativity*. In 2015, this initiative became a Springer journal— retaining the requirement of maintenance—and has been joined by similar efforts in solar physics and computational astrophysics; the idea also reached biology in 2014 [15]. Of course, these notions of living review do not have, or even need, the possibility of an underlying formalisation, but they do strongly illustrate the need for constant integration of new results in dynamic research fields.

The seminal initiative in mathematics to build a comprehensive formalised library of knowledge is clearly the QED project, dating from the 1990s, which set out to "build a computer system that effectively represents all important mathematical knowledge and techniques" [1]. Despite an active mailing list and two international workshops (held in 1994 and 1995), QED never proceeded beyond the stage of initial plans; the reasons for this have been discussed elsewhere— see, for example, [19,27]—but seem mainly to have stemmed from the difficulty either of choosing a single logic upon which to found everything or, alternatively, of providing for the inter-operability of different logics.

While it seems unlikely that a consensus will ever be reached on the first question, considerable progress has subsequently been made on the second— see, for example, [2,19] and the MMT system—and the success of the Twenty Years of the QED Manifesto workshop, held as part of the 2014 Vienna Summer of Logic, proves that the original vision remains intact. Indeed, many other projects now exist such as Kerodon and Stacks or, with a more general intention, Xena or ForMath. Although the notion of living monograph has different, albeit overlapping, motivations to these various approaches, we hope that it will fulfil a useful rôle within the larger ecosystem of formalised mathematics.

References

1. The QED manifesto. In: Bundy, A. (ed.) Automated Deduction - CADE-12. CADE 1994. LNCS, vol. 814, pp. 238–251. Springer, Berlin, Heidelberg (1994)
2. Alama, J., Brink, K., Mamane, L., Urban, J.: Large formal wikis: issues and solutions. In: Davenport, J.H., Farmer, W.M., Urban, J., Rabe, F. (eds.) CICM 2011. LNCS (LNAI), vol. 6824, pp. 133–148. Springer, Heidelberg (2011). https://doi.org/10.1007/978-3-642-22673-1_10
3. Arias, E.J.G., Pin, B., Jouvelot, P.: jsCoq: towards hybrid theorem proving interfaces. In: Proceedings of UITP 2016, vol. 239, pp. 15–27. EPTCS (2017)

4. Arsac, S.: Coq formalization of graph transformation. Master's thesis, Université Paris-Cité (2022). https://www.samuelarsac.fr/rapport22.pdf
5. Baresi, L., Heckel, R.: Tutorial introduction to graph transformation: a software engineering perspective. In: Corradini, A., Ehrig, H., Kreowski, H.-J., Rozenberg, G. (eds.) ICGT 2002. LNCS, vol. 2505, pp. 402–429. Springer, Heidelberg (2002). https://doi.org/10.1007/3-540-45832-8_30
6. Behr, N., Harmer, R., Krivine, J.: Concurrency theorems for non-linear rewriting theories. In: Gadducci, F., Kehrer, T. (eds.) ICGT 2021. LNCS, vol. 12741, pp. 3–21. Springer, Cham (2021). https://doi.org/10.1007/978-3-030-78946-6_1
7. Behr, N., Harmer, R., Krivine, J.: Fundamentals of compositional rewriting theory (2022). https://arxiv.org/abs/2204.07175
8. Bertot, Y., Kahn, G., Théry, L.: Proof by pointing. In: Hagiya, M., Mitchell, J.C. (eds.) TACS 1994. LNCS, vol. 789, pp. 141–160. Springer, Heidelberg (1994). https://doi.org/10.1007/3-540-57887-0_94
9. Bertot, Y., Castéran, P.: Interactive theorem proving and program development. In: Bertot, Y., Castéran, P. (eds.) Texts in Theoretical Computer Science. An EATCS Series, Springer, Berlin, Heidelberg (2004). https://doi.org/10.1007/978-3-662-07964-5
10. Claus, V., Ehrig, H., Rozenberg, G.: Graph-grammars and their application to computer science and biology. LNCS, vol. 73. Springer, Berlin, Heidelberg (1979). https://doi.org/10.1007/BFb0025713
11. Ehrig, H., Ehrig, K., Prange, U., Taentzer, G.: Fundamentals of algebraic graph transformation. In: Ehrig, H., Ehrig, K., Prange, U., Taentzer, G. (eds.) Texts in Theoretical Computer Science. An EATCS Series, Springer, Berlin, Heidelberg (2006). https://doi.org/10.1007/3-540-31188-2
12. Ehrig, H., Engels, G., Kreowski, H.J., Rozenberg, G. (eds.): Handbook of Graph Grammars and Computing by Graph Transformation, vol. 2: Applications, Languages and Tools. World Scientific, Singapore (1999)
13. Ehrig, H., Kreowski, H.J., Montanari, U., Rozenberg, G. (eds.): Handbook of Graph Grammars and Computing by Graph Transformation, vol. 3: Concurrency, Parallelism, and Distribution. World Scientific, Singapore (1999)
14. Ehrig, H., Nagl, M., Rozenberg, G., Rosenfeld, A. (eds.): Graph Grammars 1986. LNCS, vol. 291. Springer, Heidelberg (1987). https://doi.org/10.1007/3-540-18771-5
15. Elliott, J., Turner, T., Clavisi, O., Thomas, J., Higgins, J., Mavergames, C., Gruen, R.: Living systematic reviews: an emerging opportunity to narrow the evidence-practice gap. PLoS Med. 11(2), e1001603 (2014)
16. Garner, R., Lack, S.: On the axioms for adhesive and quasiadhesive categories. Theory Appl. Categories 27(3), 27–46 (2012)
17. Harju, T., Petre, I., Rozenberg, G.: Tutorial on DNA computing and graph transformation. In: Corradini, A., Ehrig, H., Kreowski, H., Rozenberg, G. (eds.) Graph Transformation. ICGT 2002. LNCS, vol. 2505, pp. 430–434. Springer, Berlin, Heidelberg (2002)
18. Heckel, R., Taentzer, G.: Graph Transformation for Software Engineers. Springer, Cham (2020). https://doi.org/10.1007/978-3-030-43916-3
19. Kohlhase, M., Rabe, F.: QED reloaded: towards a pluralistic formal library of mathematical knowledge. J. Formalized Reasoning 9(1), 201–234 (2016)
20. Lack, S., Sobociński, P.: Adhesive and quasiadhesive categories. RAIRO - Theor. Inform. Appl. 39(3), 511–545 (2005)
21. Pfaltz, J.L., Rosenfeld, A.: Web Grammars. In: Proceedings of the 1st International Joint Conference on Artificial Intelligence, pp. 609–619 (1969)

22. Rozenberg, G. (ed.): Handbook of Graph Grammars and Computing by Graph Transformation, vol. 1: Foundations. World Scientific, Singapore (1997)

23. Schneider, H.J.: Chomsky-languages for multidimensional input-media. In: Proceedings of the International Computing Symposium, pp. 599–608 (1970)

24. Söldner, R., Plump, D.: Towards mechanised proofs in double-pushout graph transformation. In: Proceedings of GCM 2022, vol. 374, 59–75. EPTCS (2022)

25. Strecker, M.: Interactive and automated proofs for graph transformations. Math. Struct. Comput. Sci. **28**(8), 1333–1362 (2018)

26. Wheary, J., Schutz, B.: Living reviews in relativity: making an electronic journal live. J. Electr. Publish. **3**(1) (1997)

27. Wiedijk, F.: The QED manifesto revisited. Stud. Logic, Grammar Rhetoric **10**(23), 121–133 (2007)

Graph Rewriting for Graph Neural Networks

Adam Machowczyk[✉] and Reiko Heckel[✉]

University of Leicester, Leicester, UK
{amm106,rh122}@le.ac.uk

Abstract. Given graphs as input, Graph Neural Networks (GNNs) support the inference of nodes, edges, attributes, or graph properties. Graph Rewriting investigates the rule-based manipulation of graphs to model complex graph transformations. We propose that, therefore, (i) graph rewriting subsumes GNNs and could serve as formal model to study and compare them, and (ii) the representation of GNNs as graph rewrite systems can help to design and analyse GNNs, their architectures and algorithms. Hence we propose Graph Rewriting Neural Networks (GReNN) as both novel semantic foundation and engineering discipline for GNNs. We develop a case study reminiscent of a Message Passing Neural Network realised as a Groove graph rewriting model and explore its incremental operation in response to dynamic updates.

1 Introduction

Neural Networks (NNs) are among the most successful techniques for machine learning. Starting out from application data, e.g. in a relational database, NNs require an encoding of the data into discrete vectors. Graph Neural Networks (GNNs) operate on graphs instead of tabular data. According to [19] a GNN is "an optimizable transformation on all attributes of the graph (nodes, edges, global context) that preserves graph symmetries (permutation invariances)". GNNs are applied to [19] images, text, molecules, social networks, programming code [1] and even mathematical equations [14], in areas such as antibacterial discovery [22], physics simulation [18], fake news detection [17], traffic prediction [15] and recommendation systems [7]. They support supervised, semi-supervised or unsupervised learning (where data labelling is expensive or impossible) and allow links between input vectors, resembling more closely the structure of the application data. In a recommender system where engagements with social media posts are nodes, they may predict such nodes based on the history of engagement, with attributes representing the strength of engagement.

Graph rewriting is the rule-based transformation of graphs [12]. Since GNNs are essentially graph transformations, we ask if they can be modelled by graph rewriting, and what the benefits of such a representation might be. Drawing from its rich theory and tool support, we would like to explore the use of graph rewriting as a computational model for GNNs, to specify and compare GNN

M. Fernández and C. M. Poskitt (Eds.): ICGT 2023, LNCS 13961, pp. 292–301, 2023.
https://doi.org/10.1007/978-3-031-36709-0_16

variants, study their expressiveness and complexity, and as a design and proto-
typing language for GNNs, evaluating specific GNN architectures and analysing
interactions between GNNs and the software and real-world systems with which
they interact. In modern data-driven applications, data is diverse, large-scale,
evolving, and distributed, demanding incremental and distributed solutions. This
has led to a variety of GNNs [25] and a need for a unifying theory to describe
their expressiveness and cost [4], evolution and incremental updates [8].

In this short paper, we suggest that typed attributed graph rewriting with
control can address these demands. We will illustrate this claim by means of a
case study of a recommender system for posts in a social network modelled as a
Graph Rewriting Neural Network (GReNN), a controlled graph rewriting system
based on a cycle of training and inference transformations with the possibility
of dynamic data updates. Training involves the tuning of parameters such as
weights and thresholds to support the inference of graph features or properties.
The example does not follow a particular GNN architecture, but can be seen
as an elaboration of message-passing GNNs over directed heterogeneous graphs
with dynamic changes and incremental updates, a combination we have not
encountered in any existing formalism. In particular, typed attributed graphs
allow for heterogeneity, where nodes of different types have different attributes
and are connected by differently-typed edges, while the rule-based approach
supports both local inference operations of message-passing GNNs and updates
to the graph structure representing changes in the real-world data.

2   A Recommender System GReNN

We consider a social network with users, posts, and engagements, e.g., reads,
likes or retweets, modelled as a Groove graph rewriting system [9]. In the type
graph

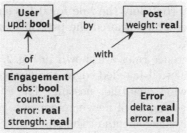

users have an *upd* attribute that determines if training is required on engagements with posts by this user. Posts have *weights* that model how representative they are of their author, i.e., how well engagements with these posts predict engagements with other posts by that user. Engagements have a *strength*
attribute to the model quality of the interaction of a user with a post, e.g. a
read may be weaker than a like or retweet. Engagements also have *error, obs*
and *count* attributes. The error represents the difference between the actual
strength of engagement and the strength inferred from engagements with similar posts. This is used for training the posts' weights by the gradient descent rule.

The Boolean *upd* and *obs* attributes signal, respectively, if an update is required to the training after a change of data affecting this node and if the strength of the node is observed from real-world engagement as opposed to being inferred. The *count* attribute, which is always 1, is used to calculate the cardinality of Engagement nodes in a multi-pattern match in Groove. To control training we add an Error node with *error* and *delta* attributes. All attributes except *strength* are part of the general mechanism while *weight* also has application-specific meaning.

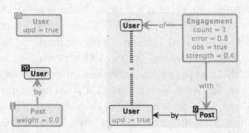

Fig. 1. Rules for dynamic updates: *newUser*, *newPost*, and *newEngagement04*. (Color figure online)

The rules come in two flavours: dynamic update rules to create users, posts and engagements, and rules for training and inference. Update rules are shown in Fig. 1 in Groove notation. As usual, elements in thin solid black outline represent read access, dashed blue elements are deleted, and thick green elements are created, while nodes and edges in dotted red represent negative conditions. The dotted red line labelled "=" means that the linked nodes are not equal. Rules to create engagements exist in three versions to create engagements of strengths 0.2, 0.4 and 0.8 (but only the rule for 0.4 is shown). When creating a new engagement, the user who authored the post has their *upd* attribute set to *true* to trigger retraining.

The training and inference rules are shown in Fig. 2. Nodes labelled ∀ or ∀⁺ represent nested multi rules, @-labelled edges associating nodes to their nesting levels and *in* edges showing quantifier nesting. The *infer* rule at the bottom calculates the expected strengths of new engagements as weighted sums of the strengths of all observed engagements of the same user with posts by a second user whose *upd* attribute is *true* (and then changes to *false*). The red-green dotted outline says that an Engagement node is only created if there is not already one, combining node creation with a negative condition. The assignment shown in green inside that Engagement node realises the following formula.

$$strength = \frac{\sum_{m \in M}(p(m).weight * e(m).strength)}{\text{card } M} \tag{1}$$

M is the set of matches of the multi pattern represented by the non-empty quantification node $\forall^>$, with $e(m)$ and $p(m)$ referencing the Engagement and

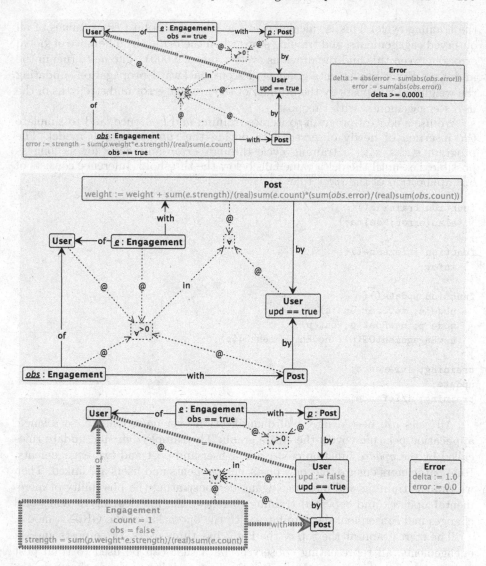

Fig. 2. Top to bottom: Rules *error*, *delta* and *infer*. (Color figure online)

Post nodes of match $m \in M$. This is nested in another $\forall$ node to apply the inference to all eligible Engagement nodes at once. Recall that the weights of Post nodes show how well a post represents the average post by that user.

The *error* rule in the top computes the inference error as the difference between the observed value of strength and the inferred value. This is typical of semi-supervised learning in GNNs where some of the nodes represent training samples while others are inferred. Structure and inference formula are similar to the *infer* rule. We use an Error node with a global *error* attribute to control the training cycle. This is calculated as the sum of absolute *error* values of all observed engagements, and training stops when the *delta*, the difference of global errors between this and the previous cycle is below 0.0001. The *delta* rule in the middle of Fig. 2 implements gradient descent backward propagation, updating the weights of the posts by the product of the average error and strengths of the user's engagements' with this post.

We use a control program to iterate training and inference, and to simulate the insertion of newly observed data during the runtime of the model. The program starts with a training cycle through *error* and *delta* for as long as possible, i.e., until the delta value falls below the threshold. Inference consists of one application of the *infer* rule:

```
function training(){
   alap{error; delta;}
}
function inference(){
   infer;
}
function update(){
   node u; newUser(out u);
   node p; newPost(u, out p);
   newEngagement02(p); newEngagement04(p);
}
training; inference;
update;
training; inference;
```

All rules use nested universal quantification over engagements, so a single application per rule covers the entire graph. The sample dynamic update rules called by the *update* function create a new user and post and two engagements. Rule parameters ensure that new posts, engagements and users are linked. Then we run the training and inference again, demonstrating the possibility of incremental updates and repeated training to reflect the interleaving of real-world changes and consequent data updates with the operation of the GReNN model.

The start graph at the top of the Fig. 3 has three users, three posts and six engagements, all representing "observed" data that can be used to train post weights. An intermediate graph, after the first training cycle, is shown in the middle, with final global error and delta values, and updated post weights. The final graph extends this by, among other things, the structure at the bottom, where the user on the right of the fragment is the same as the one in the top left of the intermediate graph. In the final graph, it has two newly observed and one newly inferred engagement with a new post by the new user on the left.

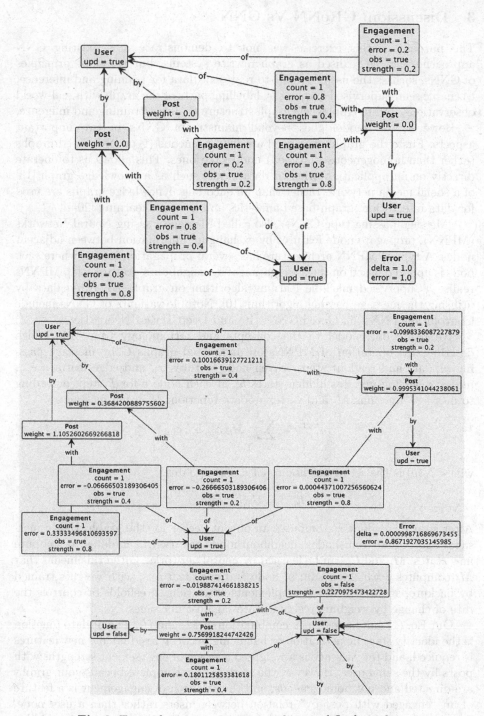

Fig. 3. Top to bottom: start, intermediate and final graphs.

3   Discussion: GReNN Vs GNN

The purpose of this exercise was not to demonstrate how existing GNN approaches can be realised as graph rewrite systems, but to adopt principles of GNNs, such as the use of graphs to represent data for training and inference, to enable semi-supervised learning by labelling parts of the graph with real-world observations, and to employ the graph structure to direct training and inference.

However, our model goes beyond mainstream GNNs in two important aspects. First, the graphs employed are heterogeneous (typed) directed graphs rather than homogeneous (untyped) undirected ones. This allows us to operate directly on an application-oriented data model such as a knowledge graph [13] of a social media network, rather than an encoding. Knowledge graphs are used for data integration, graph-based analytics and machine learning [20].

"Message-passing type GNNs, also called Message Passing Neural Networks (MPNN), propagate node features by exchanging information between adjacent nodes. A typical MPNN architecture has several propagation layers, where each node is updated based on the aggregation of its neighbours' features" [4]. MPNNs realise a supervised machine learning algorithm on graph-structured data by utilising the message-passing algorithm [10]. Notable variants of GNNs include Convolutional NNs [6], Gated GNNs [16] and Deep Tensor NNs [21].

We relate our model to the concepts and terminology from [10], which describes the operation of MPNNs on undirected graphs G by message passing, update and readout steps. Given node features x_v and edge features e_{vu}, message passing updates hidden states h_v^t at each node v for T steps according to message functions M_t and vertex update functions U_t using messages

$$m_v^{t+1} = \sum_{u \in N(v)} M_t(h_v^t, h_u^t, e_{vu})$$

where $N(v)$ is the set of neighbours of v and updates

$$h_v^{t+1} = U_t(h_v^t, m_v^{t+1}).$$

After step T, readout R computes an output vector for the graph. Hence messages aggregated from nodes' neighbourhoods are used in updates to compute new states. M_t and U_t can be learned by neural networks. Often this means that M_t computes a weighted sum of node and edge features, with weights trained by backpropagation, while U_t implements activation thresholds or controls the rate of change by weighing residual states against massages.

Our Eq. (1) can be seen as a combination of M_t and U_t. The update function is the identity since no activation or balancing between residual and new features is required, and the message is a weighted average of engagement strengths with posts by the same user. However, the analogy is incomplete because our graphs are directed and not homogeneous, and the strength of engagement is a feature of an "engaged with post by" relation between users rather than a user node. Indeed "one could also learn edge features in an MPNN by introducing hidden states for all edges in the graph . . . and updating them analogously" [10]. The relational nature of engagement is the reason why we only use a single step of forward propagation: All "engagements with posts" between the same two users are accessible locally in one step. A complex pattern realises the relation, so we need a sum over the multi pattern to compute the average. Hence graph patterns allow querying heterogeneous graphs where their homogeneous encodings are processed by navigating along edges. Apart from supporting a direct representation of the data, graph patterns increase the expressive power of GNNs also on homogeneous graphs [3].

Like weights and features, the graph structure is usually represented by matrices, such as adjacency or incidence and degree matrices. If the graph is undirected, its adjacency matrix is symmetrical. If the graph is weighted, it contains real-valued edge weights. A GReNN model supports directed edges, class heterogeneity and incremental updates to process dynamic application data in its original form without encoding. As of now, most GNN approaches only support undirected and homogenous graphs, but there is a trend towards more general models: In [11], the authors present an example of a directed MPNN and use it in conjunction with a graph attention network. In [10] the authors state that "It is trivial to extend the formalism to directed multigraphs". Multi-class approaches are utilised in IoT applications such as [2]. Emerging support for heterogeneous graph data includes [23, 24] which present a new algorithm called Explicit Message-Passing Heterogeneous Graph Neural Network (EMP). While GReNNs support both directed and heterogeneous data at the model level, such approaches may provide a platform to implement our models efficiently.

Incremental forward and back propagation in GNNs can substantially save time and resources [8] by reducing computations to nodes affected by updates. This is especially useful for big-data applications, where batch processing is not affordable and efficient. NNs and GNNs do not easily support data updates. Hence, the second generalisation in our approach is to model the dynamic updates of the graph and control specifically where retraining is required. This includes adding new nodes, an operation that would break most matrix-based representations by changing the dimensions of the matrices involved.

GReNNs reimagine GNNs free of considerations of efficient representation of and computations on graphs. This is appropriate from a modelling perspective, but unlikely to scale to large graphs. To support an engineering approach to GNN development based on graph rewriting, a study of the possible implementations of such models in mainstream GNNs is required. First experiments show that it is possible in principle to map a model such as ours to an undirected homogeneous graph by interpreting the edges of that simpler graph as paths or patterns in our model. This is a reduction that GNN developers have to perform when they present their application data as input to a GNN, but we believe that by studying this systematically we can find standard mappings from a typed attributed graph model into graphs supported by suitable GNN technology.

The suggestion that graph rewriting can serve as a semantic foundation for GNNs requires mappings in the opposite direction, taking the various GNN approaches and modelling them as GReNNs. Then, such embeddings could be used to compare the different graph types, inference or training rules. The theory of graph grammars with its language hierarchies and corresponding rule formats may well play a role here. GNNs can be based on attributed directed or undirected, simple or multi, homogeneous or typed, binary, hyper or hierarchical graphs, with little support for reuse of algorithms or implementations [19]. The categorical theory and potentially generic implementation of graph rewriting [5] can be a model to unify such diverse approaches.

References

1. Allamanis, M., Brockschmidt, M., Khademi, M.: Learning to represent programs with graphs. arXiv:1711.00740 [cs], May 2018. https://arxiv.org/abs/1711.00740
2. Anjum, A., Ikram, A., Hill, R., Antonopoulos, N., Liu, L., Sotiriadis, S.: Approaching the internet of things (IoT): a modelling, analysis and abstraction framework. Concurrency Comput.: Pract. Experience 27(8), 1966–1984 (2013)
3. Bronstein, M.: Using subgraphs for more expressive GNNs. Medium, December 2021. https://towardsdatascience.com/using-subgraphs-for-more-expressive-gnns-8d06418d5ab
4. Bronstein, M.: Beyond message passing: a physics-inspired paradigm for graph neural networks. Gradient (2022). https://thegradient.pub/graph-neural-networks-beyond-message-passing-and-weisfeiler-lehman
5. Brown, K., Patterson, E., Hanks, T., Fairbanks, J.P.: Computational category-theoretic rewriting. In: Behr, N., Strüber, D. (eds.) Graph Transformation - 15th International Conference, ICGT 2022, Held as Part of STAF 2022, Nantes, France, July 7–8, 2022, Proceedings. Lecture Notes in Computer Science, vol. 13349, pp. 155–172. Springer, Cham (2022). https://doi.org/10.1007/978-3-031-09843-7_9
6. Duvenaud, D., et al.: Convolutional networks on graphs for learning molecular fingerprints. In: Cortes, C., Lawrence, N.D., Lee, D.D., Sugiyama, M., Garnett, R. (eds.) Advances in Neural Information Processing Systems 28: Annual Conference on Neural Information Processing Systems, December 2015, pp. 7–12 (2015). Montreal, Quebec, Canada, pp. 2224–2232 (2015). https://proceedings.neurips.cc/paper/2015/hash/f9be311e65d81a9ad8150a60844bb94c-Abstract.html
7. Eksombatchai, C., et al.: Pixie: a system for recommending 3+ billion items to 200+ million users in real-time. arXiv:1711.07601 [cs], November 2017. https://arxiv.org/abs/1711.07601
8. Galke, L., Vagliano, I., Scherp, A.: Incremental training of graph neural networks on temporal graphs under distribution shift. CoRR abs/2006.14422 (2020). https://arxiv.org/abs/2006.14422
9. Ghamarian, A.H., de Mol, M., Rensink, A., Zambon, E., Zimakova, M.: Modelling and analysis using GROOVE. Int. J. Softw. Tools Technol. Transf. 14(1), 15–40 (2012). https://doi.org/10.1007/s10009-011-0186-x
10. Gilmer, J., Schoenholz, S.S., Riley, P.F., Vinyals, O., Dahl, G.E.: Neural message passing for quantum chemistry. CoRR abs/1704.01212 (2017). http://arxiv.org/abs/1704.01212
11. Han, X., Jia, M., Chang, Y., Li, Y., Wu, S.: Directed message passing neural network (D-MPNN) with graph edge attention (GEA) for property prediction of biofuel-relevant species. Energy AI 10, 100201 (2022). https://www.sciencedirect.com/science/article/pii/S2666546822000477

12. Heckel, R., Taentzer, G.: Graph Transformation for Software Engineers. Springer International Publishing, Cham (2020). http://graph-transformation-for-software-engineers.org/
13. Hogan, A., et al.: Knowledge Graphs. No. 22 in Synthesis Lectures on Data, Semantics, and Knowledge. Morgan & Claypool (2021). https://kgbook.org/
14. Lample, G., Charton, F.: Deep learning for symbolic mathematics. arXiv:1912.01412 [cs], December 2019. https://arxiv.org/abs/1912.01412
15. Lange, O., Perez, L.: Traffic prediction with advanced graph neural networks, September 2020. https://www.deepmind.com. https://www.deepmind.com/blog/traffic-prediction-with-advanced-graph-neural-networks
16. Li, Y., Tarlow, D., Brockschmidt, M., Zemel, R.S.: Gated graph sequence neural networks. In: Bengio, Y., LeCun, Y. (eds.) 4th International Conference on Learning Representations, ICLR 2016, San Juan, Puerto Rico, May 2–4, 2016, Conference Track Proceedings (2016). http://arxiv.org/abs/1511.05493
17. Monti, F., Frasca, F., Eynard, D., Mannion, D., Bronstein, M.M.: Fake news detection on social media using geometric deep learning. arXiv:1902.06673 [cs, stat], February 2019. https://arxiv.org/abs/1902.06673
18. Sanchez-Gonzalez, A., Godwin, J., Pfaff, T., Ying, R., Leskovec, J., Battaglia, P.: Learning to simulate complex physics with graph networks, November 2020. https://proceedings.mlr.press/v119/sanchez-gonzalez20a.html
19. Sanchez-Lengeling, B., Reif, E., Pearce, A., Wiltschko, A.: A gentle introduction to graph neural networks. Distill **6**(8), August 2021
20. Schad, J.: Graph powered machine learning: Part 1. ML Conference Berlin, October 2021. https://mlconference.ai/ml-summit/
21. Schütt, K.T., Arbabzadah, F., Chmiela, S., Müller, K.R., Tkatchenko, A.: Quantum-chemical insights from deep tensor neural networks. Nat. Commun. **8**(1), January 2017. https://doi.org/10.1038%2Fncomms13890
22. Stokes, J.M., et al.: A deep learning approach to antibiotic discovery. Cell **180**(4), 688–702.e13, February 2020. https://www.cell.com/cell/fulltext/S0092-8674(20)30102-1
23. Wang, Z., et al.: Heterogeneous relational message passing networks for molecular dynamics simulations. CoRR abs/2109.00711 (2021). https://arxiv.org/abs/2109.00711
24. Xu, L., He, Z.Y., Wang, K., Wang, C.D., Huang, S.Q.: Explicit message-passing heterogeneous graph neural network. IEEE Trans. Knowl. Data Eng., 1–13 (2022)
25. Zhou, J., et al.: Graph neural networks: a review of methods and applications. AI Open **1**, 57–81 (2020). https://www.sciencedirect.com/science/article/pii/S2666651021000012

Author Index

© The Editor(s) (if applicable) and The Author(s), under exclusive license
to Springer Nature Switzerland AG 2023
M. Fernández and C. M. Poskitt (Eds.): ICGT 2023, LNCS 13961, p. 303, 2023.
https://doi.org/10.1007/978-3-031-36709-0

Printed in the United States
by Baker & Taylor Publisher Services